IRMA Lectures in Mathematics and Theoretical Physics 23

Edited by Christian Kassel and Vladimir G. Turaev

Institut de Recherche Mathématique Avancée
CNRS et Université de Strasbourg
7 rue René Descartes
67084 Strasbourg Cedex
France

IRMA Lectures in Mathematics and Theoretical Physics
Edited by Christian Kassel and Vladimir G. Turaev

This series is devoted to the publication of research monographs, lecture notes, and other material arising from programs of the Institut de Recherche Mathématique Avancée (Strasbourg, France). The goal is to promote recent advances in mathematics and theoretical physics and to make them accessible to wide circles of mathematicians, physicists, and students of these disciplines.

Previously published in this series:

6 *Athanase Papadopoulos*, Metric Spaces, Convexity and Nonpositive Curvature
7 Numerical Methods for Hyperbolic and Kinetic Problems, *Stéphane Cordier, Thierry Goudon, Michaël Gutnic and Eric Sonnendrücker* (Eds.)
8 AdS/CFT Correspondence: Einstein Metrics and Their Conformal Boundaries, *Oliver Biquard* (Ed.)
9 Differential Equations and Quantum Groups, *D. Bertrand, B. Enriquez, C. Mitschi, C. Sabbah and R. Schäfke* (Eds.)
10 Physics and Number Theory, *Louise Nyssen* (Ed.)
11 Handbook of Teichmüller Theory, Volume I, *Athanase Papadopoulos* (Ed.)
12 Quantum Groups, *Benjamin Enriquez* (Ed.)
13 Handbook of Teichmüller Theory, Volume II, *Athanase Papadopoulos* (Ed.)
14 *Michel Weber*, Dynamical Systems and Processes
15 Renormalization and Galois Theory, *Alain Connes, Frédéric Fauvet and Jean-Pierre Ramis* (Eds.)
16 Handbook of Pseudo-Riemannian Geometry and Supersymmetry, *Vicente Cortés* (Ed.)
17 Handbook of Teichmüller Theory, Volume III, *Athanase Papadopoulos* (Ed.)
18 Strasbourg Master Class on Geometry, *Athanase Papadopoulos* (Ed.)
19 Handbook of Teichmüller Theory, Volume IV, *Athanase Papadopoulos* (Ed.)
20 Singularities in Geometry and Topology. Strasbourg 2009, *Vincent Blanlœil and Toru Ohmoto* (Eds.)
21 Faà di Bruno Hopf Algebras, Dyson–Schwinger Equations, and Lie–Butcher Series, *Kurusch Ebrahimi-Fard and Frédéric Fauvet* (Eds.)
22 Handbook of Hilbert Geometry, *Athanase Papadopoulos and Marc Troyanov* (Eds.)

Volumes 1–5 are available from De Gruyter (www.degruyter.de)

Sophus Lie and Felix Klein: The Erlangen Program and Its Impact in Mathematics and Physics

Lizhen Ji
Athanase Papadopoulos
Editors

European Mathematical Society

Editors:

Lizhen Ji
Department of Mathematics
University of Michigan
530 Church Street
Ann Arbor, MI 48109-1043
USA

Athanase Papadopoulos
Institut de Recherche Mathématique Avancée
CNRS et Université de Strasbourg
7 Rue René Descartes
67084 Strasbourg Cedex
France

2010 Mathematics Subject Classification: 01-00, 01-02, 01A05, 01A55, 01A70, 22-00, 22-02, 22-03, 51N15, 51P05, 53A20, 53A35, 53B50, 54H15, 58E40

Key words: Sophus Lie, Felix Klein, the Erlangen program, group action, Lie group action, symmetry, projective geometry, non-Euclidean geometry, spherical geometry, hyperbolic geometry, transitional geometry, discrete geometry, transformation group, rigidity, Galois theory, symmetries of partial differential equations, mathematical physics

ISBN 978-3-03719-148-4

The Swiss National Library lists this publication in The Swiss Book, the Swiss national bibliography, and the detailed bibliographic data are available on the Internet at http://www.helveticat.ch.

Contact address:

European Mathematical Society Publishing House
Seminar for Applied Mathematics
ETH-Zentrum SEW A27
CH-8092 Zürich
Switzerland

Phone: +41 (0)44 632 34 36
Email: info@ems-ph.org
Homepage: www.ems-ph.org

Typeset using the authors' TEX files: le-tex publishing services GmbH, Leipzig, Germany
Printing and binding: Beltz Bad Langensalza GmbH, Bad Langensalza, Germany
∞ Printed on acid free paper
9 8 7 6 5 4 3 2 1

Preface

The Erlangen program provides a fundamental point of view on the place of transformation groups in mathematics and physics. Felix Klein wrote the program, but Sophus Lie also contributed to its formulation, and his writings are probably the best example of how this program is used in mathematics. The present book gives the first modern historical and comprehensive treatment of the scope, applications and impact of the Erlangen program in geometry and physics and the roles played by Lie and Klein in its formulation and development. The book is also intended as an introduction to the works and visions of these two mathematicians. It addresses the question of what is geometry, how are its various facets connected with each other, and how are geometry and group theory involved in physics. Besides Lie and Klein, the names of Bernhard Riemann, Henri Poincaré, Hermann Weyl, Élie Cartan, Emmy Noether and other major mathematicians appear at several places in this volume.

A conference was held at the University of Strasbourg in September 2012, as the 90th meeting of the periodic *Encounter between Mathematicians and Theoretical Physicists*, whose subject was the same as the title of this book. The book does not faithfully reflect the talks given at the conference, which were generally more specialized. Indeed, our plan was to have a book interesting for a wide audience and we asked the potential authors to provide surveys and not technical reports.

We would like to thank Manfred Karbe for his encouragement and advice, and Hubert Goenner and Catherine Meusburger for valuable comments. We also thank Goenner, Meusburger and Arnfinn Laudal for sending photographs that we use in this book.

This work was supported in part by the French program ANR Finsler, by the GEAR network of the National Science Foundation (GEometric structures And Representation varieties) and by a stay of the two editors at the Erwin Schrödinger Institute for Mathematical Physics (Vienna).

Lizhen Ji and Athanase Papadopoulos
Ann Arbor and Strasbourg, March 2015

Contents

Sophus Lie.

Felix Klein.

Introduction

The Erlangen program is a perspective on geometry through invariants of the automorphism group of a space. The original reference to this program is a paper by Felix Klein which is usually presented as the exclusive historical document in this matter. Even though Klein's viewpoint was generally accepted by the mathematical community, its re-interpretation in the light of modern geometries, and especially of modern theories of physics, is central today. There are no books on the modern developments of this program. Our book is one modest step towards this goal.

The history of the Erlangen program is intricate. Klein wrote this program, but Sophus Lie made a very substantial contribution, in promoting and popularizing the ideas it contains. The work of Lie on group actions and his emphasis on their importance were certainly more decisive than Klein's contribution. This is why Lie's name comes first in the title of the present volume. Another major figure in this story is Poincaré, and his role in highlighting the importance of group actions is also critical.

Thus, groups and group actions are at the center of our discussion. But their importance in mathematics had already been crucial before the Erlangen program was formulated.

From its early beginning in questions related to solutions of algebraic equations, group theory is merged with geometry and topology. In fact, group actions existed and were important before mathematicians gave them a name, even though the formalization of the notion of a group and its systematic use in the language of geometry took place in the 19th century. If we consider group theory and transformation groups as an abstraction of the notion of symmetry, then we can say that the presence and importance of this notion in the sciences and in the arts was realized in ancient times.

Today, the notion of group is omnipresent in mathematics and, in fact, if we want to name one single concept which runs through the broad field of mathematics, it is the notion of group. Among groups, Lie groups play a central role. Besides their mathematical beauty, Lie groups have many applications both inside and outside mathematics. They are a combination of algebra, geometry and topology.

Besides groups, our subject includes geometry.

Unlike the word "group" which, in mathematics has a definite significance, the word "geometry" is not frozen. It has several meanings, and all of them (even the most recent ones) can be encompassed by the modern interpretation of Klein's idea. In the first version of Klein's Erlangen program, the main geometries that are emphasized are projective geometry and the three constant curvature geometries (Euclidean, hyperbolic and spherical), which are considered there, like affine geometry, as part of projective geometry. This is due to the fact that the transformation groups of all these geometries can be viewed as restrictions to subgroups of the transformation group of projective geometry. After these first examples of group actions in geometry, the stress shifted to Lie transformation groups, and it gradually included many new notions, like Riemannian manifolds, and more generally spaces equipped with affine connections. There is a wealth of geometries which can be described by transformation groups in the spirit of the Erlangen program. Several of these geometries were studied by Klein and Lie; among them we can mention Minkowski geometry, complex geometry, contact geometry and symplectic geometry. In modern geometry, besides the transformations of classical geometry which take the form of motions, isometries, etc., new notions of transformations and maps between spaces arose. Today, there is a wealth of new geometries that can be described by transformation groups in the spirit of the Erlangen program, including modern algebraic geometry where, according to Grothendieck's approach, the notion of morphism is more important than the notion of space.[1] As a concrete example of this fact, one can compare the Grothendieck–Riemann–Roch theorem with the Hirzebruch–Riemann–Roch. The former, which concerns morphisms, is much stronger than the latter, which concerns spaces.

Besides Lie and Klein, several other mathematicians must be mentioned in this venture. Lie created Lie theory, but others' contributions are also immense. About two decades before Klein wrote his Erlangen program, Riemann had introduced new geometries, namely, in his inaugural lecture, *Über die Hypothesen, welche der Geometrie zu Grunde liegen* (On the hypotheses which lie at the bases of geometry) (1854). These geometries, in which groups intervene at the level of infinitesimal transformations, are encompassed by the program. Poincaré, all across his work, highlighted the importance of groups. In his article on the Future of mathematics[2], he wrote: "Among the words that exerted the most beneficial influence, I will point out the words group and invariant. They made us foresee the very essence of mathematical reasoning. They showed us that in numerous cases the ancient mathematicians considered groups without knowing it, and how, after thinking that they were far away from each other, they suddenly ended up close together without understanding why." Poincaré stressed several times the importance of the ideas of Lie in the theory of group transformations. In his analysis of his own works,[3] Poincaré declares: "Like Lie, I believe that the notion, more or less unconscious, of a continuous group is the unique logical basis of our geometry." Killing, É. Cartan, Weyl, Chevalley and many others refined the structures of Lie theory and they developed its global aspects and applications to homogeneous spaces. The generalization of the Erlangen program to these new spaces uses the notions of connections and gauge groups, which were

[1]See A. Grothendieck, Proceedings of the International Congress of Mathematicians, 14–21 August 1958, Edinburgh, ed. J.A. Todd, Cambridge University Press, p. 103–118. In that talk, Grothendieck sketched his theory of cohomology of schemes.

[2]H. Poincaré, *L'Avenir des mathématiques*, Revue générale des sciences pures et appliquées 19 (1908) p. 930–939. [Parmi les mots qui ont exercé la plus heureuse influence, je signalerai ceux de groupe et d'invariant. Ils nous ont fait apercevoir l'essence de bien des raisonnements mathématiques ; ils nous ont montré dans combien de cas les anciens mathématiciens considéraient des groupes sans le savoir, et comment, se croyant bien éloignés les uns des autres, ils se trouvaient tout à coup rapprochés sans comprendre pourquoi.]

[3]*Analyse de ses travaux scientifiques, par Henri Poincaré.* Acta Mathematica, 38 (1921), p. 3–135. [Comme Lie, je crois que la notion plus ou moins inconsciente de groupe continu est la seule base logique de notre géométrie]; p. 127. There are many similar quotes in Poincaré's works.

closely linked to new developments in physics, in particular, in electromagnetism, phenomena related to light, and Einstein's theory of general relativity.

Today, instead of the word "geometry" we often use the expression "geometric structure", and there is a wealth of geometric structures which can be described by transformation groups in the spirit of the Erlangen program. We mention in particular the notion of (G, X) structure introduced by Charles Ehresmann in the 1930s, which is of paramount importance. Here X is a homogeneous space and G a Lie group acting transitively on G. A (G, X) structure on a manifold M is then an atlas whose charts are in X and whose coordinate changes are restrictions of elements of G acting on X. Ehresmann formulated the notions of developing map and of holonomy transformations, which are basic objects in the study of these structures and their moduli spaces. (G, X) structures have several variants and they have been developed and adapted to various settings by Haefliger, Kuiper, Benzécri, Thurston, Goldman and others to cover new structures, including foliations and singular spaces. The most spectacular advancement in this domain is certainly Thurston's vision of the eight geometries in dimension three, his formulation of the geometrization conjecture and the work around it, which culminated in the proof of the Poincaré conjecture by Perelman.

We talked about mathematics, but the Erlangen program also encompasses physics. In fact, geometry is closely related to physics, and symmetry is essential in modern physics. Klein himself investigated the role of groups in physics, when he stressed the concept of geometric invariants in his description of Einstein's theories of special and general relativity, in particular by showing the importance of the Lorentz group, and also in his work on the conservation laws of energy and momentum in general relativity. Another milestone that led to conceptual clarifications and made it possible to systematically exploit the notion of symmetry in physics was E. Noether's work that related symmetries of physical systems to conserved quantities.

In conclusion, the central questions that are behind the present volume are:

- What is geometry?
- What is the relation between geometry and physics?
- How are groups used in physics, especially in contemporary physics?

Let us now describe briefly the content of this volume.

Chapters 1 and 2, written by Lizhen Ji, are introductions to the lives and works on Lie and Klein. Even though Klein was a major mathematician, surprisingly enough, there is no systematic English biography of him. The author's aim is to fill this gap to a certain extent. Besides providing convenient short biographies of Lie and Klein, the author wishes to convince the reader of the importance of their works, especially those which are in close relation with the Erlangen program, and also to show how close the two men were in their ideas and characters. They both learned from each other and they had a profound influence on each other. This closeness, their ambitiousness, the competition among them and their disputes for priority of some discoveries were altogether the reasons that made them split after years of collaboration and friendship. The conflict between them is interesting and not so well known. The author describes this conflict, also mentions the difficulties that these two men encountered in their professional lives and in their relations with other mathematicians. Both of them experienced nervous breakdowns.[4] The chapter on Lie also contains an outline of his important theories as well as statements of some of his most significant theorems. In particular, the author puts forward in modern language and comments on three fundamental theorems of Lie. Concerning Klein, it is more difficult to pick out individual theorems, because Klein is known for having transmitted ideas rather than specific results. The author explains how Klein greatly influenced people and the world around him through his lectures and conversations, his books, the journals he edited, and he also recalls his crucial influence in shaping up the university of Göttingen to be the world's most important mathematics center. In these surveys, the author also mentions several mathematicians who were closely related in some way or another to Lie and Klein, among them Hilbert, Hausdorff, Engel, Plücker, Sylow, Schwarz and Poincaré. The chapter on Lie also reviews other aspects of Lie's work besides Lie groups.

Chapter 3, by Jeremy Gray, is a historical commentary on the Erlangen program. The author starts by a short summary of the program manifesto and on the circumstances of its writing, mentioning the influence of several mathematicians, and the importance of the ideas that originate from projective geometry (specially those of von Staudt). He then brings up the question of the impact of this program on the views of several mathematicians, comparing the opinions of Birkhoff and Bennet and of Hawkins.

In Chapter 4, Hubert Goenner presents a critical discussion of the general impact and of the limitations of the Erlangen program in physics. He starts by recalling that the influence of the Erlangen program in physics was greatly motivated by the geometrization of special relativity by H. Minkowski, in which the Lorentz group appears as one of the main objects of interest, but he stresses the fact that the notion of *field* defined on a geometry – and not the notion of geometry itself – is then the central element. He comments on the relation of Lie transformations with theories of conservation laws and the relations of the Erlangen program with symplectic geometry, analytical mechanics, statistical physics, quantum field theories, general relativity, Yang–Mills theory and supergravity. The paper has a special section where the author discusses supersymmetry. In a final section, the author mentions several generalizations of the notion of Lie algebra.

In Chapter 5, Norbert A'Campo and Athanase Papadopoulos comment on the two famous papers of Klein, *Über die sogenannte Nicht-Euklidische Geometrie* (On the so-called non-Euclidean geometries), *I* and *II*. The two papers were written respectively one year and a few months before the Erlangen program, and they contain in essence the main ideas of this program. We recall that the 19th century saw the birth of non-Euclidean geometry by Lobachevsky, Bolyai and Gauss, and at the same time, the development of projective geometry by Poncelet, Plücker, von Staudt and others, and also of conformal geometry by Liouville and others. Groups made the first link between all these geometries, and also between geometry and algebra. Klein, in the

[4] Klein's nervous breakdown was probably due to overwork and exhaustion, caused in part by his rude competition with Poincaré on Fuchsian functions, whereas Lie's nervous breakdown was the consequence of a chronic illness, pernicious anemia, related to a lack in vitamin B12, which at that time was incurable.

papers cited above, gives models of the three constant-curvature geometries (hyperbolic, Euclidean and spherical) in the setting of projective geometry. He defines the distance functions in each of these geometries by fixing a conic (the "conic at infinity") and taking a constant multiple of the logarithm of the cross ratio of four points: the given two points and the two intersection points of the line joining them with the conic at infinity. The hyperbolic and spherical geometries are obtained by using real and complex conics respectively, and Euclidean geometry by using a degenerate one. The authors in Chapter 5 comment on these two important papers of Klein and they display relations with works of other mathematicians, including Cayley, Beltrami, Poincaré and the founders of projective geometry.

Klein's interaction with Lie in their formative years partly motivated Lie to develop Lie's version of Galois theory of differential equations and hence of Lie transformation theory.[5] In fact, a major motivation for Lie for the introduction of Lie groups was to understand differential equations. This subject is treated in Chapter 6 of this volume. The author, Alexandre Vinogradov, starts by observing that Lie initiated his work by transporting the Galois theory of the solvability of algebraic equations to the setting of differential equations. He explains that the major contribution of Lie in this setting is the idea that symmetries of differential equations are the basic elements in the search for their solutions. One may recall here that Galois approached the problem of solvability of polynomial equations through a study of the symmetries of their roots. This is based on the simple observation that the coefficients of a polynomial may be expressed in terms of the symmetric functions of their roots, and that a permutation of the roots does not change the coefficients of the polynomial. In the case of differential equations, one can naively define the symmetry group to be the group of diffeomorphisms which preserve the space of solutions, but it is not clear how such a notion can be used. There is a *differential Galois theory* which is parallel to the Galois theory of polynomial equations. In the differential theory, the question "what are the symmetries of a (linear or nonlinear, partial or ordinary) differential equation?" is considered as the central question. Chapter 6 also contains reviews of the notions of jets and jet spaces and other constructions to explain the right setup for formulating the question of symmetry, with the goal of providing a uniform framework for the study of nonlinear partial differential equations. The author is critical of the widely held view that each nonlinear partial differential equation arising from geometry or physics is special and often requires its own development. He believes that the general approach based on symmetry is the right one.

The author mentions developments of these ideas that were originally formulated by Lie and Klein in works of E. Noether, Bäcklund, É. Cartan, Ehresmann and others. A lot of questions in this domain remain open, and this chapter will certainly give the reader a new perspective on the geometric theory of nonlinear partial differential equations.

In Chapter 7, Charles Frances surveys the modern developments of geometric structures on manifolds in the lineage of Klein and Lie. The guiding idea in this

[5] Lie has had a course at Oslo by Sylow on Galois and Abel theory before he meets Klein, but it is clear that Klein also brought some of his knowledge to Lie.

chapter is the following question: When is the automorphism group of a geometric structure a Lie group, and what can we say about the structure of such a Lie group?

The author considers the concept of *Klein geometry*, that is, a homogeneous space acted upon by a Lie group, and a generalization of this notion, leading to the concept of a *Cartan geometry*. (Cartan used the expression *espace généralisé*.) Besides the classical geometries, like constant curvature spaces (Euclidean, Lobachevsky and spherical) as well as projective geometry which unifies them, the notion of Cartan geometry includes several differential-geometric structures. These notions are defined using fiber bundles and connections. They describe spaces of variable curvature and they also lead to pseudo-Riemannian manifolds, conformal structures of type (p, q), affine connections, CR structures, and the so-called *parabolic geometries*. The author presents a series of important results on this subject, starting with the theorem of Myers and Steenrod (1939) saying that the isometry group of any Riemannian manifold is a Lie group, giving a bound on its dimension, and furthermore, it says that this group is compact if the manifold is compact. This result gave rise to an abundance of developments and generalizations. The author also explains in what sense pseudo-Riemannian manifolds, affine connections and conformal structures in dimensions ≥ 3 are rigid, symplectic manifolds are not rigid, and complex manifolds are of an intermediate type.

Thus, two general important questions are addressed in this survey:

- What are the possible continuous groups that are the automorphism groups of a geometry on a compact manifold?
- What is the influence of the automorphism group of a structure on the topology or the diffeomorphism type of the underlying manifold?

Several examples and recent results are given concerning Cartan geometries and in particular pseudo-Riemannian conformal structures.

Chapter 8, by Norbert A'Campo and Athanase Papadopoulos, concern transitional geometry. This is a family of geometries which makes a continuous transition between hyperbolic and spherical geometry, passing through Euclidean geometry. The space of transitional geometry is a fiber space over the interval $[-1, 1]$ where the fiber above each point t is a space of constant curvature t^2 if $t > 0$ and of constant curvature $-t^2$ if $t > 0$. The fibers are examples of *Klein geometries* in the sense defined in Chapter 7. The elements of each geometry are defined group-theoretically, in the spirit of Klein's Erlangen program. Points, lines, triangles, trigonometric formulae and other geometric properties transit continuously between the various geometries.

In Chapter 9, by Athanase Papadopoulos and Sumio Yamada, the authors introduce a notion of cross ratio which is proper to each of the three geometries: Euclidean, spherical and hyperbolic. This highlights the relation between projective geometry and these geometries. This is in the spirit of Klein's view of the three constant curvature geometries as part of projective geometry, which is the subject of Chapter 5 of the present volume.

Chapter 10, by Yuri Suris, concerns the Erlangen program in the setting of discrete differential geometry. This is a subject which recently emerged, whose aim is to develop a theory which is the discrete analogue of classical differential geometry.

It includes discrete versions of the differential geometry of curves and surfaces but also higher-dimensional analogues. There are discrete notions of line, curve, plane, volume, curvature, contact elements, etc. There is a unifying transformation group approach in discrete differential geometry, where the discrete analogues of the classical objects of geometry become invariants of the respective transformation groups. Several classical geometries survive in the discrete setting, and the author shows that there is a discrete analogue of the fact shown by Klein that the transformation groups of several geometries are subgroups of the projective transformation group, namely, the subgroup preserving a quadric.

Examples of discrete differential geometric geometries reviewed in this chapter include discrete line geometry and discrete line congruence, quadrics, Plücker line geometry, Lie sphere geometry, Laguerre geometry and Möbius geometry. Important notions such as curvature line parametrized surfaces, principal contact element nets, discrete Ribeaucour transformations, circular nets and conical nets are discussed. The general underlying idea is that the notion of transformation group survives in the discretization process. Like in the continuous case, the transformation group approach is at the same time a unifying approach, and it is also related to the question of "multi-dimensional consistency" of the geometry, which says roughly that a 4D consistency implies consistency in all higher dimensions. The two principles – the transformation group principle and consistency principle – are the two guiding principles in this chapter.

Chapter 11 by Catherine Meusburger is an illustration of the application of Klein's ideas in physics, and the main example studied is that of three-dimensional gravity, that is, Einstein's general relativity theory[6] with one time and two space variables. In three-dimensions, Einstein's general relativity can be described in terms of certain domains of dependence in thee-dimensional Minkowski, de Sitter and anti de Sitter space, which are homogeneous spaces. After a summary of the geometry of spacetimes and a description of the gauge invariant phase spaces of these theories, the author discusses the question of quantization of gravity and its relation to Klein's ideas of characterizing geometry by groups.

Besides presenting the geometrical and group-theoretical aspects of three-dimensional gravity, the author mentions other facets of symmetry in physics, some of them related to moduli spaces of flat connections and to quantum groups.

Chapter 12, by Jean-Bernard Zuber, is also on groups that appear in physics, as group invariants associated to a geometry. Several physical fields are mentioned, including crystallography, piezzoelectricity, general relativity, Yang–Mills theory, quantum field theories, particle physics, the physics of strong interactions, electromagnetism, sigma-models, integrable systems, superalgebras and infinite-dimensional algebras. We see again the work of Emmy Noether on group invariance principles in variational problems. Representation theory entered into physics through quantum mechanics, and the modern theory of quantum group is a by-product. The author comments on Noether's celebrated paper which she presented at the occasion of Klein's academic Jubilee. It contains two of her theorems on conservation laws.

[6] We recall by the way that Galileo's relativity theory is at the origin of many of the twentieth century theories.

Today, groups are omnipresent in physics, and as Zuber puts it: "To look for a group invariance whenever a new pattern is observed has become a second nature for particle physicists".

We hope that the various chapters of this volume will give to the reader a clear idea of how group theory, geometry and physics are related to each other, the Erlangen program being a major unifying element in this relation.

Lizhen Ji and Athanase Papadopoulos

Chapter 1

Sophus Lie, a giant in mathematics

Lizhen Ji

Contents

1 Introduction

There are very few mathematicians and physicists who have not heard of Lie groups or Lie algebras and made use of them in some way or another. If we treat discrete or finite groups as special (or degenerate, zero-dimensional) Lie groups, then almost every subject in mathematics uses Lie groups. As H. Poincaré told Lie [25] in October 1882, "all of mathematics is a matter of groups." It is clear that the importance of groups comes from their actions. For a list of topics of group actions, see [17].

Lie theory was the creation of Sophus Lie, and Lie is most famous for it. But Lie's work is broader than this. What else did Lie achieve besides his work in Lie theory? This might not be so well known. The differential geometer S. S. Chern wrote in 1992 that "Lie was a great mathematician even without Lie groups" [7]. What did and can Chern mean? We will attempt to give a summary of some major contributions of Lie in §6.

One purpose of this chapter is to give *a glimpse of Lie's mathematical life* by recording several things which I have read about Lie and his work. Therefore, it is *short and emphasizes only a few things* about his mathematics and life. For a fairly detailed account of his life (but not his mathematics), see the full length biography [27].

We also provide some details about the unfortunate conflict between Lie and Klein and the famous quote from Lie's preface to the third volume of his books on transformation groups, which is usually only quoted without explaining the context. The fruitful collaboration between Engel and Lie and the publication of Lie's collected works are also mentioned.

We hope that this chapter will be interesting and instructive to the reader of this book and might serve as a brief introduction to the work and life of Lie discussed in this book.

2 Some general comments on Lie and his impact

It is known that Lie's main work is concerned with understanding how continuous transformation groups provide an organizing principle for different areas of mathematics, including geometry, mechanics, and partial differential equations. But it might not be well known that Lie's collected works consist of 7 large volumes of the total number of pages about 5600. (We should keep in mind that a substantial portion of these pages are commentaries on his papers written by the editors. In spite of this, Lie's output was still enormous.) Probably it is also helpful to keep in mind that Lie started to do mathematics at the age of 26 and passed away at 57. Besides many papers, he wrote multiple books, which total over several thousands of pages. According to Lie, only a part of his ideas had been put down into written form. In an autobiographic note [9, p. 1], Lie wrote:

> My life is actually quite incomprehensible to me. As a young man, I had no idea that I was blessed with originality, Then, as a 26-year-old, I suddenly realized that I could create. I read a little and began to produce. In the years 1869–1874, I had a lot of ideas which, in the course of time, I have developed only very imperfectly.
>
> In particular, it was group theory and its great importance for the differential equations which interested me. But publication in this area went woefully slow. I could not structure it properly, and I was always afraid of making mistakes. Not the small inessential mistakes ... No, it was the deep-rooted errors I feared. I am glad that my group theory in its present state does not contain any fundamental errors.

Lie was a highly original and technically powerful mathematician. The recognition of the idea of Lie groups (or transformation groups) took time. In 1870s, he wrote in a letter [26, p. XVIII]:

> If I only knew how to get the mathematicians interested in transformation groups and their applications to differential equations. I am certain, absolutely certain in my case, that these theories in the future will be recognized as fundamental. I want to form such an impression now, since for one thing, I could then achieve ten times as much.

In 1890, Lie was confident and wrote that he strongly believed that his work would stand through all times, and in the years to come, it would be more and more appreciated by the mathematical world.

Eduard Study was a privatdozent (lecturer) in Leipzig when Lie held the chair in geometry there. In 1924, the mature Eduard Study summarized Lie as follows [26, p. 24]:

> Sophus Lie had the shortcomings of an autodidact, but he was also one of the most brilliant mathematicians who ever lived. He possessed something which is not found very often and which is now becoming even rarer, and he possessed it in abundance: creative imagination. Coming generations will learn to appreciate this visionary's mind better than the present generation, who can only appreciate the mathematicians' sharp intellect. The all-encompassing scope of this man's vision, which, above all, demands recognition, is nearly completely lost. But, the coming generation [...] will understand the importance of the theory of transformation groups and ensure the scientific status that this magnificent work deserves.

What Lie studied are infinitesimal Lie groups, or essentially Lie algebras. Given what H. Weyl and É. Cartan contributed to the global theory of Lie groups starting around the middle of 1920s and hence made Lie groups one of the most basic and essential objects in modern (or contemporary) mathematics, one must marvel at the above visionary evaluation of Lie's work by Study. For a fairly detailed overview of the historical development of Lie groups with particular emphasis on the works of Lie, Killing, É. Cartan and Weyl, see the book [14].

Two months after Lie died, a biography of him appeared in the American Mathematical Monthly [12]. It was written by George Bruce Halsted, an active mathematics educator and a mathematician at the University of Texas at Austin, who taught famous mathematicians like R. L. Moore and L. E. Dickson. Reading it more than one hundred years later, his strong statement might sound a bit surprising but is more justified than before, "[...] the greatest mathematician in the world, Sophus Lie, died [...] His work is cut short; his influence, his fame, will broaden, will tower from day to day."

Probably a more accurate evaluation of Lie was given by Engel in a memorial speech on Lie [9, p. 24] in 1899:

> If the capacity for discovery is the true measure of a mathematician's greatness, then Sophus Lie must be ranked among the foremost mathematicians of all time. Only extremely few have opened up so many vast areas for mathematical research and created such rich and wide-ranging methods as he [...] In addition to a capacity for discovery, we expect a mathematician to posses a penetrating mind, and Lie was really an exceptionally gifted mathematician [...] His efforts were based on tackling problems which are important, but solvable, and it often happened that he was able to solve problems which had withstood the efforts of other eminent mathematicians.

In this sense, Lie was a giant for his deep and original contribution to mathematics, and is famous not for other reasons. (One can easily think of several mathematicians, without naming them, who are famous for various things besides mathematics). Incidentally, he was also a giant in the physical sense. There are some vivid descriptions of Lie by people such as É. Cartan [1, p. 7], Engel [27, p. 312], and his physics colleague Ostwald at Leipzig [27, p. 396]. See also [27, p. 3]. For some interesting discussions on the relations between giants and scientists, see [11, pp. 163–164, p. 184] and [22, pp. 9–13].

3 A glimpse of Lie's early academic life

Lie was born on December 17, 1842. His father, Johann Herman Lie, was a Lutheran minister. He was the youngest of the six children of the family. Lie first attended school in the town of Moss in South Eastern Norway and on the eastern side of the Oslo Fjord. In 1857 he entered Nissen's Private Latin School in Christiania, which became Oslo in 1925. At that time, he decided to pursue a military career, but his poor eyesight made this impossible, and he entered University of Christiania to pursue a more academic life.

During his university time, Lie studied science in a broad sense. He took mathematics courses and attended lectures by teachers of high quality. For example, he attended lectures by Sylow in 1862.[1]

Though Lie studied with some good mathematicians and did well in most courses, on his graduation in 1865, he did not show any special ability for mathematics or any particular liking for it. Lie could not decide what subject to pursue and he gave some private lessons and also volunteered some lectures for a student union while trying to make his decision. He knew he wanted an academic career and thought for a while that astronomy might be the right topic. He also learnt some mechanics, and wondered about botany, zoology or physics. Lie reached the not-so-tender age of 26 in 1868 and was still not sure what he should pursue as a career. But this year was a big turning point for him.

In June 1868, the Tenth Meeting of Scandinavian Natural Sciences was held in Christiania. It attracted 368 participants. Lie attended many lectures and was particularly influenced by the lecture of a former student of the great French geometer Michel Chasles, which referred to works of Chasles, Möbius, and Plücker.

It seems that the approaching season, the autumn of 1868, became one long continuing period of work for Sophus Lie, with his frequent borrowing of books from the library. In addition to Chasles, Möbius and Plücker, Lie discovered the Frenchman Poncelet, the Englishman Hamilton, and the Italian Cremona, as well as others who had made important contributions to algebraic and analytic geometry.

Lie plowed through many volumes of the leading mathematical journals from Paris and Berlin, and in the Science Students Association he gave several lectures during the spring of 1869 on what he called his "Theory of the imaginaries", and on how information on real geometric objects could be transferred to his "imaginary objects."

[1]Ludwig Sylow (1832–1918) was Norwegian, like Lie. He is now famous and remembered for the Sylow subgroups. At that time, he was not on the permanent staff of the university of Christiania, but he was substituting for a regular faculty member and taught a course. In this course, he explained Abel's and Galois' work on algebraic equations. But it seems that Lie did not understand or remember the content of this course, and it was Klein who re-explained these theories to him and made a huge impact on Lie's mathematical life.

Sophus Lie wrote a paper on his discovery. The paper was four pages long and it was published at his own expenses.[2] After this paper was translated into German, it was published in the leading mathematics journal of the time, *Crelle's Journal*.[3] With this paper, he applied to the Collegium for a travel grant and received it. Then he left for Berlin in September 1869 and begun his glorious and productive mathematical career.

There were several significant events in Berlin for Lie on this trip. He met Felix Klein and they immediately became good friends. They shared common interests and common geometric approaches, and their influence on each other was immense. Without this destined (or chance) encounter, Lie and Klein might not have been the people we know.

Lie also impressed Kummer by solving problems which Kummer was working on. This gave him confidence in his own power and originality. According to Lie's letter to his boyfriend [26, p. XII]:

> Today I had a triumph which I am sure you will be interested to hear about. Professor Kummer suggested that we test our powers on a discussion of all line congruences of the 3rd degree. Fortunately, a couple of months ago, I had already solved a problem which was in a way special of the above, but was nevertheless much more general. ... I regard this as a confirmation of my good scientific insight that I, from the very first, understood the value of findings. That I have shown both energy and capability in connection with my findings; that I know.

In the summer of 1970, Lie and Klein visited Paris and met several important people such as Jordan and Darboux. The interaction with Jordan and Jordan's new book on groups had a huge influence on both of them. This book by Jordan contained more than an exposition of Galois theory and can be considered as a comprehensive discussion of how groups were used in all subjects up to that point. For Klein and Lie, it was an eye opener. Besides learning Galois theory, they started to realize the basic and unifying role groups would play in geometry and other parts of mathematics. In some sense, the trips with Klein sealed the future research direction of Lie. Klein played a crucial role in the formative years of Lie, and the converse is also true. We will discuss their interaction in more detail in §8 and §9.

At the outbreak of the Franco–Prussian war in July, Klein left, and Lie stayed for one more month and then decided to hike to Italy. But he was arrested near Fontainebleau as he was suspected of being a spy and spent one month in jail. Darboux came and freed him. In [8], Darboux wrote:

> True, in 1870 a misadventure befell him, whose consequences I was instrumental in averting. Surprised at Paris by the declaration of war, he took refuge at

[2] The publication of this paper is unusual also by today's standards. According to [2], "His first published paper appeared in 1869. It gives a new representation of the complex plane and uses ideas of Plücker. But Lie had difficulties in getting these ideas published by the Academy in Christiania. He was impatient. Professor Bjerknes asked for more time to look at the paper, but Professor Broch returned it after two days – saying he had understood nothing! However, three other professors – who probably understood the material even less – supported publication. This happened as a result of influence by friends of Lie."

[3] The German version of this paper is still only 8 pages long, but in his collected works edited by Engel and Heegaard, there are over 100 pages of commentaries devoted to it.

> Fontainebleau. Occupied incessantly by the ideas fermenting in his brain, he would go every day into the forest, loitering in places most remote from the beaten path, taking notes and drawing figures. It took little at this time to awaken suspicion. Arrested and imprisoned at Fontainebleau, under conditions otherwise very comfortable, he called for the aid of Chasles, Bertrand, and others; I made the trip to Fontainebleau and had no trouble in convincing the procureur impérial; all the notes which had been seized and in which figured complexes, orthogonal systems, and names of geometers, bore in no way upon the national defenses.

Afterwards, Lie wrote to his close friend [26, p. XV], "except at the very first, when I thought it was a matter of a couple of days, I have taken things truly philosophically. I think that a mathematician is well suited to be in prison."

In fact, while he was in prison, he worked on his thesis and a few months later, he submitted his thesis, on March 1871. He received his doctorate degree in July 1872, and accepted a new chair at the university of Christiania set up for him by the Norwegian National Assembly. It was a good thesis, which dealt with the integration theory of partial differential equations. After his thesis, Lie's mathematics talent was widely recognized and his mathematical career was secured.

When Lie worked on his thesis with a scholarship from the University of Christiania, he needed to teach at his former grammar school to supplement his income. With this new chair, he could devote himself entirely to mathematics. Besides developing his work on transformation groups and working with Klein towards the formulation of the Erlangen program, Lie was also involved in editing with his former teacher Ludwig Sylow the collected works of Abel. Since Lie was not familiar with algebras, especially with Abel's works, this project was mainly carried out by Sylow. But locating and gathering manuscripts of Abel took a lot of effort to both men, and the project took multiple years.

In his personal life, Lie married Anna Birch in 1874, and they had two sons and a daughter.

Lie published several papers on transformation groups and on the applications to integration of differential equations and he established a new journal in Christiania to publish his papers, but these papers did not receive much attention. Because of this, Lie started to work on more geometric problem such as minimal surfaces and surfaces of constant curvature.

Later in 1882 some work by French mathematicians on integration of differential equations via transformation groups motivated Lie to go back to his work on integration of differential equations and theories of differential invariants of groups.

4 A mature Lie and his collaboration with Engel

There were two people who made, at least contributed substantially to make, Lie the mathematician we know today. They were Klein and Engel. Of course, his story with Klein is much better known and dramatic and talked about, but his interaction with Engel is not less important or ordinary.

In the period from 1868 to 1884, Lie worked constantly and lonely to develop his theory of transformation groups, integration problems, and theories of differential invariants of finite and infinite groups. But he could not describe his new ideas in an understandable and convincing way, and his work was not valued by the mathematics community. Further, he was alone in Norway and no one could discuss with him or understand his work.

In a letter to Klein in September 1883 [9, p. 9], Lie wrote that "It is lonely, frightfully lonely, here in Christiania where nobody understands my work and interests."

Realizing the seriousness of the situation of Lie and the importance of summarizing in a coherent way results of Lie and keeping him productive, Klein and his colleague Mayer at Leipzig decided to send their student Friedrich Engel to assist Lie. Klein and Mayer knew that without help from someone like Engel, Lie could not produce a coherent presentation of his new novel theories.

Like Lie, Engel was also a son of a Lutheran minister. He was born in 1861, nineteen years younger than Lie. He started his university studies in 1879 and attended both the University of Leipzig and the University of Berlin. In 1883 he obtained his doctorate degree from Leipzig under Mayer with a thesis on contact transformations. After a year of military service in Dresden, Engel returned to Leipzig in the spring of 1884 to attend Klein's seminar in order to write a Habilitation. At that time, besides Klein, Mayer was probably the only person who understood Lie's work and his talent. Since contact transformations form one important class in Lie's theories of transformation groups, Engel was a natural candidate for this mission. Klein and Mayer worked together to obtain a stipend from the University of Leipzig and the Royal Society of Sciences of Saxony for Engel so that he could travel to work with Lie in Christiania.

In June 1884, Lie wrote a letter to Engel [9, p. 10],

> From 1871–1876, I lived and breathed only transformation groups and integration problems. But when nobody took any interest in these things, I grew a bit weary and turned to geometry for a time. Now just in the last few years, I have again taken up these old pursuits of mine. If you will support me with the further development and editing of these things, you will be doing me a great service, especially in that, for once, a mathematician finally has a serious interest in these theories. Here in Christiania, a specialist like myself is terribly lonely. No interest, no understanding.

According to a letter of Engel in the autumn of 1884 after meeting Lie [27, p. 312]:

> The goal of my journey was twofold: on one hand, under Lie's own guidance, I should become immersed in his theories, and on the other, I should exercise a sort of pressure on him, to get him to carry on his work for a coherent presentation of one of his greater theories, with which I should help him apply his hand.

Lie wanted to write a major comprehensive monograph on transformation groups, not merely a simple introduction to his new theory. It "should be a systematic and strict-as-possible account that would retain its worth for a long time" [9, p. 11].

Lie and Engel met twice every day, in the morning at the apartment of Engel and in the afternoon at Lie's apartment. They started with a list of chapters. Then Lie dictated an outline of each chapter and Engel would supply the detail. According to Engel [9, p. 11],

> Every day, I was newly astonished by the magnificence of the structure which Lie had built entirely on his own, and about which his publications, up to then, gave only a vague idea. The preliminary editorial work was completed by Christmas, after which Lie devoted some weeks to working through all of the material in order to lay down the final draft. Starting at the end of January 1885, the editorial work began anew; the finished chapters were reworked and new ones were added. When I left Christiania in June of 1885, there was a pile of manuscripts which Lie figured would eventually fill approximately thirty printer's sheets. That it would be eight years before the work was completed and the thirty sheets would become one hundred and twenty-five was something neither of us could have imagined at that time.

Lie and Engel worked intensively over the nine month period when Engel was there. This collaboration was beneficial to both parties. To Engel, it was probably the best introduction to Lie theories and it served his later mathematical research well. According to Kowalewski, a student of Lie and Engel, [9, p. 10], "Lie would never have been able to produce such an account by himself. He would have drowned in the sea of ideas which filled his mind at that time. Engel succeeded in bringing a systematic order to this chaotic mass of thought."

After returning to Leipzig, Engel finished his Habilitation titled "On the defining equations of the continuous transformation groups" and became a Privatdozent.

In 1886, Klein moved to Göttingen for various reasons. (See [18] for a brief description of Klein's career). Thanks to the efforts of Klein, Lie moved to Leipzig in 1886 to take up a chair in geometry. More description of this will be given in §8 below.

When Lie visited Leipzig in February 1886 to prepare for his move, he wrote to his wife excitedly [27, p. 320], "to the best of my knowledge, there have been no other foreigners, other than Abel and I, appointed professor at a German University. (The Swiss are out of the calculation here.) It is rather amazing. In Christiania I have often felt myself to be treated unfairly, so I have truly achieved an unmerited honor."

Leipzig was the hometown of the famous Leibniz and a major culture and academic center. In comparison to his native country, it was an academic heaven for Lie.

In April 1886, Lie became the Professor of Geometry and Director of the Mathematical Seminar and Institute at the University of Leipzig. Lie and Engel resumed to work intensively on their joint book again. In 1888, the leading German scientific publisher Teubner, based in Leipzig, published the first volume of *Theory of transformation groups*, which was 632 pages long.

In that year, Engel also became the assistant to Lie after Friedrich Schur left. When Lie went to a nerve clinic near the end of 1889, Engel gave Lie's lectures for him.

The second volume of their joint book was published in 1890 and was 555 pages long, and the third volume contained 831 pages and was published in 1893.

The three big volumes of joint books with Engel would not see the light of day or even start without the substantial contribution of Engel.

It was a major piece of work. In a 21-page review of the first volume [9, p. 16], Eduard Study wrote,

> The work in question gives a comprehensive description of an extensive theory which Mr. Lie has developed over a number of years in a large number of individual articles in journals ... Because most of these articles are not well known, and because of their concise format, the content, in spite of its enormous value, has remained virtually unknown to the scientific community. But by the same token we can also be thankful that the author has had the rare opportunity of being able to let his thoughts mature in peace, to form them in harmony and think them through independently, away from the breathless competitiveness of our time. We do not have a *textbook* written by a host of authors who have worked together to introduce their theories to a wider audience, but rather the creation of one man, an *original work* which, from beginning to end, deals with completely new things [...] We do not believe that we are saying too much when we claim that there are few areas of mathematical science which will not be enriched by the fundamental ideas of this new discipline.

It is probably interesting to note that Engel's role and effort in this massive work were not mentioned here. Maybe the help of a junior author or assistant was taken for granted in the German culture at that time.

In the preface to the third volume, Lie wrote [9, p. 15]:

> For me, Professor Engel occupies a special position. On the initiative of F. Klein and A. Mayer, he traveled to Christiania in 1884 to assist me in the preparation of a coherent description of my theories. He tackled this assignment, the size of which was not known at that time, with the perseverance and skill which typifies a man of his caliber. He has also, during this time, developed a series of important ideas of his own, but has in a most unselfish manner declined to describe them here in any great detail or continuity, satisfying himself with submitting short pieces to *Mathematische Annalen* and, particularly, *Leipziger Berichte*. He has, instead, unceasingly dedicated his talent and free time which his teaching allowed him to spend, to work on the presentation of my theories as fully, as completely and systematically, and above all, as *precisely* as is in any manner possible. For this selfless work which has stretched over a period of nine years, I, and, in my opinion, the entire scientific world owe him the highest gratitude.

Lie and Engel formed a team both in terms of writing and teaching. Some students came to study with both Lie and Engel. Engel also contributed to the success of Lie's teaching. For example, a major portion of students who received the doctoral degree at Leipzig was supervised under Lie. Lie also thanked Engel for this in the preface of the third volume.

But this preface also contained a description of some conflict with Klein, and hence Engel's academic future suffered due to this. See §9 for more detail.

5 Lie's breakdown and a final major result

After his move to Leipzig, Lie worked hard and was very productive. While Leipzig was academically stimulating to Lie, it was not stress-free for him, and relations with others were complicated too. "The pressure of work, problems of collaboration, and domestic anxieties made him sleepless and depressed, and in 1889 he had a complete breakdown" [27, p. 328]. Lie had to go to a nerve clinic and stayed there for seven months. He was given opium, but the treatment was not effective, and he decided to cure the problem himself.[4] He wrote to his friend [26, XXIII]:

> In the end I began to sleep badly and finally did not sleep at all. I had to give up my lecturing and enter a nerve clinic. Unfortunately I have been an impossible patient. It has always been my belief that the doctors did not understand my illness. I have been treated with opium, in enormous dose, to calm my nerves, but it did not help. Also sleeping draughts.
>
> Three to four weeks ago I got tired of staying at the nerve clinic. I decided to try to see what I could do myself to regain my equilibrium and the ability to sleep. I have now done what the doctors say no one can endure, that is to say I have completely stopped taking opium. It has been a great strain. But now, on a couple of days, against the doctors' advice, I have taken some exercise.
>
> I hope now that in a week's time I shall have completely overcome the harmful effects of the opium cure. I think myself that the doctors have only harmed me with opium.
>
> My nerves are very strained, but my body has still retained its horsepower. I shall cure myself on my own. I shall walk from morning to evening (the doctors say it is madness). In this way I shall drive out all the filth of the opium, and afterwards my natural ability to sleep will gradually return. That is my hope.

Finally he thought that he had recovered, and was released. Actually he was not cured at the time of release. Instead, in the reception book of the clinic, his condition at that time was recorded as "a Melancholy not cured" [27, p. 328]. His friends and colleagues found changes in Lie's attitudes towards others and his behaviors: mistrusting and accusing others for stealing his ideas. Indeed, according to Engel [27, p. 397], Lie did recover his mathematical ability, but "not as a human being. His mistrust and irritability did not dissipate, but rather they grew more and more with the years, such that he made life difficult for himself and all his friends. The most painful thing was that he never allowed himself to speak openly about the reasons for his despondency."

When he was busy teaching and working out his results, he did not have much time to pick up new topics. While at the nerve clinic, he worked again on the so-called Helmholtz problem on the axioms of geometry[5] and wrote two papers about it.

---

[4]According to the now accepted theory, Lie suffered from the so-called pernicious anemia psychosis, an incurable disease at that time. People also believe that his soured relations with Klein and others were partly caused by this disease. See the section on the period 1886–1898 in [9] and the reference [29] there.

[5]Lie's work on the Helmholtz problem was apparently well known at the beginning of the 20th century. According to [5, Theorem 16.7], the Lie–Helmholtz Theorem states that spaces of constant curvature, i.e., the Euclidean space, the hyperbolic space and the sphere, can be characterized by abundance of isometries: for every

Lie had thought of and worked on this problem for a long time and had also criticized the work of Helmholtz and complained to Klein about it. According to [27, p. 380–381],

> Very early on, Lie was certainly clear that the transformation theory he was developing was related to non-Euclidean geometry, and in a letter to Mayer as early as 1875, Lie had pointed out that von Helmhotz's work on the axioms of geometry from 1868, were basically and fundamentally an investigation of a class of transformation groups: "I have long assumed this, and finally had it verified by reading his work."
>
> Klein too, in 1883, has asked Lie what he thought of von Helmhotz's geometric work. Lie replied immediately that he found the results correct, but that von Helmhotz operated with a division between the real and imaginary that was hardly appropriate. And a little later, after having studied the treatise more thoroughly, he communicated to Klein that von Helmhotz's work contained "substantial shortcomings", and he thought it positively impossible to overcome these shortcomings by means of the elementary methods that von Helmhotz had applied. Lie went on to complete and simplify von Helmhotz's spatial theory [...]

In 1884, Lie wrote to Klein [26, XXVI]:

> If I ever get as far as to definitely complete my old calculations of all groups and point transformations of a three-dimensional space I shall discuss in more detail Helmholtz's hypothesis concerning metric geometry from a purely analytical aspect.

According to [27, p. 381],

> Lie did further work with von Helmhotz's space problem, and confided to Klein in April 1887, that the earlier works on the problem had now come to a satisfying conclusion – at least when one was addressing finite dimensional transformation groups, and therein, a limited number of parameters. What remained was to deduce some that extended across the board such that infinite-dimensional groups could be included.

Lie's work on the Helmholtz problem led him to being awarded the first Lobatschevsky prize in 1897. Klein wrote a very strong report on his work, and this report was the determining factor for this award.

6 An overview of Lie's major works

As mentioned before, Lie was very productive and he wrote many thousands of pages of papers and multiple books. His name will be forever associated with Lie groups and Lie algebras and several other dozen concepts and definitions in mathematics

two congruently ordered triples of points, there is an isometry of the space that moves one triple to the other, where two ordered triples of points (v_1, v_2, v_3), (v'_1, v'_2, v'_3) are *congruent* if the corresponding distances are equal, $d(v_i, v_j) = d(v'_i, v'_j)$ for all pairs i, j. References to papers of Weyl and Enriques on this theorem were given in [5].

(almost all of them involve Lie groups or Lie algebras in various ways). One natural question is what *exactly Lie had achieved in Lie theory.* The second natural question is: besides Lie groups and Lie algebras, *what else Lie had done.*

It is not easy to read and understand Lie's work due to his writing style. In a preface to a book of translations of some papers of Lie [21] in a book series *Lie Groups: History, Frontiers and Applications*, which contain also some classical books and papers by É. Cartan, Ricci, Levi-Civita and also other more modern ones, Robert Hermann wrote:

> In reading Lie's work in preparation for my commentary on these translations, I was overwhelmed by the richness and beauty of the geometric ideas flowing from Lie's work. Only a small part of this has been absorbed into mainstream mathematics. He thought and wrote in grandiose terms, in a style that has now gone out of fashion, and that would be censored by our scientific journals! The papers translated here and in the succeeding volumes of our translations present Lie in his wildest and greatest form.

We nevertheless try to provide some short summaries. Though articles in Encyclopedia Britannica are targeting the educated public, articles about mathematicians often give fairly good summaries. It might be informative and interesting to take a look at such an article about Lie *before the global theory of Lie groups were developed by Weyl and Cartan.* An article in Encyclopedia Britannica in 1911 summarized Lie's work on Lie theory up to that time:

> Lie's work exercised a great influence on the progress of mathematical science during the later decades of the 19th century. His primary aim has been declared to be the advancement and elaboration of the theory of differential equations, and it was with this end in view that he developed his theory of transformation groups, set forth in his *Theorie der Transformationsgruppen* (3 vols., Leipzig, 1888–1893), a work of wide range and great originality, by which probably his name is best known. A special application of his theory of continuous groups was to the general problem of non-Euclidean geometry. The latter part of the book above mentioned was devoted to a study of the foundations of geometry, considered from the standpoint of B. Riemann and H. von Helmholtz; and he intended to publish a systematic exposition of his geometrical investigations, in conjunction with Dr. G. Scheffers, but only one volume made its appearance (Geometrie der Berührungstransformationen, Leipzig, 1896).

The writer of this article in 1911 might not have imagined the wide scope and multifaceted applications of Lie theory. From what I have read and heard, a list of topics of major work of Lie is as follows:

1. Line complexes. This work of Lie was the foundation of Lie's future work on differential equations and transformation groups, and hence of Lie theory [13]. It also contains the origin of toric varieties.
2. Lie sphere geometry and Lie contact structures. Contact transformations are closely related to contact geometry, which is in many ways an odd-dimensional counterpart of symplectic geometry, and has broad applications in physics. Relatively recently, it was applied to low-dimensional topology.

3. The integration theory of differential equations. This subject has died and recovered in a strong way in connection with integrable systems and hidden symmetries.
4. The theory of transformation groups (or Lie groups). This has had a huge impact through the development, maturing and applications of Lie groups. The theory of transformation groups reached its height in the 1960–1970s. But the theory of Lie groups is becoming more important with the passage of time and will probably stay as long as mathematics is practiced.
5. Infinitesimal transformation groups (or Lie algebras). Lie algebras are simpler than Lie groups and were at first used as tools to understand Lie groups, but they are important in their own right. For example, the infinite-dimensional Kac–Moody Lie algebras are natural generalizations of the usual finite dimensional Lie groups, and their importance and applications are now well-established. Though they also have the corresponding Kac–Moody Lie groups, it is not clear how useful they can be.
6. Substantial contribution to the Erlangen program, which was written and formally proposed by Felix Klein and whose success and influence was partially responsible for the breakup of the friendship between Lie and Klein. Lie contributed to the formulation and also the development of this program, and his role has been recognized more and more by both historians of mathematics and practicing mathematicians.
7. The Helmholtz space problem: determine geometries whose geometric properties are determined by the motion of rigid bodies. See footnote 4. The solution of this problem led Lie to be awarded the Lobachevsky prize. Lie's work on this problem also had a big impact on Poincaré's work on geometry.
8. Minimal surfaces. In 1878, building on the work of Monge on integration of the Euler–Lagrange equations for minimal surfaces, Lie assigned each minimal surface a complex-analytic curve. This was the starting point of a fruitful connection between minimal surfaces and analytic curves. Together with the work of Weierstrass, Riemann, Schwarz, and others, this introduced the wide use of methods and results of complex function theory in the theory of minimal surfaces at the end of the 19th century.

7 Three fundamental theorems of Lie in the Lie theory

When people talk about Lie's work, they often mention three fundamental theorems of Lie. His second and third fundamental theorems are well known and stated in many textbooks on Lie theories. On the other hand, the first fundamental theorem is not mentioned in most books on Lie groups and Lie algebras. The discussion below will explain the reasons:

1. It addresses a basic problem in transformation group theory rather than a problem in abstract Lie theory.
2. It is such a basic result that people often take it for granted.

We will first discuss these theorems in the original setup of transformation groups and later summarize all three theorems in the modern language.

The first theorem says that a local group action on a manifold is determined by the induced vector fields on the manifold. Now the space of vector fields of the manifold forms a Lie algebra. So the study of Lie group actions is reduced to the study of Lie algebras.

This is a deep insight of Lie and is one of the reasons for people to say that *Lie reduced the study of Lie groups to Lie algebras, and hence reduced a nonlinear object to a linear one.*

In the case of a one-parameter group of local diffeomorphisms of a manifold, the action is determined by one vector field on the manifold. Conversely, given a vector field, the existence of the corresponding local solution should have been well known in Lie's time. The proper definition of manifold was not known then, but no notion of manifolds was needed since the action of a Lie group in Lie's work is local and hence can be considered on $\mathbb{R}^n$.

In Lie's statement, the key point is to show how the vector fields on a manifold M associated with a Lie group G is determined by a homomorphism from $\mathfrak{g} = T_e G$ to the space $\chi(M)$ of vector fields on M. (One part of the theorem is that $\mathfrak{g} = T_e G$.)

Lie's second theorem says that given a Lie algebra homomorphism $\mathfrak{g} = T_e G \to \chi(M)$, then there is a local action. One important point is that there is already a Lie group G whose Lie algebra is $\mathfrak{g}$.

One special case of Lie's Fundamental Theorems 1 and 2 is that a one-parameter group of diffeomorphisms φ_t of a manifold M amounts to a vector field X on the manifold. This has two components:

1. The family φ_t induces a vector field X by taking the derivative, and φ_t is uniquely determined by X. The uniqueness follows from the fact that φ_t satisfies an ODE.
2. Given a vector field X, there is an one-parameter family of local diffeomorphisms φ_t which induces X. If M is compact, then the diffeomorphisms φ_t are global. This amounts to integrating a vector field on a manifold into a flow.

The third theorem says that given any abstract Lie algebra $\mathfrak{g}$, and a Lie algebra homomorphism $\mathfrak{g} \to \chi(M)$, then there is a local Lie group (or the germ of a Lie group) G and an action of G on M which induces the homomorphism $\mathfrak{g} \to \chi(M)$.

Lie was interested in Lie group actions. Now people are more interested in the theory of abstract Lie groups and usually reformulate these results in terms of **abstract** Lie groups and Lie algebras.

If we generalize and put these three fundamental theorems in the *modern language* of Lie theory, then they can be stated as follows and can be found in most books on Lie groups and algebras:

1. The **first theorem** should be stated as: *a Lie group homomorphism is determined locally by a Lie algebra homomorphism.*
2. The **second theorem** says that *any Lie group homomorphism induces a Lie algebra homomorphism. Conversely, given a Lie algebra homomorphism, there is a local group homomorphism between corresponding Lie groups. If the domain of the locally defined map is a simply connected Lie group, then there is a global Lie group homomorphism.*
3. The **third theorem** says that *given a Lie algebra* $\mathfrak{g}$, *there exists a Lie group G whose Lie algebra is equal to* $\mathfrak{g}$. (Note that there is no group action here and hence this statement is different from the statements above.)

8 Relation with Klein I: the fruitful cooperation

There are many differences and similarities between Lie and Klein. Lie was a good natured, sincere great mathematician. For example, he gave free lectures in the summer to USA students to prepare them for his later formal lectures. He went out of his way to help his Ph.D. students. He was not formal, and his lectures were not polished and could be messy sometimes.

Klein was a good mathematician with a great vision and he was also a powerful politician in mathematics. He was a noble, strict gentleman. His lectures were always well-organized and polished.

Lie and Klein first met in Berlin in the winter semester of 1869–1870 and they became close friends. It is hard to overestimate the importance of their joint work and discussions on their mathematical works and careers. For example, it was Klein who helped Lie to see the analogy between his work on differential equations and Abel's work on the solvability of algebraic equations, which motivated Lie to develop a general theory of differential equations that is similar to the Galois theory for algebraic equations, which lead to Lie theory. On the other hand, it was Lie who provided substantial evidence to the general ideas in the Erlangen program of Klein that were influential on the development of that program.

Klein also helped to promote Lie's work and career in many ways. For example, when Klein left Leipzig, he secured the vacant chair for Lie in spite of many objections. Klein drafted the recommendation of the Royal Saxon Ministry for Cultural Affairs and Education in Dresden to the Philosophical Faculty of the University of Leipzig, and the comment on Lie run as follows [9, p. 12]:

> Lie is the only one who, by force of personality and in the originality of his thinking, is capable of establishing an independent school of geometry. We received proof of this when Kregel von Sternback's scholarship was to be awarded. We sent a young mathematician – our present Privatdocent, Dr. Engel – to Lie in Chtristiania, from where he returned with a plethora of new ideas.

It is also helpful to quote here a letter written by Weierstrass at that time [9, p. 12]:

> I cannot deny that Lie has produced his share of good work. But neither as a scholar nor as a teacher is he so important that there is a justification in preferring him, a foreigner, to all of those, our countrymen, who are available. It now seems that he is being seen as a second Abel who must be secured at any cost.

One particular fellow countryman Weierstrass had in mind was his former student Hermann Schwarz, who was also a great mathematician.

Another crucial contribution of Klein to Lie's career was to send Engel to help Lie to write up his deep work on transformation groups. Without Engel, Lie's contributions might not have been so well known and hence might not have had the huge impacts on mathematics and physics that they have now. It is perhaps sad to note that Engel was punished by Klein in some way for being a co-author of Lie after the breakup between Lie and Klein. One further twist was that Klein made Engel edit Lie's collected works carefully after Lie passed away.

9 Relation with Klein II: conflicts and the famous preface

The breakup between Lie and Klein is famous for one sentence Lie put down in the preface of the third volume of his joint book with Engel on Lie transformation groups published in 1893: "*I am not a student of Klein, nor is the opposite the case, even if it perhaps comes closer to the truth.*"

This is usually the only sentence that people quote and say. It sounds quite strong and surprising, but there are some reasons behind it. The issue is about the formulation and credit of ideas in the Erlangen program, which was already famous at that time. It might be helpful to quote more from the foreword of Lie [9, p. 19]:

> F. Klein, whom I kept abreast of all my ideas during these years, was occasioned to develop similar viewpoints for discontinuous groups. In his Erlangen Program, where he reports on his and on my ideas, he, in addition, talks about groups which, according to my terminology, are neither continuous or discontinuous. For example, he speaks of the group of all Cremona transformations and of the group of distortions. The fact that there is an essential difference between these types of groups and the groups which I have called continuous (given the fact that my continuous groups can be defined with the help of differential equations) is something that has apparently escaped him. Also, there is almost no mention of the important concept of a differential invariant in Klein's program. Klein shares no credit for this concept, upon which a general invariant theory can be built, and it was from me that he learned that each and every group defined by differential equations determines differential invariants which can be found through integration of complete systems.
>
> I feel these remarks are called for since Klein's students and friends have repeatedly represented the relationship between his work and my work wrongly. Moreover, some remarks which have accompanied the new editions of Klein's interesting program (so far, in four different journals) could be taken the wrong way. I am no

student of Klein and neither is the opposite the case, though the latter might be closer to the truth.

By saying all this, of course, I do not mean to criticize Klein's original work in the theory of algebraic equations and function theory. I regard Klein's talent highly and will never forget the sympathetic interest he has taken in my research endeavors. Nonetheless, I don't believe he distinguishes sufficiently between induction and proof, between a concept and its use.

According to [27, p. 317], in the same preface,

> Lie's assertion was that Klein did not clearly distinguish between the type of groups which were presented in the Erlangen Programme – for example, Cremonian transformations and the group of rotations, which in Lie's terminology were neither continuous nor discontinuous – and the groups Lie had later defined with the help of differential equations:
>
> "One finds almost no sign of the important concept of differential invariants in Klein's programme. This concept, which first of all a common invariant theory could be build upon, is something Klein has no part of, and he has learned from me that every group that it defined by means of differential equations, determines differential invariants, which can be found through the integration of integrable systems."
>
> ... Lie continued, in their investigations of geometry's foundation, Klein, as well as von Helmhotz, de Tilly, Lindemann, and Killing, all committed gross errors, and this could largely be put down to their lack of knowledge of group theory.

Maybe some explanations are in order to shed more light on these strong words of Lie. According to [26, pp. XXIII–XXIV],

> Sophus Lie gradually discovered that Felix Klein's support for his mathematical work no longer conformed with his own interests, and the relationship between the two friends became more reserved. When, in 1892, Felix Klein wanted to republish the *Erlangen program* and explain its history, he sent the manuscript to Sophus Lie for a comment. Sophus Lie was dismayed when he saw what Felix Klein has written, and got the impression that his friend now wanted to have his share of what Sophus Lie regarded as his life's work. To make things quite clear, he asked Felix Klein to let him borrow the letters he had sent him before the *Erlangen program* was written. When he learned that these letters no longer existed, Sophus Lie wrote to Felix Klein, November 1892.

The letter from Lie to Klein in November 1892 goes as follows [9, p. XXIV]:

> I am reading through your manuscript very thoroughly. In the first place, I am afraid that you, on your part, will not succeed in producing a presentation that I can accept as correct. Even several points which I have already criticized sharply are incorrect, or at least misleading, in your current presentation. I shall try as far as possible to concentrate my criticism on specific points. If we do not succeed in reaching agreement, I think that it is only right and reasonable that we each present our views independently, and the mathematical public can then form their own opinion.
>
> For the time being I can only say how sorry I am that you were capable of burning my so significant letters. In my eyes this was vandalism; I had received your specific promise that you would take care of them.

> I have already told you that my period of naiveness is now over. Even if I still firmly retain good memories from the years 1869–1872, I shall nevertheless try to keep myself that which I regard as my own. It seems that you sometimes believe that you have shared my ideas by having made use of them.

The comprehensive biography of Lie [27, p. 371] gives other details on the origin of this conflict:

> The relationship between them [Lie and Klein] had certainly cooled over recent years, although they continued to exchange letters the same way, although not as frequently as earlier. But it was above all professional divergencies that were central to the fact that Lie now broke off relationship. Following Lie's publication of the first volume of his *Theorie der Transformationsgruppen*, Klein judged that there was sufficient interest to have his Erlangen Programme been republished. But before Klein's text from 1872 was printed anew, Klein had contacted Lie to find out how the working relationship and exchange of ideas between them twenty years earlier should be presented. Lie had made violent objections to the way in which Klein had planned to portray the ideas and the work. But Klein's Erlangen Programme was printed, and it came out in four different journals, in German, Italian, English and French – without taking into account Lie' commentary on his assistance in formulating this twenty-year-old programme. More and more in mathematical circles, Klein's Erlangen Programme was spoken of as central to the paradigm shift in geometry that occurred in the previous generation. A large part of the third volume of Lie's great work on transformation groups was devoted to a deepening discussion on the hypotheses or axioms that ought to be set down as fundamental to a geometry, that – whether or not it accepted Euclid's postulates – satisfactorily clarified classical geometry as well as the non-Euclidean geometry of Gauss, Lobachevsky, Bolyai, and Riemann.
>
> The information that spread regarding the relationship between Klein's and Lie's respective work, was, according to Lie, both wrong and misleading. Lie considered he had been side-lined but was eager to "set things right", and grasped the first and best opportunity. In front of the professional substance of his work he placed his twenty-page foreword. The power-charge that liquidated their friendship and sent shock-waves through the mathematical milieu was short, if not sweet.

Klein was the king of German mathematics and probably also of the European mathematics at that time. What was people's reaction to the strong preface of Lie? Maybe a letter from Hilbert to Klein in 1893 will explain this [26, p. XXV]: "In his third volume, his megalomania spouts like flames."

Lie probably did not suffer too much professionally from this conflict with Klein since he had the chair at Leipzig. But this was not the case with Engel. Since Engel's name also appeared on the book, he had to pay for this. Engel was looking for a job, and a position of professorship was open at that time at the University of Königsberg, the hometown and home institution of Hilbert where he held a chair in mathematics, and this open position was a natural and likely choice for Engel. In the same letter to Klein, Hilbert continued [26, p. XXV], "I have excluded Engel completely. Although he himself has not made any comment in the preface, I hold him to some extent co-responsible for the incomprehensible and totally useless personal animosity which the third volume of Lie's work on transformation groups is full off."

Engel could not get an academic job for several years,[6] and Klein arranged Engel to edit the collected works of Grassmann and then later the collected works of Lie; on the latter he worked for several decades.

Another consequence of this conflict with Klein was that Lie could not finish another proposed joint book with Engel on applications of transformation groups to differential equations. According to [27, p. 390–391], after the publication of these three volumes,

> The next task that Lie saw for himself was to make refinements and applications of what had now been completely formulated. But this foreword [of the third volume] with its sharp accusations against Klein, caused hindrances to the further work. Because Lie in the same foreword had praised Engel to the skies for his "exact" and "unselfish activity", it now became difficult for Engel to continue to collaborate with Lie – consequently as well, nothing came of the announced work on, among other things, differential invariants and infinite-dimensional continuous groups. As for Engel, his career outlook certainly now lay in other directions than Lie's. According to Lie's German student, Gerhard Kowalewski, relations between Lie and Engel gradually became so cool that they were seldom to be seen in the same place.

It should be pointed out that relations between Lie and Engel had some hard time before this foreword came out. It was caused by the fight between Lie and Killing due to some overlap in their work on Lie theories, in particular, Lie algebras. For some reason, at the initial stage, Killing communicated with Engel and cited some papers of Engel instead of Lie's, and Lie felt than Engel betrayed his trust. For a more detailed discussion on this issue, see [27, pp. 382–385, p. 395].[7] After Lie's death, Engel continued and carried out his mentor's work in several ways. See §11 for example. He was a faithful disciple and was justly awarded with the Norwegian Order of St Olaf and an honorary doctorate from the University of Oslo.

Maybe there is one contributing factor to these conflicts.[8] It is the intrinsic madness of *all people* who are devoted to research and are doing original work, in partic-

---

[6]On the other hand, all things ended well with Engel. In 1904, he accepted the chair of mathematics at the University of Greifswald when his friend Eduard Study resigned, and in 1913, took the chair of mathematics at the University of Giessen. Engel also received a Lobachevsky Gold Medal. The Lobachevsky medal is different from the Lobatschevsky prize won by his mentor Lie and his fellow countryman Wilhelm Killing. The medal was given on a few occasions to the referee of a person nominated for the prize. For instance, Klein also received, in 1897, a the Gold medal, for his report on Lie. See [28].

[7]Manfred Karbe pointed out that in his autobiography [20, pp. 51–52], Kowalewski speculates about the mounting alienation of Lie and Engel, and reports about Lie's dislike of his three-volume Transformation Groups. When Lie needed some material of his own in his lectures or seminars, he never made use of these books but only of the papers in Math. Ann. And Kowalewski continues on page 52, line -5:

"Von hier aus kann man es vielleicht verstehen, dass die Abneigung gegen das Buch sich auf den Mitarbeiter bertrug, dem er doch so sehr zu Dank verpflichtet war." (From this one may perhaps understand that the aversion to the book is transferred to the collaborator to whom he was so much indebted.)

[8]There has been an explanation of Lie's behavior in this conflict with Klein by establishing a relation between genius and madness. According to [27, p. 394], after Lie's death, "In Göttingen, Klein made a speech that gave rise to much rumor, not least because here, in addition to all his praise for his old friend, Klein suggested that the close relationship between genius and madness, and that Lie had certainly been struck by a mental condition that was tinged with a persecution complex – at least, by assessing the point from notes that Klein made for his speech, it seems that this was the expression he used."

ular mathematicians and scientists. According to a comment of Lie's nephew, Johan Vogt, a professor at the University of Oslo in economics, and also a translator, writer and editor, made in 1930 on his uncle [27, p. 397],

> We shall avail ourselves of a popular picture. Every person has within himself some normality and some of what may be called madness. I believe that most of my colleagues possess ninety-eight percent normality and two percent madness. But Sophus Lie certainly had appreciably more of the latter. The merging of a pronounced scientific gift and an impulsiveness that verges on the uncontrollable, would certainly describe many of the greatest mathematicians. In Sophus Lie this combination was starkly evident.

It might be helpful to point out that later at the request of the committee of Lobatschevsky prize, Klein wrote a very strong report about the important work of Lie contained in this third volume on transformation groups, and this report was instrumental in securing the inaugural Lobatschevsky prize for Lie.

It might also be helpful to quote from Klein on Lie's work related to this conflict. The following quote of Klein [19, pp. 352–353], its translation and information about it were kindly provided by Hubert Goenner:

> I will now add a personal remark. The already mentioned Erlangen Program is about an outlook which – as already stated in the program itself – I developed in personal communication with Lie (now professor in Leipzig, before in Christiana). Lie who has been engaged particularly with transformation groups, created a whole theory of them, which finds its account in a larger œuvre "Theory of Transformation Groups", edited by Lie and Engel, Vol. I 1888, Vol. II 1890. In addition, a third volume will appear, supposedly in not all too distant a time. Obviously, we cannot think about responding now to the contents of Lie's theories [...]. My remark is limited to having called attention to Lie's theories.

The above comment was made by Klein in the winter of 1889 or at the beginning of 1890, but Klein backed its publication until 1893, the year of the ill-famed preface to the third volume by Lie and Engel.

Further details about this unfortunate conflict and the final reconciliation between these two old friends are also given in [27, pp. 384–394]. See also the article [18] for more information about Klein and on some related discussion on the relationship between Lie and Klein.

The above discussion showed that the success of the Erlangen program was one cause for the conflict between Lie and Klein. A natural question is how historians of mathematics have viewed this issue. Given the fame and impacts of the Erlangen program, it is not surprising that there have been many historical papers about it. Two papers [15] [3] present very different views on the contributions of Klein and Lie to the success and impacts of this program. The paper [15] argues convincingly that Lie's work in the period 1872–1892 made the Erlangen program a solid program with substantial results, while the paper [3] was written to dispute this point of view. It seems that the authors are talking about slightly different things. For example, [3] explains the influence of Klein and the later contribution of Study, Cartan and Weyl, but most of their contributions were made after 1890. The analysis of the situation in [10, p. 550] seems to be fair and reasonable:

> It seems that the Erlangen Program met with a slow reception until the 1890s, by which time Klein's status as a major mathematician at the University of Göttingen had a great deal to do with its successful re-launch. By that time too a number of mathematicians had done considerable work broadly in the spirit of the programme, although the extent to which they were influenced by the programme, or were even aware of it, is not at all clear [...] Since 1872 Lie had gone on to build up a vast theory of groups of continuous transformations of various kinds; but however much it owned to the early experiences with Klein, and however much Klein may have assisted Lie in achieving a major professorship at Leipzig University in 1886, it is doubtful if the Erlangen Program had guided Lie's thoughts. Lie was far too powerful and original a mathematician for that.

10 Relations with others

As mentioned, both Klein and Engel played crucial roles in the academic life of Lie. Another important person to Lie is Georg Scheffers, who obtained his Ph.D. in 1890 under Lie. Lie thought highly of Scheffers. In a letter to Mittag–Leffler [27, p. 369], Lie wrote "One of my best pupils (Scheffers) is sending you a work, which he has prepared before my eyes, and who has taken his doctorate here in Leipzig with a dissertation that got the best mark ... Scheffers possesses an unusually evident talent and his calculations are worked through with great precision, and bring new results."

After the collaboration between Lie and Engel unfortunately broke off, Scheffers substituted for Engel and edited two of Lie's lecture notes in the early 1890s. They are *Lectures on differential equations with known infinitesimal transformations* of 568 pages, and *Lectures on Continuous groups* of 810 pages. Later in 1896, they also wrote a book together, *Geometry of contact transformations* of 694 pages. All these book projects of Lie with others indicate that Lie might not have been able to efficiently write up books by himself. For example, he only wrote by himself a book of 146 pages and a program for a course in Christiania in 1878.

In 1896 Scheffers became docent at the Technical University of Darmstadt, where he was promoted to professor in 1900. The collaboration with Lie stopped after this move. From 1907 to 1935, when he retired, Scheffers was a professor at the Technical University of Berlin.

According to a prominent American mathematician, G. A. Miller, from the end of the nineteenth century, "The trait of Lie's character which impressed me most forcibly when I first met him in the summer of 1895 was his extreme openness and lack of effort to hide ignorance on any subject."

Though he was motivated by discontinuous groups (or rather finite groups) taught by Sylow and kept on studying a classical book on finite substitution groups by Jordan, he could never command the theory of finite groups. Miller continued, "In fact he frequently remarked during his lectures that he always got stuck when he entered upon the subject of discontinuous groups."

When Lie first arrived in Leipzig, teaching was a challenge for him due to both lack of students and the amount of time needed for preparation. In a letter to a friend from the youth, Lie wrote [26, p. XXI], "While, in Norway, I hardly spent five minutes a day on preparing the lectures, in Germany I had to spend an average of about 3 hours. The language is always a problem, and above all, the competition implies that I had to deliver 8–10 lectures a week."

When Lie and his assistant Engel decided to present their own research on transformation groups, students from all over the world poured in, and the Ecole Normale sent its best students to study with Lie. It was a big success. According to the recollection of a student of Lie [26, p. XXVII]:

> Lie liked to teach, especially when the subject was his own ideas. He had vivid contact with his students, who included many Americans, but also Frenchmen, Russians, Serbs and Greeks. It was his custom to ask us questions during the lectures and he usually addressed each of us by name.
>
> Lie never wore a tie. His full beard covered the place where the tie would have been, so even the most splendid tie would not have shown to advantage. At the start of a lecture he would take off his collar with a deft movement, saying: "I love to be free", and he would then begin his lecture with the words "Gentlemen, be kind enough to show me your notes, to help me remember what I did last time." Someone or other on the front bench would immediately stand up and hold out the open notebook, whereupon Lie, with a satisfied nod of the bead, would say "Yes, now I remember." In the case of difficult problems, especially those referring to Lie's complex integration theories, it could happen that the great master, who naturally spoke without any kind of preparation whatsoever, got into difficulties and, as the saying goes, became stuck. He would then ask one of his elite students for help.

Lie had many students. Probably one of the most famous was Felix Hausdorff. Lie tried to convince Hausdorff to work with him on differential equations of the first order without success. Of course, Hausdorff became most famous for his work on topology. See [27, p. 392–393] for a description of Hausdorff and his interaction with Lie.

Throughout his life, Lie often felt that he was under-recognized and under-appreciated. This might be explained by his late start in mathematics and his early isolation in Norway. He paid careful attention to other people's reaction to his work. For example, Lie wrote about Darboux in a letter to Klein [25] in October 1882:

> Darboux has studied my work with remarkable thoroughness. This is good insofar as he has given gradually more lectures on my theories at the Sorbonne, for example on line and sphere geometry, contact transformation, and first-order partial differential equations. The trouble is that he continually plunders my work. He makes inessential changes and then publishes these without mentioning my name.

11 Collected works of Lie: editing, commentaries and publication

Since Lie theories are so well known and there are many books on different aspects of Lie groups and Lie algebras, Lie's collected works are not so well known to general mathematicians and students. The editing and publication of Lie's collected works are both valuable and interesting to some people. In view of this, we include some relevant comments.

Due to his death at a relatively young age, the task of editing Lie's work completely fell on others. It turned out that editing and printing the collected work of Sophus Lie was highly nontrivial and a huge financial burden on the publisher. The situation is well explained in [6]:

> Twenty-three years after the death of Sophus Lie appears the first volume to be printed of his collected memoirs. It is not that nothing has been done in the meantime towards making his work more readily available. A consideration of the matter was taken up soon after his death but dropped owing to the difficulties in the way of printing so large a collection as his memoirs will make. An early and unsuccessful effort to launch the enterprise was made by the officers of Videnskapsselskapet i Kristiania; but plans did not take a definite form till 1912; then through the Mathematisch-physische Klasse der Leipziger Akademie and the publishing firm of B. G. Teubner steps were taken to launch the project. Teubner presented a plan for raising money by subscription to cover a part of the cost of the work and a little later invitations to subscribe were sent out. The responses were at first not encouraging; from Norway, the homeland of Lie, only three subscriptions were obtained in response to the first invitations.
>
> In these circumstances, Engel, who was pressing the undertaking, resorted to an unusual means. He asked the help of the daily press of Norway. On March 9, 1913, the newspaper TIDENS TEGN of Christiania carried a short article by Engel with the title *Sophus Lies samlede Afhandlinger* in which was emphasized the failure of Lie's homeland to respond with assistance in the work of printing his collected memoirs. This attracted the attention of the editor and he took up the campaign: two important results came from this, namely, a list of subscriptions from Norway to support the undertaking and an appropriation by the Storthing to assist in the work. By June the amount of support received and promised was sufficient to cause Teubner to announce that the work could be undertaken; and in November the memoirs for the first volume were sent to the printer, the notes and supplementary matter to be supplied later.
>
> The Great War so interfered with the undertaking that it could not be continued, and by the close of the war circumstances were so altered that the work could not proceed on the basis of the original subscriptions and understandings and new means for continuing the work had to be sought. Up to this time the work had been under the charge of Engel as editor. But it now became apparent that the publication of the memoirs would have to become a Norwegian undertaking. Accordingly, Poul Heegaard became associated with Engel as an editor. The printing of the work became an enterprise not of the publishers but of the societies which support them in this undertaking. Under such circumstances the third volume of the series, but the first one to be printed, has now been put into our hands. "The printing of further vol-

> umes will be carried through gradually as the necessary means are procured; more I cannot say about it," says Engel, "because the cost of printing continues to mount incessantly."

The first volume was published in 1922, and the sixth volume was published in 1937. The seventh volume consisted of some unpublished papers of Lie and was published only in 1960 due to the World War II and other issues. This was certainly a major collected work in the last 100 years.

The collected works of Lie are very well done with the utmost dedication and respect thanks to the efforts of Engel and Heegaard. This can be seen in the editor's introduction to volume VI of Lie's Collected Works,

> If one should go through the whole history of mathematics, I believe that he will not find a second case where, from a few general thoughts, which at first sight do not appear promising, has been developed so extensive and wide-reaching a theory. Considered as an edifice of thought Lie's theory is a work of art which must stir up admiration and astonishment in every mathematician who penetrates it deeply. This work of art appears to me to be a production in every way comparable with that [...] of a Beethoven [...] It is therefore entirely comprehensible if Lie [...] was embittered that 'deren Wesen, ja Existenz, den Mathematikern fort-während unbekannt zu sein scheint' (p. 680). This deplorable situation, which Lie himself felt so keenly, exists no longer, at least in Germany. In order to do whatever lies in my power to improve the situation still further, [...] I have sought to clarify all the individual matters (Einzelheiten) and all the brief suggestions in these memoirs.

Each volume contains a substantial amount of notes, commentaries and supplementary material such as letters of Lie, and "This additional material has been prepared with great care and with the convenience of the reader always in mind." For example, as mentioned before, Lie's first paper was only 8 pages long, but the commentary consisted of over 100 pages. According to [4],

> Although Engel was himself an important and productive mathematician he has found his place in the history of mathematics mainly because he was the closest student and the indispensable assistant of a greater figure: Sophus Lie, after N. H. Abel the greatest Norwegian mathematician. Lie was not capable of giving to the ideas that flowed inexhaustibly from his geometrical intuition the overall coherence and precise analytical form they needed in order to become accessible to the mathematical world [...] Lie's peculiar nature made it necessary for his works to be elucidated by one who knew them intimately and thus Engel's "Annotations" completed in scope with the text itself.

Acknowledgments. I would like to thank Athanase Papadopoulos for carefully reading preliminary versions of this article and his help with references on the Lie–Helmholtz Theorem, and Hubert Goenner for several constructive and critical suggestions and references. I would also like to thank Manfred Karbe for the reference [20] and a summary of a possible explanation in [20] about the alienation between Lie and Engel.

Bibliography

[1] M. A. Akivis and B. A. Rosenfeld, *Elie Cartan (1869–1951)*. Translated from the Russian manuscript by V. V. Goldberg. Translations of Mathematical Monographs, 123. American Mathematical Society, Providence, RI, 1993.

[2] N. Baas, Sophus Lie. *Math. Intelligencer* 16 (1994), no. 1, 16–19.

[3] G. Birkhoff and M. K. Bennett, Felix Klein and his "Erlanger Programm". *History and philosophy of modern mathematics* (Minneapolis, MN, 1985), 145–176, Minnesota Stud. Philos. Sci., XI, Univ. Minnesota Press, Minneapolis, MN, 1988.

[4] H. Boerner, *Friedrich Engel, Complete Dictionary of Scientific Biography*. Vol. 4. Detroit: Charles Scribner's Sons, 2008. p. 370–371.

[5] H. Busemann, Local metric geometry. *Trans. Amer. Math. Soc.* 56 (1944). 200–274.

[6] R. Carmichael, Book Review: Sophus Lie's Gesammelte Abhandlungen. *Bull. Amer. Math. Soc.* 29 (1923), no. 8, 367–369; 31 (1925), no. 9–10, 559–560; 34 (1928), no. 3, 369–370; 36 (1930), no. 5, 337.

[7] S. S. Chern, Sophus Lie and differential geometry. *The Sophus Lie Memorial Conference* (Oslo, 1992), 129–137, Scand. Univ. Press, Oslo, 1994.

[8] G. Darboux, Sophus Lie. *Bull. Amer. Math. Soc.* 5 (1899), 367–370.

[9] B. Fritzsche, Sophus Lie: a sketch of his life and work. *J. Lie Theory* 9 (1999), no. 1, 1–38.

[10] J. Gray, Felix Klein's Erlangen Program, 'Comparative considerations of recent geometrical researches'. In *Landmark Writings in Western Mathematics 1640–1940*, Elsevier, 2005, 544–552.

[11] J. Gribbin, *The Scientists: A History of Science Told Through the Lives of Its Greatest Inventors*. Random House Trade Paperbacks, 2004.

[12] G. Halsted, Sophus Lie. *Amer. Math. Monthly* 6 (1899), no. 4, 97–101.

[13] T. Hawkins, The birth of Lie's theory of groups. *The Sophus Lie Memorial Conference* (Oslo, 1992), 23–50, Scand. Univ. Press, Oslo, 1994.

[14] T. Hawkins, *Emergence of the theory of Lie groups. An essay in the history of mathematics 1869–1926*. Sources and Studies in the History of Mathematics and Physical Sciences. Springer-Verlag, New York, 2000.

[15] T. Hawkins, The Erlanger Programm of Felix Klein: reflections on its place in the history of mathematics. *Historia Math.* 11 (1984), no. 4, 442–470.

[16] S. Helgason, Sophus Lie, the mathematician. *The Sophus Lie Memorial Conference* (Oslo, 1992), 3–21, Scand. Univ. Press, Oslo, 1994.

[17] L. Ji, A summary of topics related to group actions. *Handbook of group actions*, ed. L. Ji, A. Papadopoulos and S.-T. Yau, Vol.1, Higher Education Press, Beijing and International Press, Boston, 2015, 33–187.

[18] L. Ji, Felix Klein: his life and mathematics. In *Sophus Lie and Felix Klein: The Erlangen program and its impact in mathematics and physics*, ed. L. Ji and A. Papadopoulos, European Mathematical Society Publishing House, Zürich, 2015, 27–58.

[19] F. Klein, *Nicht-Euklidische Geometrie, I. Vorlesung, Wintersemester 1889/90*. Ausgearbeitet von F. Schilling, Göttingen, 1893.

[20] G. Kowalewski, *Bestand und Wandel. Meine Lebenserinnerungen zugleich ein Beitrag zur neueren Geschichte der Mathematik.* Verlag Oldenbourg, München 1950.

[21] S. Lie, Sophus Lie's 1880 transformation group paper. In part a translation of "Theorie der Transformations-gruppen"'by S. Lie [*Math. Ann.* 16 (1880), 441–528]. Translated by Michael Ackerman. Comments by Robert Hermann. Lie Groups: History, Frontiers and Applications, Vol. I. Math. Sci. Press, Brookline, Mass., 1975.

[22] R. Merton, *On the Shoulders of Giants: A Shandean Postscript*. University of Chicago Press, 1993.

[23] G. Miller, Some reminiscences in regard to Sophus Lie. *Amer. Math. Monthly* 6 (1899), no. 8–9, 191–193.

[24] C. Reid, *Hilbert*. Reprint of the 1970 original. Copernicus, New York, 1996.

[25] D. Rowe, Three letters from Sophus Lie to Felix Klein on Parisian mathematics during the early 1880s. Translated from the German by the author. *Math. Intelligencer* 7 (1985), no. 3, 74–77.

[26] E. Strom, Sophus Lie. *The Sophus Lie Memorial Conference* (Oslo, 1992), Scand. Univ. Press, Oslo, 1994.

[27] A. Stubhaug, *The mathematician Sophus Lie. It was the audacity of my thinking.* Translated from the 2000 Norwegian original by Richard H. Daly. Springer-Verlag, Berlin, 2002.

[28] A. Vassilief, Prox Lobachevsky (premier concours), *Nouv. Ann. Math.*, 3e sér. 17 (1898), 137–139.

Chapter 2

Felix Klein: his life and mathematics

Lizhen Ji

Contents

1 Introduction

Felix Klein was not only one of the great mathematicians but also one of the great educators and scientific writers of the nineteenth and the early twentieth century. He was a natural born leader and had a global vision for mathematics, mathematics education and mathematical development. He had the required ambition, drive and ability to remove obstacles on his way and to carry out his plans. He was a benign and noble dictator, and was the most kingly mathematician in the history of mathematics. He had a great deal of influence on German mathematics and the world mathematics community. Indeed, it was Klein who turned Göttingen into the leading center of mathematics in the world.

I have heard of Felix Klein for a long time but have not really tried to look up information about him. On the other hand, when I became interested in him and wanted to learn more about him, for example, the exact mathematical content of his competition with Poincaré, his relation with Lie and their conflicts, and his contribution towards the Erlangen program, there was no biography of Klein in English which was easily available. Later we found some extensive writings about him in obituary notices [18], and in books [13] [12]. After reading various sources about Klein which were available to me in English, I found him an even more interesting person than I thought. One purpose in writing this chapter was to share with the reader a summary of what I read about this incredible mathematician with some emphasis on his lectures in the USA around 1893 and their influence on the development of the American mathematics community, and some thoughts which occurred to me during this reading process. Another purpose is to give a brief outline of the rich life of Klein and to supplement other more scientific papers in this book.[1]

Probably, Klein is famous to different groups of people for different reasons. For example, for people working in Lie theories and geometry, the Erlangen program he proposed had far-reaching consequences; for people working on discrete subgroups of Lie groups and automorphic forms, Kleinian groups and his famous books with Fricke on discrete groups of linear fractional transformations (or Möbius transformations) and automorphic forms have had a huge impact; many other people enjoy his books on elementary mathematics and history of mathematics, and of course high school and college teachers also appreciate his books and points of view on education. A special feature of Klein is his broad command of and vision towards mathematics. We will discuss more on this last point from different aspects.

In terms of mathematical contribution, Klein was probably not of the same rank as his predecessors Gauss or Riemann and his younger colleagues Hilbert and Weyl in Göttingen, but he also made real contributions and was a kingly mathematician, more kingly than any of the others mentioned above. He could command respect from others like a king or even like a god. In some sense, Klein *earned* this. On the influence of Klein on the mathematics in Göttingen, Weyl [7, p. 228] said that "Klein ruled mathematics there like a god, but his godlike power came from the force of his personality, his dedication, and willingness to work, and his ability to get things done."

But kings are kings and can be tough and remote from the ordinary people. In 1922, the eminent analyst Kurt O. Friedrichs was young and visited Klein [7, p. 229]: "I was amazed, simply swept off my feet by Klein's grace and charm. ... He could be very charming and gentlemanly when all went his way but with anyone who crossed him he was a tyrant."

Klein made important original contributions to mathematics early in his career. But his research was cut short due to exhaustion and the nervous breakdown caused by his competition with Poincaré on Fuchsian groups and the uniformization of Riemann surfaces.[2] He fell down but picked it up. Very few people can do this. Since he could not do mathematical research anymore, he devoted his time and energy to

[1] Most mathematicians have heard of Klein and have been influenced by his mathematics. On the other hand, various writings and stories about him are scattered in the literature, and we feel that putting several snapshots from various angles could convey a vague global picture of the man and his mathematics. It might be helpful to quote [18, p. i]: "After his death there appeared, one after another, a number of sketches of the man and his work from the pens of many of his pupils. But, just as a photograph of a man of unusual personality, or of a place of striking beauty, conveys little to one not personally acquainted with the original, so it is, and so it must be, with these sketches of Klein." For the reader now, these many sketches are probably not easily accessible.

[2] It is interesting to see a different explanation of the breakdown of Klein in [18, p. vii]: "His breakdown was probably accelerated by the antagonism he experienced at Leipzig. He was much younger than his colleagues, and they resented his innovating tendencies. In particular, there was opposition to his determination to avail himself of the vaunted German "Lehrfreiheit", and to interpret the word "Geometry" in its widest sense, beginning his lectures with a course on the Geometric Theory of Functions."

education and mathematical writing, and more importantly to building and providing a stimulating environment for others. For example, he brought Hilbert to Göttingen and turned this small college city into the leading center of mathematics, which attracted people from all over the world. His lecture tours in the USA near the end of the nineteenth century also played a pivotal role in the emergence of the US mathematics community as one of the leading ones in the world. In this sense, Klein was also a noble mathematician and had a far-reaching and long-lasting impact.

His approach to mathematics emphasizes the big picture and connections between different subjects without paying too much attention to details or work to substantiate his vision. He valued results and methods which can be applied to a broad range of topics and problems. For example, he proposed the famous Erlangen program but did not really work on it to substantiate it. Instead, his friend Lie worked to make it an important concrete program. As Courant once commented, Klein tended to soar above the terrain that occupied ordinary mathematicians, taking in and enjoying the vast view of mathematics, but it was often difficult for him to land and to do the hard boring work. He had no patience for thorny problems that require difficult technical arguments. What counted for him was the big pictures and the general pattern behind seemingly unrelated results. According to Courant [13, p. 179], Klein had certainly understood "that his most splendid scientific creations were fundamentally gigantic sketches, the completion of which he had to leave to other hands."

Klein was a master writer and speaker. In some sense, he was a very good salesman of mathematics, but he drew a lot of criticism for this. Because of this, he had a lot of influence on mathematics and the mathematics community. He is a unique combination of an excellent mathematician, master teacher and efficient organizer.

People's responses to these features of Klein were not all positive. According to a letter from Mittag-Leffler to Hermite in 1881 [7, p. 224]:

> You asked me what are the relations between Klein and the great Berliners ... Weierstrass finds that Klein is not lacking in talent but is very superficial and even sometimes rather a charlatan. Kronecker finds that he is quite simply a charlatan without real merit. I believe that is also the opinion of Kummer.

In 1892, when the faculty of Berlin University discussed the successor to Weierstrass, they rejected Klein as a "dazzling charlatan" and a "complier." His longtime friend Lie's opinion was also harsh but more concrete [7, p. 224]:

> I rank Klein's talents highly and shall never forget the ready sympathy with which he always accompanied me in my scientific attempts; but I opine that he does not, for instance, sufficiently distinguish induction from proof, and the introduction of a concept from its exploitation.

In spite of all these criticisms and reservations, Klein was the man who was largely responsible for restoring Göttingen's former luster and hence for initiating a process that transformed the whole structure of mathematics at German universities and also in some other parts of the world. Among people from Europe, he had the most important influence on the emergence of mathematics in the USA. There is no question of the fact as to he was the most dynamic mathematical figure in the world in the last quarter of the nineteenth century.

Klein represented an ideal German scholar in the nineteenth century. He strove for and attained an extraordinary breadth of knowledge, much of which he acquired in his active interaction with other mathematicians. He also freely shared his insights and knowledge with his students and younger colleagues, attracting people towards him from around the world.

Unlike most mathematicians who only affected certain parts of mathematics and the mathematics community, the influence of Klein is also global and foundational. The legacy of Klein lives on as it is witnessed in the popularity of his books, the continuing and far-reaching influence of the general philosophy of the Erlangen program, interaction between mathematics and physics, and the theory of Kleinian groups. In some sense, Klein did not belong to his generation but was ahead of time.

2 A nontrivial birth

It is customary that kings are born at special times and places. Maybe their destiny gives them something extra to start with. The mostly kingly mathematician, Felix Klein, was born in the night of April 25, 1849, in Düsseldorf in the Rhineland when [18, p. i]

> there was anxiety in the house of the secretary to the Regierungspräsident. Without, the canon thundered on the barricades raised by the insurgent Rhinelanders against their hated Prussian rulers. Within, although all had been prepared for flight, there was no thought of departure; [...] His birth was marked by the final crushing of the revolution of 1848; his life measured the domination of Prussia over Germany, typifies all that was best and nobest in that domination.

Shortly after his birth, his hometown and the nearby region became the battleground of the last war of the 1848 Revolution in the German states.

In the twenty years that followed Klein's birth, Prussia became a major power in Europe, and there were almost constantly conflicts and turmoils, culminating in the Franco–Prussian war, with a crushing victory over France.

Later Klein served in the voluntary corps of emergency workers and he witnessed firsthand the battle sites of Metz and Sedan, where the Empire of Louis-Napoléon Bonaparte (Napoléon III) finally collapsed and was replaced by the Third Republic. In Germany the Second German Empire began, with Otto von Bismarck as a powerful first chancellor.[3]

[3]Otto von Bismarck was a conservative German statesman who dominated European affairs from the 1860s to his dismissal in 1890. After a series of short victorious wars he unified numerous German states into a powerful German Empire under Prussian leadership, then created a "balance of power" that preserved peace in Europe from 1871 until 1914.

Klein's academic life practically coincided with the rise and fall of the second Reich. All these historical events influenced Klein's character and his perspective on mathematics and the mathematics community.

3 Education

Overall, Klein had a fairly normal life and uninterrupted education. He attended the gymnasium in Düsseldorf, and did not find the Latin and Greek classics exciting.

Klein entered the University of Bonn in 1865 at the age of 16 and found the courses there, with emphasis on natural sciences, ideally suited to him. His university education at Bonn contributed significantly to his universalist outlook with a wide variety of subjects including mathematics, physics, botany, chemistry, zoology, and mineralogy, and he participated in all five sections of the Bonn Natural Sciences Seminar.

In mathematics, Klein took some courses with the distinguished analyst Rudolf Lipschitz, including analytic geometry, number theory, differential equations, mechanics and potential theory. But Lipschitz was just an ordinary teacher to Klein.

When he entered the University of Bonn, Klein aspired to become a physicist and studied with Plücker, a gifted experimental physicist and geometer. Plücker picked Klein to be an assistant for the laboratory courses in physics when Klein was only in his second semester. The interaction with Plücker probably had the most important influence on Klein in his formative years.

By the time Klein met Plücker in 1866, Plücker's interests had returned to geometry after having worked exclusively on physics for nearly twenty years, and he was writing a two-volume book on line geometry titled "Neue Geometrie des Raumes." When he died unexpectedly in May 1868, Plücker had only finished the first volume. As a student of Plücker, the death of Plücker provided a uniquely challenging opportunity for Klein: to finish the second volume and edit the work of his teacher.

Originally, the rising and inspiring geometer Clebsch at Göttingen was responsible for completing the book of Plücker. But he delegated this task to Klein. This seemingly impossible task changed the life of Klein in many ways.

First, it gave Klein a good chance to learn line geometry solidly, which played an important role in his future work with Lie and eventually in the Erlangen program. Second, it also brought him into close contact with Clebsch and his school which included many distinguished mathematicians such as Gordan, Max Noether, Alexander von Brill, etc. Through them, Klein learned and worked on Riemann's theory of functions, which eventually became Klein's favorite subject. He also became Clebsch's natural successor in many other ways. For example, he took over many students of Clebsch and the journal *Mathematische Annalen* started by Clebsch.

Klein obtained his Ph.D. degree in December 1868 with Rudolf Lipschitz as a joint (or nominal) advisor.

4 Three people who had most influence on Klein

There are three people who played a crucial role in the informative years of Klein.

The first person was Plücker, his teacher during his college years. Physics and the interaction between mathematics and physics had always played an important role in the mathematical life of Klein. It is reasonable to guess that this might have something to do with the influence of Plücker. For most mathematicians, Plücker is well known for Plücker coordinates in projective geometry. But he started as a physicist. In fact, In 1836 at the age of 35, he was appointed professor of physics at the University of Bonn and he started investigations of cathode rays that led eventually to the discovery of the electron. Almost 30 years later, he switched to and concentrated on geometry.

Klein had written many books, some of which are still popular. Probably the most original book by him is "Über Riemann's Theorie der Algebraischen Functionen und ihre Integrale", published in 1882, where he tried to explain and justify Riemann's work on functions on Riemann surfaces, in particular, the Dirichlet principle, using ideas from physics. Klein wrote [12, p. 178],

> in modern mathematical literature, it is altogether unusual to present, as occurs in my booklet, general physical and geometrical deliberations in naive *anschaulicher* form which later find their firm support in exact mathematical proofs. [...] I consider it unjustifiable that most mathematicians suppress their intuitive thoughts and only publish the necessary, strict (and mostly arithmetical) proofs [...] I wrote my work on Riemann precisely as a physicist, unconcerned with all the careful considerations that are usual in a detailed mathematical treatment, and, precisely because of this, I have also received the approval of various physicists.

In a biography of Klein [6], Halsted wrote: "The death of Plücker on May 22nd 1868 closed this formative period, of which the influence on Klein cannot be overestimated. So mighty is the power of contact with the living spirit of research, of taking part in original work with a master, of sharing in creative authorship, that anyone who has once come intricately in contact with a producer of first rank must have had his whole mentality altered for the rest of his life. The gradual development, high attainment, and then continuous achievement of Felix Klein are more due to Plücker than to all other influences combined. His very mental attitude in the world of mathematics constantly recalls his great maker."

The second person was Alfred Clebsch, another important teacher of Klein. Clebsch could be considered as a postdoctor mentor of Klein. After obtaining his Ph.D. in 1868, Klein went to Göttingen to work under Clebsch for eight months. When Klein first met him, Clebsch was only 35 years old and was already a famous teacher and leader of a new school in algebraic geometry.

Clebsch made important contributions to algebraic geometry and invariant theory. Before Göttingen, he taught in Berlin and Karlsruhe. His collaboration with Paul Gordan in Giessen led to the introduction of the Clebsch–Gordan coefficients for spherical harmonics, which are now widely used in representation theory of compact Lie groups and in quantum mechanics, and to find the explicit direct sum decomposition of the tensor product of two irreducible representations into irreducible representations. Together with Carl Neumann at Göttingen, Clebsch founded the mathematical research journal *Mathematische Annalen* in 1868.

When *Mathematische Annalen* was started, the leading journal in the world was Crelle's Journal, which is formally called *the Journal für die reine und angewandte Mathematik* and was founded by August Leopold Crelle in 1826 in Berlin. This journal is famous for many important, high-quality papers it has published. For example, seven of the legendary Abel great papers were published in the first volume of Crelle's Journal.

After Clebsch died, Klein took over the *Mathematische Annalen.* This allowed Klein to promote the mathematics he liked and he turned the journal into the leading journal in the world, overtaking the role of Crelle's Journal. Both journals are still high quality journals and well regarded by mathematicians, in spite of hundreds of mathematics journals around the world.

The interaction of Klein with Clebsch and his students such as Max Noether exposed Klein to the functional ideas of Riemann, and Klein's works on Riemann surfaces and geometric function theory were probably his deepest *concrete* contributions. According to Courant [14, p. 178], "If today we are able to build on the work of Riemann, it is thanks to Klein."

The third person who had a great influence on Klein was Lie, who might had the most important impact on Klein's mathematics among the three people. Against the advice of Clebsch, Klein went to Berlin in the winter semester of 1869–1870. Berlin was the center of mathematics at that time, dominated by Weierstrass, Kummer and Kronecker. There were also impressive students in Berlin at that time, which included Cantor, Frobenius, Killing and Mittag–Leffler. Klein did not benefit much from the lectures of the masters and did not stand out among these distinguished students. But he met Sophus Lie and they became lifelong friends. In a letter written to his mother, Klein wrote [7, p. 221]:

> Among the young mathematicians I have made the acquaintance of someone who appeals to me very much. He is Lie, a Norwegian whose name I already knew from an article he had published in Christiania. We have especially busied ourselves with similar things, so there is no lack of material for conversation. Yet we are not only united by this common love, but also a certain repulsion to the art and manner in which mathematics here asserts itself over and against the accomplishments of others, particularly foreigners.

Though Lie was almost six years older than Klein, in their interaction Klein often played the role of a more senior partner. This was also often the way viewed by others. As we will discuss later, this was also one cause of their unfortunate later conflict.

After his visit to Berlin, Klein and Lie were together in Paris in the spring of 1870. When the Franco–Prussian war broke out, Klein was forced to return to Germany for military service. They met again at many other occasions later. The interaction and cooperation was beneficial to both of them. It is an inspiring story with sad components: their breakup and reconciliation.

By all accounts, Klein is now best known for the Erlangen program. But it is fair to say that without learning Lie groups from Lie and working with Lie, Klein might not have proposed the Erlangen program, and without the work of Lie and his school towards it, the Erlangen program would not have been so famous now with such a huge impact.

5 Academic career

Klein is famous for being a great teacher. But at his first job as a professor at the University of Erlangen, only two students signed up for his class. After the first lecture, only one of the two remained.

In 1872 Klein was appointed full professor at the University of Erlangen at the age of 23. Clebsch had nominated him and strongly supported him, believing that Klein would be the star of his generation. In connection with his appointment as a professor, he gave a speech on mathematical education and *separately* wrote a booklet titled "Vergleichende Betrachtungen über neuere geometrische Forschungen", which became the famous Erlangen program.[4]

Klein was not too successful as a teacher in Erlangen, and it seems that only a small number of students attended his lectures. But he was successful as a researcher during his three years there. In 1875, Klein was given a chair at the Technical University of Munich. His lectures were very popular and his first semester course was attended by over 200 students. Together with Brill, he founded a laboratory for the construction of mathematical models. These models became very popular and were bought by many mathematics departments around the world. On the second floor of the Mathematics Institute at Göttingen, one can still see a large collection of such models. For more discussion about these models by Klein, see [4, §3.3].

During his years in Munich, both his research and his role as an editor of the *Mathematische Annalen* went really well. Klein worked on the interface between algebra and the theory of complex variables and he developed a geometric approach to Galois theory and a unified theory of elliptic modular functions, which prepared him for his deep and original contribution to an entirely new field: the theory of automorphic functions. In short, he found a field of research which was well suited to his mathematical taste and perspective: a mixture of group theory, algebraic equations and function theory. Without any question, he was seen as the rising star of German mathematics. In 1881, he accepted a new chair in geometry at the University of Leipzig, which was a leading university in Germany but with a weak mathematics department.

Klein reached his height in terms of mathematical creativity at the University of Leipzig. He started to work on automorphic functions and uniformization of Riemann surfaces in competition with Poincaré. Poincaré was 4 years younger than Klein. In 1881, Klein was world famous, and Poincaré was an unknown young mathematician.

[4]It seems that some people confused Klein's written report with his inaugural oral speech.

But Poincaré quickly caught up and surpassed Klein. According to a letter written from Paris to Klein [15], Lie wrote: "Poincaré said that at first it was hard for him to read your work, but now it goes very easily. A number of mathematicians, such as Darboux and Jordan, say that you make great demands on the reader in that you often do not supply proofs."

Poincaré made a steady and substantial progress which was difficult for Klein to keep up with. Klein tried his best in order not to lose the competition. After publishing his results in October 1882 on the uniformization of Riemann surfaces of finite type, Klein collapsed and experienced depression. This uniformization theorem was a natural capping result for the earlier works of Klein and Poincaré on Fuchsian groups and Kleinian groups. But Klein's paper was only an announcement with a sketch of ideas. In fact, it could not be completed without the much later work of Brouwer on topology. Poincaré announced a similar result around the same time, also without enough details. The uniformization theorem of Riemann surfaces was eventually proved independently by Poincaré and Koebe in 1907.[5]

During Klein's stay at Leipzig, he turned its mathematics department into a major mathematics department in the world and he built a school of geometry. Realizing that his research career was over, he started to think about ways to retain his role as a head of an important mathematics school.

In 1886, the opportunity which Klein had been waiting for arose. He received an offer from Göttingen and he accepted it immediately. It was an ideal position for him. Göttingen had a famous tradition and was a leading university in Germany. Probably more importantly, a comrade of his from the time he served the military dealt with university appointments at the Ministry of Culture in Berlin. Since the Ministry had a policy of strengthening science departments at Göttingen, Klein was able to build up his power base at Göttingen. Eventually he did take advantage of these opportunities and hired Hilbert. The rest is the well-known history, the legendary Göttingen school of mathematics of Hilbert and Klein.

Of course, Göttingen had a long tradition of famous mathematicians before Klein's arrival. It started with the legendary Gauss, who studied, worked and taught there for over 50 years. But Gauss was like an eighteenth century scholar. He did not write and publish all his results, and did not incorporate his research into his teaching. In fact, he almost never taught advanced mathematics courses (he only taught elementary aspects of mathematical astronomy). Instead, he preferred to carry out an extensive scientific correspondence with a few friends and peers. The immediate successor of Gauss was Dirichlet. Dirichlet was a great mathematician and a great teacher, and he spent most of his academic career in Berlin. He died rather unexpectedly only three years after he moved to Göttingen. The successor of Dirichlet was Riemann. But Riemann had a poor health and took extended leaves in the sunny Northern Italy. Af-

[5] One dispute in the correspondence between Klein and Poincaré was concerned with the naming of Fuchsian groups, which Klein felt was not fair. It is interesting to quote a letter from Lie to Klein [15] on this issue, when Lie was visiting Paris: "Tell me when you get a chance whether Fuchs has answered your last remarks and, if so, where, I have no doubts that the mathematical world will give the essential work you've done prior to Poincaré's discoveries its just due. With all of your students you have an army that represents a mighty force ..."

ter his death in 1866, the natural academic successor was Dedekind.[6] But Dedekind was quiet and withdrawn, and did not interact much with others. He was not suitable to maintain and develop the tradition of Göttingen. After him, Clebsch established a major school in algebraic geometry in Göttingen, but only for a short period from 1868 to 1872. His unexpected death made this just one more flicker in the history of Göttingen. Though Klein was the youngest member of the Clebsch circle, he was the most dynamic and assumed much of the burden for developing further his mentor's research and publication programs. Göttingen remained quiet until the arrival of Klein and Hilbert, and Klein was the man largely responsible for restoring Göttingen's glorious history.

It is perhaps helpful to point out that Klein and Hilbert formed a perfect pair since they were so different and they complemented each other so well. Klein was a worldly man and good at politics and enjoyed it, and Hilbert was a rather "simple" man who was really (or only) interested in mathematics but did not care for power or politics; Klein was formal and kept a distance with students and younger colleagues, and Hilbert was casual and mingled with students and others as well. If both of them were like Klein, there might not have been peace at Göttingen (or in the German mathematics community) and Göttingen could not have been at the cutting edge of mathematics and its interaction with physics; and if both were like Hilbert, they might not have gotten things done and attracted people from all over the world, for example, there might not have been the new idea of an open book mathematics library for students or more chairs of mathematics in Göttingen.

Hilbert was an academic grandson of Klein, and it is interesting to see how the relations between Klein and Hilbert evolved with time by quoting from the biography of Hilbert by C. Reid. As mentioned before, Klein collapsed after his competition with Poincaré on the uniformization of Riemann surfaces in 1882. Hilbert met Klein in Leipzig in 1885 [13, p. 19]:

> But the Klein whom Hilbert now met in Leipzig in 1885 was not his same dazzling prodigy...
>
> Hilbert attended Klein's lectures and took part in a seminar. He could not have avoided being impressed. Klein was a tall, handsome, dark haired and darken beard man with shining eyes, whose mathematical lectures were universally admired and circulated even as far as America...
>
> As for Klein's reaction to the young doctor from Königsberg – he carefully preserved the Vortrag, or lecture, which Hilbert presented to the seminar and he later said: "When I heard this Vortrag, I knew immediately that he was the coming man in mathematics."

In March 1886, Hilbert went to Paris on the recommendation of Klein [13, p. 22].

> As soon as Hilbert was settled, he wrote to Klein. The letter shows how important he considered the professor. It was carefully drafted out with great attention to the

[6]Though Dedekind received his Ph.D. and Habilitation in Göttingen, he never worked in Göttingen. Instead, he was a professor in Zürich from 1858 to 1862, and then in Braunschweig for the rest of his life. One of his major contributions to Göttingen mathematics was his substantial editing and publication of Dirichlet's lecture notes "Vorlesungen über Zahlentheorie". In this sense, he was a major figure in Göttingen mathematics.

> proper, elegant wording, then copied over in a large, careful Roman script rather than the Gothic which he continued to use in his letters to Hurwitz...

At the end of April, 1886, Klein wrote to Hilbert [13, p. 25]:

> "Not as much about mathematics as I expected." Klein commented disapprovingly to Hilbert. He then proceeded to fire off half a dozen questions and comments which had occurred to him while glancing through the most recent number of the Comptes Rendus: "Who is Sparre? The so-called Theorem of Sparre is already in a Munich dissertation (1878, I think). Who is Stieltjes? I have an interest in this man, I have come across an earlier paper by Humbert – it would be very interesting if you could check on the originality of his work (perhaps via Halphen?) and find out for me a little more about his personality. It is strange that the geometry in the style of Veronese–Segre happens to be coming back in fashion again ..."
>
> "Hold it always before your eyes", he admonished Hilbert, "that the opportunity you have now will never come again."

In December 1894, Klein wrote to Hilbert [13, p. 45]:

> I want to inform you that I shall make every effort to see that no one other than you is called here.
>
> You are the man whom I need as my scientific complement because of the direction of your work and the power of your mathematical thinking and the fact that you are still in the middle of your productive years. I am counting on it that you will give a new inner strength to the mathematical school here, which has grown continuously and, as it seems, will grow even more — and that perhaps you will even exercise a rejuvenating effect upon me ...
>
> I can't know whether I will prevail in the faculty. I know even less whether the offer will follow from Berlin as we propose it. But this one thing you must promise me, even today: that you will not decline the call if you receive it!

Hilbert replied [13, p. 46]: "Without any doubt I would accept a call to Göttingen with great joy and without hesitation."

Klein had some reservation about working with Hilbert if Hilbert came [13, p. 46]:

> As for Klein's feelings –already it was clear that Hilbert questioned any authority, personal or mathematical, and went his own way... When in the faculty meeting his colleagues accused him of wanting merely a comfortable younger man, he replied, "I have asked the most difficult person of all."

Both Klein and Hilbert knew that the other party might not be easy to work with, but they felt that it was beneficial for everyone. According to [13, p.46],

> Hilbert worked very hard on his reply to Klein's letter, crossing out and rewriting extensively to get exactly the effect he wanted....
>
> "Your letter has surprised me in the happiest way," he began. "It has opened up a possibility for the realization of which I might have hoped at best in the distant future and as the final goal of all my efforts [...] Decisive for me above all would

> be the scientific stimulation which would come from you and the greater sphere of influence and the glory of your university. Besides, it would be a fulfillment of mine and my wife's dearest wish to live in a smaller university town, particular one which is so beautifully situated in Göttingen."
>
> Upon receiving this letter from Hilbert, Klein proceeded to plan out a campaign.
>
> "I have already told Hurwitz that we will not propose him this time so that we will be more successful in proposing you. We will call Minkowski in second place. I have discussed this with Althoff [the person in the Ministry of Culture in charge of faculty appointments.] and he thinks that that will make it easier then for Minkowski to get your place in Königsberg."
>
> Within a week he was writing triumphantly to Hilbert:
>
> "This has been just marvelous, much faster than I ever dared to hope it could be. Please accept my heartiest welcome!"

6 As a teacher and educator

Throughout his life, Klein was always interested in education and teaching. In some sense, he was more like a scholar than a research professor.

Klein was a very successful teacher after the initial difficult time at the University of Erlangen and he perfected his lecture style in Göttingen. According to [13, p. 48],

> Klein's lectures were deservedly recognized as classic. It was his custom often to arrive as much as an hour before the students in order to check the encyclopedic list of references which he had had his assistant prepared. At the same time he smoothed out any roughness of expression or thought which might still remain in his manuscript. Before he began his lecture, he had mapped out in his mind an arrangement of formulas, diagrams and citations. At the conclusion the board contained a perfect summary of the presentation, every square inch being appropriately filled and logically ordered.
>
> It was Klein's theory that students should work out proofs for themselves. He gave them only a general sketch of the method. The result was that a student had to spend at least four hours outside class for every hour spent in class if he wished to master the material. Klein's forte was the comprehensive view.

One of the assistants of Klein who later became rather famous was Sommerfeld. For a detailed description of interaction between Klein and Sommerfeld, see [4].

Another successful aspect of his teaching is that he had 57 Ph.D. students. Some became very distinguished mathematicians, such as Adolf Hurwitz, Ludwig Bieberbach, Maxime Bocher, Frank Cole, C. L. Ferdinand Lindemann (who proved the transcendence of π), Alexander Ostrowski, and Axel Harnack (as in the Harnack inequality). Many of his successful books were based on his lectures.

Klein wrote many books, especially later in his life after he stopped doing research. We will describe in detail his book "Lectures on mathematics" [9] in § 9 below, and we briefly mention here several other great books:

1. (joint with Robert Fricke) *Vorlesungen über die Theorie der automorphen Funktionen. Band 1: Die gruppentheoretischen Grundlagen. Band II: Die funktionentheoretischen Ausführungen und die Andwendungen.*
2. (joint with Robert Fricke) *Vorlesungen über die Theorie der elliptischen Modulfunktionen. Band I: Grundlegung der Theorie. Band II: Fortbildung und Anwendung der Theorie.*

   These four volume books were written with his student Fricke, based on Klein's lectures. They set the foundation for the modern theory of discrete subgroups of Lie groups, arithmetic subgroups of algebraic groups and automorphic forms. They are classical in the sense that people know that they are important, but few people actually read them.
3. (joint with Arnold Sommerfeld) *The theory of the top*, 3 volumes.

In the 1890s, Klein turned to mathematical physics, a subject from which he had never strayed far. These books were originally presented by Klein as an 1895 lecture at Göttingen University that was broadened in scope and clarified as a result of his collaboration with Sommerfeld. They are still in print and the standard references in the subject.

After he left mathematics research, Klein started around 1900 to take an interest in mathematical instruction in schools below college level. For example, in 1905, he played a decisive role in formulating a plan which recommended that differential and integral calculus and the function concept be taught in high schools. This recommendation was gradually implemented in many countries around the world, and now this is standard. This contribution of Klein is typical of him: they tend to be foundational and natural retrospectively.

Klein contributed to the development of comprehensive mathematics courses for teachers and engineers, strengthening the connection between mathematics and sciences and engineering.

In 1908, Klein was elected chairman of the International Commission on Mathematical Instruction at the Rome International Congress of Mathematicians. Under his guidance, the German branch of the Commission published many volumes on the teaching of mathematics at all levels in Germany.

In 2000, The International Commission on Mathematical Instruction decided to create the Felix Klein Award recognizing outstanding lifetime achievement in mathematics education research. According to the description of the award, it

> serves not only to encourage the efforts of others, but also to contribute to the development of high standards for the field through the public recognition of exemplars.

Klein also wrote several books for the general reader, high school students and school teachers, for example, several volumes of *Elementary Mathematics from an Advanced Standpoint* on Geometry, Arithmetic, Algebra and Analysis. These books have been translated into several languages and are still in print.

7 Main contributions to mathematics

To both mathematicians and physicists, the best known contribution of Klein is the Erlangen program, which is not a concrete theorem or a theory, but rather a point of view. But other mathematicians such as Lie worked to make it more substantial and concrete.

Unlike most other great mathematicians, Klein's contribution to mathematics cannot be measured by theorems proved or conjectures solved.

But his contributions can be felt at various levels. They have become so much a part of our mathematical thinking that it is sometimes hard for us to appreciate their novelty.

The motivation for Klein to propose this grand program is to answer a simple question: What is geometry? At that time, there were many different aspects of geometric studies such as Plücker's line geometry, spherical geometry by the French school, the hyperbolic geometry by Lobachevsky and Bolyai, elliptic geometry by Riemann, projective geometry in the tradition of Cayley and Salmon, the birational geometry of Riemann and Clebsch, the work of Grassmann on the geometry of vector spaces, the synthetic approach to the foundations of geometry by von Staudt, Lie's work on contact transformations and their applications to systems of partial differential equations, etc. Klein wanted to extract the common and basic principle of all these works.

According to Klein, the answer is that geometry is the study of invariants of symmetry (or transformation) groups. This is the essential point of the Erlangen program he proposed in 1872. For example, the elliptic, hyperbolic and parabolic geometries correspond to different subgroups of the group of projective transformations, and hence they are related to each other by viewing them through projective geometry.

This important particular conclusion was reached by Klein before he proposed the Erlangen program. In 1871, he published two papers titled *Ueber die sogenannte Nicht-Euklidische Geometrie* (*On the So-called Non-Euclidean Geometry*) which showed that Euclidean and non-Euclidean geometries could be considered as special cases of the projective geometry. This gave a new proof of the fact that non-Euclidean geometry is consistent if and only if Euclidean geometry is, thus putting Euclidean and non-Euclidean geometries on the same footing, and helping end all controversy surrounding non-Euclidean geometry. See the commentary [1] in this volume. This work made Klein famous and secured his job at the University of Erlangen. If one wants to name a single important theorem of Klein, this might be it. But it should be pointed out that even in this case, he made essential use of the work of Cayley. Therefore, this model of the hyperbolic plane is often called the Cayley–Klein, or the Beltrami–Cayley–Klein model, because Beltrami also worked it out.

The Erlangen program was not widely known in the first two decades after it was written and was only available as a booklet. It was finally published in 1893 in the leading journal, *Mathematische Annalen.* An authorized English translation was published in [11].

In the meantime, others including Poincaré arrived at similar ideas. In a letter from Paris to Klein in the early 1880s [15], Lie wrote: "Poincaré mentioned on one occasion that all mathematics is a matter of groups. I told him of your Erlangen program, which he did not know about."

Lie and his school actually worked on the Erlangen program and made clear the role of groups as the great unifying principle of twentieth century mathematics, which was not clear to Klein in the original program.

The impact of the Erlangen program, or rather its point of view, was huge and included many branches of mathematics and sciences.

In the narrow sense, the Erlangen program considers homogeneous manifolds. In the broad sense, it emphasizes the importance of understanding invariants under automorphism groups. For example, for Riemannian manifolds, we need to understand invariants (both local and global) under isometries. For differentiable manifolds, we search for invariants under diffeomorphisms. And for topological manifolds, we can study invariants under either homeomorphisms or homotopy equivalences. For example, homotopy groups, homology and cohomology groups of topological spaces can be considered as such invariants. A similar principle applies to other sciences, in particular physics.

Besides the Erlangen program, Klein's work on Fuchsian groups, Kleinian groups and automorphic functions, his emphasis on the connection between mathematics and physics is still of huge current interest. This can be seen by many recent results on three-dimensional hyperbolic manifolds. Another result which might not be so well-known is that the modern notion of manifold which was introduced by Herman Weyl in a book on Riemann surfaces was motivated by Klein's work and his point of view on Riemann surfaces in his book "On Riemann's Theory of Algebraic Functions and their Integrals."

According to an article on Klein by Burau and Schoeneberg [3] in *the Dictionary of Scientific Biography*,

> Klein considered his work in function theory to be the summit of his work in mathematics. He owed some of his greatest successes to his development of Riemann's ideas and to the intimate alliance he forged between the later and the conception of invariant theory, of number theory and algebra, of group theory, and of multidimensional geometry and the theory of differential equations, especially in his own fields, elliptic modular functions and automorphic functions.

This assessment of Klein's work is consistent with that in Reid's biography of Hilbert [13, p. 19]: Klein had

> a strong drive to break down barriers between pure and applied science. His mathematical interest was all-inclusive. Geometry, number theory, group theory, invariant theory, algebra – all had been combined for his master work, the development and completion of the great Riemannian ideas on geometric function theory. The crown of this work had been his theory of automorphic functions.

For a more systematic and chronological summary of the work of Klein, see [5].

8 The Evanston Colloquium Lectures and the resulting book

Klein wrote many books at different levels and on different topics. Perhaps one book which reflects best his broad knowledge of mathematics and his perspective on the importance of mathematical topics is his book *Lectures on mathematics*, a survey of mathematics in the previous quarter of a century. It is still in print and can serve as a good model for surveys in mathematics.

This book was closely connected to the International Mathematical Congress organized in conjunction with the World's Fair: the Columbian Exposition in 1893 in Chicago.

This International Mathematical Congress in Chicago [2] was a major event and 4 years ahead of the first International Congress of Mathematicians held in 1897 in Zurich. It acted as a harbinger of a new era of international cooperation in mathematics.

Klein's book *Lectures on mathematics* consisted of the lecture notes on his two week long Evanston Colloquium Lectures which he gave at Northwestern University right after the congress. We need to view this book in the context of the fair and the status of mathematics in the USA at that time.

The World's Columbian Exposition was to celebrate the 400th anniversary of Christopher Columbus's arrival in the New World in 1492. Chicago won over New York City, Washington D. C., and St. Louis the honor of hosting this special fair. The fair had a profound effect on architecture, arts, Chicago's self-image, and American industrial optimism. For the mathematical community, the International Mathematical Congress in Chicago and the lectures of Klein represented the beginning of emergence of American mathematics.

Klein had been waiting for 10 years for a good opportunity to visit and give lectures in the USA, and the Columbia exposition finally gave him a chance. The Evanston Colloquium Lectures were highlights of his visit.

According to an arrangement with the Prussian Ministry of Culture, Klein would attend the International Mathematical Congress as an official representative of his government. Klein viewed mathematics as a great chance for the German Reich to demonstrate its growing dominance and also as a golden opportunity for him to solidify his position (or reputation) as a leading mathematician in Germany.

Besides his talk at the Congress, he also offered to give a two-week long lecture series at Northwestern University in Evanston after the Congress. The enthusiasm of the local organizers can be clearly seen from the letter from Henry White of Northwestern University to Klein [12, p. 306]:

> your scheme for a comprehensive survey of recent mathematics is the most useful, though not the least onerous. If you have strength to carry it out, it will be in every way a worthy contribution to a noble object.

E. H. Moore of Chicago University wrote to Klein [12, p. 306]:

> you cannot easily understand, or rather imagine, how much pleasure your two letters brought to me personally and to all of us interested in the success of the Congress.

> We are deeply grateful to you and your government [...] As to the September Colloquium: I esteem it a great privilege to be one of the circle of those to profit by the inspiration of your leadership through the domains of modern mathematics.

The importance of the Mathematics Congress and the Evanston Colloquium could not be over estimated for mathematicians in the Chicago area. They provided a unique opportunity to put Midwestern mathematics on the map. (Note that the traditional institutions of higher learning of USA were located on the East Coast.)

On the first day of the Congress, Klein launched upon a sweeping survey of modern mathematics titled "*The Present State of Mathematics*" [10], which reflected very much his philosophical point of view:

> When we contemplate the development of mathematics in this nineteenth century, we find something similar to what has taken place in other sciences. The famous investigators of the preceding period, Lagrange, Laplace, Gauss, were each great enough to embrace all branches of mathematics and its applications. In particular, astronomy and mathematics were in their time regarded as inseparable.
>
> With the succeeding generation, however, the tendency to specialization manifests itself. Not unworthy are the names of its early representatives: Abel, Jacobi, Galois and the great geometers from Poncelet on, and not inconsiderable are their individual achievements. But the developing science departs at the same time more and more from its original scope and purpose and threatens to sacrifice its earlier unity and to split into diverse branches.

He also pointed out that

> the attention bestowed upon it by the general scientific public diminishes. It became almost the custom to regard modern mathematical speculation as something having no general interest or importance, and the proposal has often been made that, at least for purpose of instruction, all results be formulated from the same standpoints as in the earlier period. Such conditions were unquestionably to be regretted.

Then he continued:

> This is a picture of the past. I wish on the present occasion to state and to emphasize that in the last two decades a marked improvement from within has asserted itself in our science, with constantly increasing success [...]. This unifying tendency, originally purely theoretical, comes in envitably to extend to the applications of mathematics in other sciences, and on the other hand is sustained and reinforced in the development and extension of these latter.

Klein also emphasized the importance of the team work and cooperation:

> Speaking, as I do, under the influence of our Göttingen tradition, and dominated somewhat, perhaps, by the great name of Gauss, I may be pardoned if I characterize the tendency that has been outlined in these remarks as a *return to the general Gaussian programme.* A distinction between the present and the earlier period lies evidently in this: that what was formerly begun by a single mastermind, we now must seek to accomplish by united efforts and cooperation.

All these comments made more than 100 years ago are still valid and require the attention of all mathematicians.

When Klein talked about the internal unity of mathematics, the basis comes from the concept of group and the notion of analytic functions of complex variables. To show the usefulness of mathematics in sciences, one example he gave was the application of group theory to the classification of crystallographic structures, which had been studied by various people many years earlier but were only completed by Federov and Schoenflies in 1880s. On the last day of the Congress, Klein also gave a lecture titled "Concerning the Development of the Theory of Groups during the Last Twenty Years."

The Mathematics Congress started on August 21 and ended on August 26, and the Evanston Colloquium Lectures by Klein started on August 28. It was much more than a usual mathematics lecture series. These lectures were Klein's first serious attempt to sketch an overview of some major trends in mathematics. Lectures of such nature would have appeared presumptuous to the German mathematics community, and would also have clashed with elitist ideals of Wissenschaft,[7] which rejected anything related to popularization. Because of all these factors, Klein chose to deliver his "Lectures on Mathematics" outside Germany.

These lectures gave an overview of the major development of mathematics and of some of the principal mathematicians in the proceeding 25 years. They were full of strong opinions and novel insights of a highly influencing mathematician, and Klein was the ideal person to give such lectures at an ideal place and time. Klein rarely presented even slightly complicated arguments in his lectures and almost always focused on the big picture of a theory and interconnection between different subjects. The Evanston lectures were especially so. Even for people who had studied with Klein earlier in Göttingen, these lectures were a new experience.

For 12 days, Klein delivered one lecture on each day. He spoke slowly in English and posed questions to the audience as he went along. Each lecture was followed by lengthy informal discussion. This book "Lectures on Mathematics" was based on the lecture notes by Alexander Ziwet, the chair of the mathematics department at the University of Michigan. They were revised later by Ziwet together Klein for publication in January 1894. This book probably gives a rather condensed record of what Klein said in his lectures, and it is not clear how much, or if any, of the topics discussed at the informal discussions was incorporated into the book.[8] Klein said himself that this book by itself gives no sense of the atmosphere on this historical occasion. But it is probably the best approximation to listening to a lecture by such a master speaker.

---

[7]Wissenschaft is the German word for any study or science that involves systematic research and teaching. It implies that knowledge is a dynamic process discoverable for oneself, rather than something that is handed down, hence the popularization in the form of superficial survey is against its spirit. Wissenschaft was the official ideology of German Universities during the nineteenth century. It emphasized the unity of teaching and individual research or discovery for the student. It suggests that education is a process of growing and becoming.

[8]Ziwet explained in the preface that "in reading the lectures here published, it should be kept in mind that they followed immediately upon the adjournment of the Chicago meeting, and were addressed to members of the Congress. This circumstance, in addition to the limited time and the informal character of the colloquium, must account for the incompleteness with which the various subjects are treated."

9 A summary of the book "Lectures on mathematics" and Klein's conflicts with Lie

This book "Lectures on mathematics" was published in 1894, almost 120 years ago, and mathematics now is very different from mathematics at that time. In some sense, this book characterizes Klein's perspective towards mathematics: a global vision and interconnections between different subjects of mathematics.

A natural question is why should one read it now. Or maybe a better question is how should one read it.

Some obvious answers include (1) we can read it to understand an overview of the mathematics written 120 years ago, and to understand how mathematics has evolved and new results and theories arose. Unlike any other subject, mathematics has a strong historical continuity: what was considered important in the past is also important and relevant now. Thus many topics discussed by Klein are still relevant and important today; (2) we can learn from Klein on how to study and understand mathematics, and how to take a global approach to it.

A good answer to the question is contained in a preface written by William Osgood for a reprint of this book in 1910 [9]:

> To reproduce after a lapse of seventeen years lectures which at the time they were delivered were in such close contact with the most recent work of that day, may well call for a word of justification. Has mathematics not advanced since then, and are the questions here treated of the first importance at the present time? I reply by asking: What *is* important in the development of mathematics? Is it solely the attainment of new results of potential value, or must not an essential part of the best scientific efforts of each age be devoted to possessing itself of the heritage of the age that has just preceeded it?
>
> [...] at a time when the contributions of the immediate past were so rich and so unrelated, Klein was able to uncover the essential bonds that connect them and to discern the fields to whose development the new methods were best adapted.
>
> His instinct for that which is vital in mathematics is sure, and the light with which his treatment illumines the problems here considered may well serve as a guide for the youth who is approaching the study of the problems of a later day.

Though this is not a book to learn systematically some classical mathematics, its informal style allows one to understand mathematics from a different perspective. Given that this book was written by such a major figure of the nineteenth century mathematics community, a master speaker and writer, we can learn from him on how to pick out good mathematics, and how to understand some histories and stories behind mathematics, i.e., the book is an eyewitness account from an active participant during that period. One may wonder about the experience of attending Klein's lectures. This book probably gives one a bit of flavor. Naturally, there are many other things or advices which we can learn from Klein via this book. For example, page after page, Klein used concrete examples to show the importance of geometric intuition and interconnections between different subjects.

We now briefly comment on the content of this book and point out some special features along the way.

Klein started the lectures by classifying mathematicians into three categories: (1) logicians, (2) formalists, and (3) intuitionists.

This is rather different from another current popular classification: theory builders and problem solvers.

He devoted his **first lecture** to Clebsch, the eminent geometer whom we already mentioned and his mentor who greatly influenced his academic life. It is interesting to see how he evaluated his respected teacher:

> However great the achievement of Clebsch's in making the work of Riemann more easy of access to his contemporaries, it is my opinion that at the present time the book of Clebsch is no longer to be considered the standard work for an introduction to the study of Abelian functions. The chief objections to Clebsch's presentation are twofold: they can be briefly characterized as a lack of mathematical rigor on the one hand, and a loss of intuitiveness, of geometrical perspicuity, on the other.

One point Klein emphasized here is the important role played by Anschauung, which is basically geometrical intuition, or rather direct or immediate intuition or perception of sense data with little or no rational interpretation. It is also interesting to note that Klein puts Clebsch into both categories (2) and (3).

One important mathematical reason for which Klein was critical of Clebsch is that Klein believed in Riemann's original intuitive approach to the theory of Abelian integrals, but the Clebsch school used an algebraic approach in order to put Riemann's work on the rigorous foundation. Klein wrote,

> For these reasons, it seems to me best to begin the theory of Abelian functions with Riemann's ideas, without, however, neglecting to give later the purely algebraic developments.

Such comments and points can be applied to many other subjects and problems in modern mathematics.

As mentioned before, besides Clebsch, there were two other people, Plücker and Lie, who greatly influenced Klein. The **second and third lectures** were devoted to Lie and his work. Klein used Lie's early work on geometry and the theory of partial differential equations as examples of how great mathematics can emerge through intuition and unconscious inspiration rather than only complicated computation. His other important advice was to read the original papers, instead of the final books:

> To fully understand the mathematical genius of Sophus Lie, one must not turn to the books recently published by him in collaboration with Dr. Engel [*Theorie der transformationgruppen*, the third volume of which was about to appear], but to his earlier memoirs, written during the first years of his scientific career. There Lie shows himself the true geometer that he is, while in his later publications, finding that he was but imperfectly understood by the mathematicians accustomed to the analytical point of view, he adopted a very general analytical form of treatment that was not always easy to follow. Fortunately, I had the advantage of becoming intimately acquainted

> with Lie's ideas at a very early period, when they were still, as the chemists say, in the "nascent state," and thus most effective in producing a strong reaction.

Though the first three lectures concentrated on the work of Clebsch and Lie, other mathematicians and their works were also mentioned. For example, Plücker's work on line geometry was mentioned in connection with the work of Lie.

Lecture 4 is a good example of some classical mathematics which is still important and poorly understood. Algebraic curves and algebraic surfaces over the complex numbers are now fairly well understood, but real algebraic curves and algebraic surfaces are still less known and poorly understood, not because of lack of motivation, but rather due to lack of techniques and methods.

Klein's **fifth lecture** on hypergeometric functions is relatively short and sketchy. The **sixth lecture** of Klein is probably the best known of all 12 lectures. Intuition was always important to Klein and had been emphasized by him at many occasions. He started with a discussion of two kinds of intuition: naive and refined intuition, and their roles in developing mathematical theories:

> the naive intuition is not exact, while the refined intuition is not properly intuition at all, but arises through the logical development from axioms considered as perfectly exact.

He used historical examples to illustrate his point. For example, Klein considered Euclid's *Elements* as a prototype of refined intuition, while the work of Newton and other pioneers of the differential and integral calculus resulting from naive intuition. Klein wrote:

> It is the latter that we find in Euclid; he carefully develops his system on the basis of well-formulated axioms, is fully conscious of the necessity of exact proofs, clearly distinguishes between the commensurable and incommensurable, and so forth.
>
> The naive intuition, on the other hand, was especially active during the period of the genesis of the differential and integral calculus. Thus we see that Newton assumes without hesitation the existence, in every case, of a velocity in a moving point, without troubling himself whether there might not be continuous functions having no derivative.

Klein continued:

> this is the traditional view – that it is possible finally to discard intuition entirely, basing the whole science on the axioms alone. I am of the opinion that, certainly, for the purposes of research it is always necessary to combine the intuition with the axioms. I do not believe, for instance, that it would have been possible to derive the results discussed in my former lectures, the splendid researches of Lie, the continuity of the shape of algebraic curves and surfaces, or the most general forms of triangle, without the constant use of geometrical intuition.

This comment of Klein reminds us of the Bourbaki axiomatic approach to mathematics. Of course, Bourbaki has had a definite impact on modern mathematics, but it has also drawn many criticisms and has failed to achieve its goal.

Related to these two kinds of intuition, there is the difficulty in teaching mathematics. Klein wrote:

> a practical difficulty presents itself in the teaching of mathematics [...] The teacher is confronted with the problem of harmonizing two opposite and almost contradictory requirements. On the one hand, he has to consider the limited and as yet undeveloped intellectual grasp of his students and the fact that most of them study mathematics mainly with a view to the practical applications; on the other hand, his conscientiousness as a teacher and man of science would seem to compel him to detract in nowise from perfect mathematical rigor and therefore to introduce from the beginning all the refinements and niceties of modern abstract mathematics.

After the more philosophical 6th lecture, Klein turned to the transcendence of e and π. He started the **seventh lecture** with: "Last Saturday we discussed inexact mathematics: today we shall speak of the most exact branch of mathematical science." After Hermite proved the transcendence of e in 1873, Lindemann developed further the method of Hermite to prove transcendence of π in 1882. This fact is probably well-known to most mathematicians, but many people may not know that Lindemann was a former student of Klein. Klein wrote:

> The proof that π is a transcendental number will forever make an epoch in mathematical science. It gives a final answer to the problem of squaring the circle and settles this vexed question once for all. .... The proof of the transcendence of π will hardly diminish the number of circle-squarers, however; for this class of people has always shown an absolute distrust of mathematicians and contempt for mathematics that cannot be overcome by any amount of demonstration.

Klein's statement and explanation were to the point. They are still very relevant and true today.

The **eighth lecture** continued Klein's tour of number theory to emphasize his point that geometric methods are important in number theory.

The **ninth lecture** is concerned with solutions of higher algebraic equations. Everyone is familiar with the formula for solutions of quadratic algebraic equations and knows that there is no analogous algebraic formula for general algebraic equations of degree 5 or higher. This lecture gives a good summary of algebraic equations of higher degree. Klein started with:

> All equations whose solutions cannot be expressed by radicals were classified simply as insoluble, although it is well-known that the Galois groups belonging to such equations may be very different in character.

He continued: "*The solution of an equation will,* in the present lecture, *be regarded as consisting in its reduction to certain algebraic normal equations.*" The rest of the lecture is to explain this point by several examples, in particular, the icosahedron equation.

The **tenth lecture** dealt with hyperelliptic and abelian functions, an important subject which had been intensively studied by many people. Klein started with: "The subject of hyperelliptic and Abelian functions is of such vast dimensions that it would be impossible to embrace it in its whole extent in one lecture." Later in the same lecture, Klein wrote:

> Here, as elsewhere, there seems to resign a certain pre-established harmony in the development of mathematics, what is required in one line of research being supplied by another line, so that there appears to be a logical necessity in this, independent of our individual disposition.

This sets the tone of the **eleventh lecture** on "The most recent researches in non-Euclidean geometry." As mentioned before, Lie had a great influence on Klein, and the cooperation and friendship between Lie and Klein were beneficial. For example, it was Klein who got the chair in geometry at Leipzig for Lie after Klein left Leipzig for Göttingen. But the great friendship fell apart in 1892, and the climax was a sentence Lie put in the third volume of his monumental book *Theorie der Transformationsgruppen*: "I am not a student of Klein, nor is the opposite the case, even if it perhaps comes closer to the truth."

This was an amazing statement, considering that Klein was the dominating figure in the German mathematics community. There were many reasons for this breakup.

One reason was that Klein finally published the booklet "*The Erlangen program*" in the leading journal "*Mathematische Annalen*" in 1892. When Klein asked Lie about the idea of republishing it around 1892, Lie said that it was a good idea since he regarded it as Klein's most important work from the 1872 period and felt that it could be better understood and appreciated in 1892 than it was first circulated as a booklet in 1872. Klein tried to include some of his joint work and Lie's work in the revision, and Lie thought that Klein took more credit than he deserved. There is some truth in it (see [8] for some discussion on this point) and there are also other reasons for the breakup.

A few years before that, Lie felt that several people, in particular Wilhelm Killing, had used his ideas without giving him credit. In the early 1890s, Klein crossed the line and became an enemy of Lie in Lie's mind. When Klein published a paper on non-Euclidean geometry and a set of lecture notes, he did not mention some of Lie's results, which Lie believed were important applications of his theory of transformations groups.

In the *Evanston Colloquium Lectures*, Klein was trying to make up the situation in some sense. This might also explain why Klein devoted the second and third lectures together with a substantial part of the eleventh lecture to Lie and his work. Klein wrote:

> I have the more pleasure in placing before you the results of Lie's investigations as they are not taken into due account in my paper on the foundations of projective geometry [...] in 1890 nor in my (lithographed) lectures on non-Euclidean geometry delivered at Göttingen in 1889–1890.

These efforts of Klein did not have immediate effects on Lie. Shortly after the Evanston Colloquium Lectures, Lie wrote to Adolf Mayer [12, p. 351] and compared Klein to

> an actress, who in her youth dazzled the public with glamorous beauty but who gradually relied on ever more dubious means to attain success on third-rate stages [which

> should be interpreted as the Mathematical Congress in Chicago and the Evanston Colloquium Lectures].

The relation between Klein and Lie was complicated. Though Klein was younger than Lie, in their interaction and academic careers, Klein always played the role of a senior friend.

Their breakup was finally made up. In 1894 in the Russian city of Kazan, an international prize was set up to commemorate the great geometer Lobachevsky. The prize was to be awarded to geometric research, particularly in the development of non-Euclidean geometry. In 1897 Klein was asked by the prize committee to write a report on Lie's work, which helped Lie to receive the award in 1897, the first time this prize was given. In his report, Klein emphasized Lie's contribution in the third volume of *Theorie der Transformationsgruppen*, and Klein's report was published in the *Mathematische Annalen* one year later.

In 1898, Lie wrote to Klein and thanked him for the report. This was the first letter since their breakup in 1892. Lie also told Klein that he had resigned from his chair at Leipzig and would return to Norway soon. Unfortunately, Lie died shortly after he went back to Norway.

The final reconciliation between Lie and Klein was vividly described by a letter of Klein's wife written many years after the incident [18, p. xix]:

> One summer evening, as we came home from an excursion, there, in front of our door, sat the pale sick man. 'Lie!' we cried, in joyful surprise. The two friends shook hands, looked into one another's eyes, all that had passed since their last meeting was forgotten. Lie stayed with us one day, the dear friend, and yet changed. I cannot think of him and of his tragic fate without emotion. Soon after he died, but not until the great mathematician had been received in Norway like a king.

The conflict between Klein and Lie was natural in some ways but also very complicated and rather unfortunate. Similar things had happened before and after them and will continue to happen. Fortunately, it ended with a good reconciliation. For some related discussions and more details on this conflict, see [8] in this volume.

After discussing Lie's work, Klein moved to the classification of spaces which are locally the same as the Euclidean space and spaces of positive curvature. The work of Clifford was mentioned. The problem of classifying Clifford–Klein space forms is still not completely solved today.

The **last lecture** has a rather unusual title "The study of Mathematics in Göttingen." This should be understood in the context that at that time, almost every capable American student wanted to go to study in Göttingen, in particular to become a student of Klein. Klein started with "In this last lecture I should like to make some general remarks on the way in which the study of mathematics is organized at the university of Göttingen, with particular reference to what may be of interest to American students." One advice he gave was

> Would he not do better to spend first a year or two in one of the larger American universities? Here he would find more readily the transition to specialized studies,

> and might, at the same time, arrive at a clearer judgement of his own mathematical ability: this would save him from the severe disappointment that might result from his going to Germany.

These comments make one wonder if they can be applied to students who want to go to study abroad now, and I am thinking especially of students from China.

The comments of Klein on his Göttingen lectures also applied to the Evanston Colloquium Lectures:

> my higher lectures have frequently an encyclopedic character, comfortably to the general tendency of my programme. [...] My lectures may then serve to form the wider background on which [...] special studies are projected. It is in this way, I believe, that my lectures will prove of the greatest benefit.

Time has shown that Klein's Evanstan Colloquium Lectures have greatly influenced the rising of the American mathematical community.

10 The ambitious encyclopedia in mathematics

As briefly discussed before, Klein wrote many books, several of which are substantial.

But his most ambitious plan was to edit an encyclopedia of mathematical sciences. In 1894 he launched the idea of an encyclopedia of mathematics including its applications, which became the Encyklopädie der mathematischen Wissenschaften. According to [18, p. xiii],

> The Enzyklopädie was, from Klein's point of view, an effort to render accessible to his pupils, to himself, and to the mathematical public at large, the bulk of existing mathematics. One day in the '90's the concept of the Enzyklopädie was formulated by Klein in the presence of the writer: the progress of mathematics, he said, using a favourite metaphor, was like the erection of a great tower; sometimes the growth in height is evident, sometimes it remains apparently stationary; those are the periods of general revision, when the advance, though invisible from the outside, is still real, consisting in underpinning and strengthening. And he suggested that such was the then period. What he meant, he concluded, is a general view of the state of the edifice as it exists at present.

The first volume was published in 1898, and the last volume was published in 1933. Its total length for the six volumes is over 20000 pages, and it provided an important standard reference of enduring value. It is a pity that it appeared only in German and French editions and did not continue.

Klein's former student Walther von Dyck was the chair of the editorial board and did much of the actual work. He explained the mission of this huge project [17]: "The mission was to present a simple and concise exposition, as complete as possible, of the body of contemporary mathematics and its consequences, while indicating with a detailed bibliography the historical development of mathematical methods from the beginning of the nineteenth century."

But Klein was deeply involved in the global plan and design of the whole project. Because of this, it is also called "*Klein's encyclopedia*."

In view of Klein's perspective on mathematics, this seems to have been a befitting final project for him and is beneficial to the whole mathematics community.

11 Klein's death and his tomb

Klein was always full of energy. Even during the last two years of his life, when he laid almost helpless, he never complained and remained clear to the end, working and correcting proof-sheets. At half past eight on the evening of June 22, 1925, Klein died painlessly at the age of 75, a honorable and kingly death.

After Klein died in June 1925, Hilbert said in a speech in the next morning [13, p. 178]:

> But the event after it happened touched us all deeply and affected us painfully. Up until yesterday Felix Klein was still with us, we could pay him a visit, we could get his advice, we could see how highly interested he was in us. But that is now all over.

According to C. Reid [13, p. 178],

> Everything they saw around them in Göttingen was the work of Klein, the collection of mathematical models in the adjoining corridor, the Lesesimmer with all the books on open shelves, the numerous technical institutes that had grown up around the University, the easy relation they had with the education ministry, the many important people from business and industry who were interested in them ... They had lost a "great spirit, a strong will, and a noble character."
>
> An era had come to an end.

Courant said [14, p. 179],

> Many who knew him only as an organizer [...] found him too harsh and violent, so he produced much opposition to his ideas [...] which a gentler hand would easily have overcome.

On the other hand, according to C. Reid [14, p. 179],

> Yet his nearest relatives and colleagues and the great majority of his students had known always that behind the relentlessly naive drive, a good human being stood.

In his life, Klein always appeared as a formal, stern and efficient German professor. This formality and people's respect towards him can also be seen on his tomb stone. The inscription on the stone includes *"Felix Klein, A Friend, Sincere and Con-*

stant",[9] and was well-arranged and filled the whole stone. His tomb is close to the Chapel of the huge cemetery in Göttingen, a rather important spot there.

12 Major mathematicians and mathematics results in 1943–1993 from Klein's perspective

More than 100 years ago, Klein gave 12 lectures to summarize the status of mathematics at that time, or more precisely an overview of the major mathematicians and their main contributions in the previous 25 years. One might ask the following question: if Klein had given another series of lectures in 1993 (one hundred years later) on the most important achievements in mathematics in the previous half century, i.e., from 1943 to 1993, what mathematicians and mathematical results would he have discussed?

Probably one thing is clear: twelve lectures would not be enough. There have been many more people doing mathematics research in the twentieth century that the few mathematicians at several universities in Europe in the ninteenth century.

From the above description of Klein's life and work, it seems that he would value mathematical works that shed new lights on interconnections between different subjects, that contribute globally to multiple subjects, open up new fields and generate new problems and results, but might not emphasize some isolated theorems or solutions of major conjectures which kill the subjects. It is tempting to conjecture that Klein might talk about the following mathematicians, their works and related subjects:

1. Marston Morse (1892–1977). He is best known for his work on the calculus of variations in the large, in particular Morse theory, which is a fundamental tool in topology and geometry. Though Morse theory was the underlying theme of his work, he was very productive and made many substantial contributions to related topics.
2. Carl Siegel (1896–1981). He worked in and made fundamental contributions to both number theory and celestial mechanics. His work in number theory (both analytic and algebraic) and automorphic forms in multiple variables had far reaching consequences in analytic number theory, arithmetic number theory, and the theory of arithmetic subgroups. His contribution in celestial mechanics is also both broad and deep.
3. Andrey Nikolaevich Kolmogorov (1903–1987). He made essential contributions to many fields such as probability theory, topology, intuitionistic logic, turbulence, classical mechanics and computational complexity. Without his work, probability theory will not be like it is now. Besides his work on probability, KAM theory is one of the many deep theories he developed.

---

[9]This inscription is quoted from [13, p. 179]. On the actual tomb stone, the words "Felix Klein, A Friend" are not there. Maybe C. Reid wanted to emphasize the friendly aspect of the stern Klein.

4. André Weil (1906–1998). He did foundational work in number theory, arithmetic algebraic geometry, algebraic geometry and differential geometry. For example, the foundation of algebraic geometry, the Chern–Weil theory, and abelian varieties are some of his many deep contributions, which cover a broad range of topics such as topology, differential geometry, complex analytic geometry and Lie theories. The Weil conjecture has had a huge impact on modern mathematics, in particular arithmetic algebraic geometry. He is also one of the founding fathers, maybe the leader, of the Bourbaki group.
5. Jean Leray (1906–1998). He made foundational contributions to both partial differential equations and algebraic topology, and he combined methods from these seemingly different subjects to solve difficult problems. He is probably most famous for his introduction and work on sheave theories, and spectral sequences.
6. Hassler Whitney (1907–1989). He is one of the founders of singularity theory, and did foundational work in differential topology such as embeddings and in algebraic topology and differential geometry such as characteristic classes.
7. Lev Semenovich Pontryagin (1908–1988). He made major discoveries in many subjects such as topological groups and analysis on topological groups, algebraic topology (in particular characteristic classes) and differential topology such as cobordism theory. He also made important contributions to differential equations and control theory.
8. Claude Chevalley (1909–1984). He was a highly original and cultured person and he made important contributions to number theory, algebraic geometry, class field theory, finite group theory, and algebraic groups. Basic notions he introduced include Chevelley groups with spectacular applications to finite simple groups and adeles, which are basic in modern number theory. His books on Lie theories have had a huge impact.
9. Shiing-Shen Chern (1911–2004). He was regarded as one of the leaders in global differential geometry, which emerged as a major theory in the 20th century. His work covered all the classic fields of differential geometry and was most famous for the Chern–Weil theory and Chern classes, which are widely used in modern mathematics.
10. Oswald Teichmüller (1913–1943). Though he died at the age of 30, and some of his papers were published in 1944 in a famous (or infamous) Nazi journal and hence were relatively unknown, he introduced quasiconformal mappings and differential geometric methods into complex analysis and solved the problem of moduli asked by Riemann. The idea of rigidifying moduli problems and the resulting Teichmüller theory has had a long lasting impact on many subjects ranging from algebraic geometry to low-dimensional topology.
11. Israel Moiseevich Gelfand (1913–2009). He made major contributions to many branches of mathematics, including group theory, representation theory of non-compact Lie groups, differential equations, functional analysis, and applied

mathematics. He also educated and inspired generations of students through his legendary seminar at Moscow State University.

12. Kunihiko Kodaira (1915–1997). He made fundamental contributions in algebraic geometry and complex geometry such as the Kodaira embedding theorem, Hodge theory, the deformation theory of complex manifolds, and the classification of algebraic surfaces.
13. Kiyoshi Ito (1915–2008). He made fundamental contribution to stochastic processes. His theory is called Ito calculus and is widely applied in various fields, in particular in financial mathematics.
14. Jean-Pierre Serre (1926–). He has made fundamental contributions to the fields of algebraic topology (such as Serre spectral sequence), algebraic geometry (such as GAGA and sheaf theory), number theory and several other fields such as homological algebra and combinatorial group theory. His many books also have educated many people around the world and are models of exposition.
15. Alexandre Grothendieck (1928–2014). He was the central figure behind the creation of the modern theory of algebraic geometry and also made major contributions to many subjects such as functional analysis. In some sense, the language and landscape of mathematics changed after his work. His generalization of the classical Riemann–Roch theorem launched the study of algebraic and topological K-theory and also played an important role in general index theory, and his discovery of ℓ-adic étale cohomology was the key tool in the proof of the Weil conjectures, completed by his student Pierre Deligne.
16. John Nash (1928–). He made highly original and fundamental contributions to game theory, differential geometry and partial differential equations such as the De Giorgi–Nash–Moser theorem and the Nash embedding theorem.
17. Michael Atiyah (1929–). He laid the foundations for topological K-theory and index theories. In particular together with Singer he proved the Atiyah–Singer index theorem, which has been widely used in both mathematics and physics. He also made many fundamental contributions towards interaction between geometry and analysis.
18. Goro Shimura (1930–). He made important and extensive contributions to arithmetical geometry and automorphic forms. One key concept is the one of a Shimura variety, which is the higher-dimensional equivalent of modular curve and plays an important role in the Langlands program. He also made the important Taniyama–Shimura conjecture on modularity of elliptic curves and contributed substantially to various topics in arithmetical geometry and automorphic forms.
19. John Willard Milnor (1931–). He did pioneering work in topology by proving the existence of exotic differential structures on 7-dimensional spheres. This made people realize the subtle difference between smooth and topological structures and had a huge influence on differential topology. His other works on K-theory, Milnor fibration and his multiple books also have had great impacts on mathematics.

20. Robert Langlands (1936–). He made fundamental contributions to automorphic forms and representation theory, which has had a major effect on number theory. He proposed a collection of far-reaching and influential conjectures, called the Langlands program, that relate Galois groups in algebraic number theory to automorphic forms and representation theory of algebraic groups over local fields and adeles.
21. Jacques Tits (1930–). His best known work is the theory of Tits buildings, natural spaces for algebraic groups. It has unexpected far reaching consequences that arose in a broad range of subjects. In some sense, he made the Erlangen program concrete for algebraic groups, in particular exceptional algebraic groups over all fields.
22. William Thurston (1946–2012). He was highly original and made fundamental contributions to the study of 3-manifolds. His work and perspective completely changed the landscape of 3-dimensional topology through his geometrization conjecture.
23. Mikhail Leonidovich Gromov (1943–). He is probably known as the mathematician with the largest number of ideas. He made original and important contributions in many different areas of mathematics including differerential geometry, coarse geometry, differential equations, symplectic geometry and geometric group theory.
24. Gregori Aleksandrovich Margulis (1946–). He made fundamental and highly original contributions to structure properties and applications of lattices in semisimple Lie groups, and initiated the approach of using ergodic theory to solve diophantine problems in number theory and questions in combinatorics and measure theory. The Erlangen program is mainly concerned with Lie groups, and Margulis showed the importance of discrete subgroups of Lie groups in many contexts.
25. Shing-Tung Yau (1949–). He was one of the first persons who combined differential equations and geometry, and efficiently used analysis to solve outstanding problems in algebraic geometry, differential geometry, low dimensional topology, mathematical physics such as general relativity and string theory. His lists of problems have had major impact on the broad areas of mathematics related to geometry and analysis.
26. Edward Witten (1951–). He made contributions in mathematics and helped bridge gaps between fundamental physics and other areas of mathematics. For example, he gave a simple proof of the positive mass conjecture using the idea of supersymmetry, and interpreted Morse theory, elliptic genus and other fundamental results in mathematics via supersymmetry. Using his physics intuition, Witten also provided many stimulating problems and conjectures for generations of mathematicians. Klein always liked physics and emphasized the importance of the interaction between mathematics and physics.

Acknowledgements. I would like to thank Athanase Papadopoulos for his interest in this article, for reading this chapter carefully and for valuable suggestions, and

Hubert Goenner for carefully reading an earlier version of this chapter and his many constructive suggestions. I would also like to thank Manfred Karbe for his careful reading of this chapter and for suggesting the reference [4]. During the preparation of this chapter, I have consulted the books and articles related to Felix Klein that appear in the bibliography below, among other references. My debt to them is clear and immense.

Bibliography

[1] N. A'Campo and A. Papadopoulos, On Klein's *So-called Non-Euclidean geometry*. In *Sophus Lie and Felix Klein: The Erlangen program and its impact in mathematics and physics*, ed. L. Ji and A. Papadopoulos, European Mathematical Society Publishing House, Zürich, 2015, 91–136.

[2] O. Bolza, H. Maschke, E. Moore and H. White, *Mathematical papers read at International Congress of Mathematicians (1893: Chicago, Ill.)* New York: Macmillan, for the American Mathematical Society, 1896.

[3] W. Burau and B. Schoeneberg, Klein, Christian Felix. In *Complete Dictionary of Scientific Biography*. Vol. 7. Detroit: Charles Scribner's Sons, 2008. 396–400.

[4] M. Eckert, *Arnold Sommerfeld. Science, life and turbulent times 1868–1951.* Translated from the German original by Tom Artin. Springer, New York, 2013.

[5] G. Frei, Felix Klein (1849–1925) – A biographical sketch. In *Jahrbuch Überblicke Mathematik* 1984, Bibliographisches Institut, Mannheim, pp. 229–254.

[6] G. Halsted, Biography: Felix Klein. *Amer. Math. Monthly* 1 (1894), no. 12, 416–420.

[7] I. James, *Remarkable mathematicians. From Euler to von Neumann.* MAA Spectrum. Mathematical Association of America, Washington, DC; Cambridge University Press, Cambridge, 2002.

[8] L. Ji, Sophus Lie, a giant in mathematics In *Sophus Lie and Felix Klein: The Erlangen program and its impact in mathematics and physics*, ed. L. Ji and A. Papadopoulos, European Mathematical Society Publishing House, Zürich, 2015, 1–26.

[9] F. Klein, *Lectures on mathematics*. The Evanston Colloquium. A. Ziwet & H. S. White eds. New York and London. Macmillan and Co. (1894). A reprint of the 1910 edition with the preface by Osgood, American Mathematical Society, 2000.

[10] F. Klein, The present state of mathematics. In *Mathematical Papers read at the International Mathematical Congress*, held in Connection with the World's Columbian Exposition Chicago, New York: Macmillan, for the American Mathematical Society, 1896, pp. 133–135.

[11] F. Klein, *Erlangen program*, translated by M. W. Haskell, New York Math Soc., 2 (1892/1893), 215–249.

[12] K. Parshall and D. Rowe, *The emergence of the American mathematical research community, 1876–1900: J. J. Sylvester, Felix Klein, and E. H. Moore.* History of Mathematics, 8. American Mathematical Society, Providence, RI, London Mathematical Society, London, 1994.

[13] C. Reid, *Hilbert. With an appreciation of Hilbert's mathematical work by Hermann Weyl.* Springer-Verlag, New York-Berlin 1970.

[14] C. Reid, *Hilbert–Courant.* Springer-Verlag, New York, 1986.

[15] D. Rowe, Three letters from Sophus Lie to Felix Klein on Parisian mathematics during the early 1880s. Translated from the German by the author. *Math. Intelligencer* 7 (1985), no. 3, 74–77.

[16] D. Rowe, Felix Klein, Adolf Hurwitz, and the "Jewish question" in German academia. *Math. Intelligencer* 29 (2007), no. 2, 18–30.

[17] W. von Dyck, Enzyklopädie der mathematischen Wissenschaften. *Proceedings of the International Congress of Mathematicians*, vol. 2, pp. 123–134, 1908.

[18] W. H. Young, Christian Felix Klein, *Proceedings of the Royal Society of London. Series A.* Vol. 121, No. 788 (Dec. 1, 1928).

Chapter 3

Klein and the Erlangen Programme

Jeremy J. Gray

Contents

1 Introduction

Klein's Erlangen Programme[1] is one of the best known papers on mathematics written in the 19th century. Or, rather, it is one of handful of books and papers from the century that people still mention today, whether they have read it or not. The chapter will describe the fundamental ideas in the Programme and their significance, and then consider how the young Felix Klein came to those ideas by building on an influence not often considered: the synthetic projective geometry of von Staudt. It will only briefly discuss the impact of the Erlangen Programme on later work, which is entangled in complicated ways with the work of Sophus Lie, Henri Poincaré, and others.

2 The Erlangen Programme

The Erlangen Programme is the name given to a paper Klein circulated on the occasion of his appointment as a Professor at the University of Erlangen in 1872 under the title of "A comparative review of recent researches in geometry." Inevitably, his presentation of his ideas fell far short of modern standards of precision. It also belongs to the genre of programmes or manifestos rather than research papers, and like those today it makes claims about what can be done without providing much detail.

[1]The programme has many names, because Erlangen is the name of the place, Erlanger is the genitive of the name, and Programme has an English and American spelling (Program). Some authors quoted in this chapter refer to it as the E. P.

It seems best, therefore, to present it in a summary that sticks close to the original and only later to offer some modern comments.

The Programme opens with the claim that projective geometry now occupies the first place among advances in geometry, having incorporated metrical geometry, so that it can be said that projective geometry embraces the whole of geometry. The viewpoint that unified these two can be extended to include other geometries, such as inversive geometry and birational geometry, Klein went on, and it seems advisable to do so because (p. 216) "geometry, which is after all one in substance, has been only too much broken up in the course of its recent rapid development into a series of almost distinct theories."[2] The essential idea, said Klein, was that of a group of space-transformations. This he defined merely as a collection of transformations of a space to itself that is closed under composition. Only in the English translation of 1893 did he add that one should further specify the existence of the inverse of every transformation. He then directed attention to those transformations that leave the geometric properties of configurations unaltered.

> For geometric properties are, from their very idea, independent of the position occupied in space by the configuration in question, of its absolute magnitude, and finally of the sense [today we would say, orientation] in which its parts are arranged. The properties of a configuration remain therefore unchanged by any motions of space, by transformation into similar configurations, by transformation into symmetrical configurations with regard to a plane (reflection), as well as by any combination of these transformations. The totality of all these transformations we designate as the *principal group* of space-transformations; *geometric properties are not changed by the transformations of the principal group.* And, conversely, *geometric properties are characterized by their remaining invariant under the transformations of the principal group.* (p. 218, italics in original.)

Klein was clear that each transformation was to be understood as a map of the space to itself, not as a map of one figure to another that in some way left the space fixed. This marked an important shift in the focus of attention in geometry. Previously there had been figures to study, which to be sure were in a space, and these figures could be studied by transforming them to others of the same kind if the problem permitted. In this way projective geometry commended itself because any non-degenerate conic could be replaced by a circle, which usually simplified the problem considerably. In Klein's view there was a space, and a transformation group. The group picked out the figures, and any part of the space was to be treated on a par with all of its transformations.

Klein then passed from space to the consideration of any 'manifoldness', as he called it, and set out

> the following comprehensive problem: Given a manifoldness and a group of transformations of the same; to investigate the configurations belonging to the manifoldness with regard to such properties as are not altered by the transformations of the group. (p. 218)

[2]For convenience, all quotations are taken from the English translation by M. W. Haskell and published as [21]. Klein was scrupulous in noting later amendments to the original publication.

Or, as he promptly rephrased it (p. 219, italics in original): "*Given a manifoldness and a group of transformations of the same; to develop the theory of invariants relating to that group.*" This is the redefinition of geometry that is the first defining feature of the Erlangen Programme.

Klein next argued that one may proceed by adding to the list of configurations, in which case the group that keeps them invariant will generally be smaller than the principal group, or conversely one may seek to enlarge the group, in which case the class of invariant configurations will generally shrink. Metrical considerations are introduced in projective geometry by adding a requirement that all configurations are considered with respect to a fundamental one (for example, a conic, which becomes the points at infinity in the metrical space). Klein also noted that if one wishes to include duality as a fundamental feature of projective geometry and the transformations that exchange a figure and one of its duals then points and lines must both be considered as space elements.

The second defining feature of the Erlangen Programme is less often remembered, but it is the idea of isomorphic group actions, or, as Klein put it, the idea that if a geometry is given as a manifoldness A investigated with respect to a group B and in some way A is converted into a manifoldness (Klein was very vague here) A' then the group B can be regarded as a group B' that gives A' a geometry. In this spirit, Klein asked his readers to consider the space of binary forms. This can be regarded as the geometry on a straight line with the group of linear transformations – we would say the geometry of the projective line with the group of Möbius transformations. The projective line is in a one-to-one correspondence with a conic, and the group now becomes the group mapping this conic to itself, and so the geometry of points on a conic can be used to import geometrical ideas into the study of binary forms.

Considerations of this kind led Klein to discuss what he called Hesse's principle of transference. In this process the group is held fixed but allowed to act on different spaces. Klein gave the example of the group of Möbius transformations just mentioned now acting on the set of point-pairs on a conic. Each point-pair defines a line in the projective plane, and so the group now acts on the projective plane with the line considered as space-element. Then, because projective geometry with respect to a conic is identical with projective metrical geometry, Klein concluded that the theory of binary forms and the plane projective metrical geometry are the same. Here Klein referred to his second paper on non-Euclidean geometry that was about to appear [18].

Klein now explained how inversive geometry fitted into this framework. He said that this was a geometry not known to German mathematicians, but that reflected his youthful ignorance rather than a historical reality. Then he gave a longer exposition of some of Lie's ideas about sphere geometry, which he related to inversive geometry. Then he turned to the group of rational transformations of space and the group of birational transformations of a curve on surface in space. He commented very briefly on how analysis situs would fit into the framework, which he defined very loosely as being concerned with all transformations that create infinitesimal distortions, and on the group of all point transformations, by which he meant transformations that are linear on infinitesimal neighborhoods. After that came the groups of contact transformations that Lie had studied. He gave the example of contact transformations of three-dimensional space, so the group acts on the space of quintuples x, y, z, p, q, where $p = \frac{\partial z}{\partial x}$ and $q = \frac{\partial z}{\partial y}$ and the expression $dz - pdx - qdy = 0$ is preserved. The intended application of this theory, said Klein, was the classification of surfaces defined either by a first-order partial differential equation or by a system of such equations. Another way the Programme could be used, he said, was in the study of manifoldnesses of constant curvature.

Klein concluded with some remarks about how this approach to geometry would necessarily be coordinate-free, and would therefore be in line with the contemporary theory of invariants in algebra and even more with the "so-called" Galois theory of equations as recently presented by Serret and Jordan. There, as here, the primary object of study was the group. The manifesto ended with a series of seven notes amplifying various points made in the earlier sections. Of these the fifth prudently attempted to head off any philosophical objection to talk of non-Euclidean geometry by making the account here entirely mathematical, although it was also, he said, in his view "an indispensable prerequisite to every philosophical discussion of the subject." The seventh discussed the geometry of binary cubics and quartics from the new point of view.

2.1 Comments Klein did not pause to define what he meant by a manifoldness. He may well have had Riemann's vague and general concept in mind [27], which is usually glossed as some sort of differentiable manifold, possibly of infinite dimension, usually with a metric of some kind. He almost certainly only had in mind what we would regard as a limited range of examples, but they were novel in his day, and these are the spaces obtained from real projective space of some dimension by forming for example the set of all lines, or all conics, or all cubic curves in that projective space. Thus the space of all conics in the plane forms a five-dimensional space in which each plane conic is referred to a space-element. On the other hand, there is no reason to restrict Klein's ideas to groups acting on manifolds in the modern sense of the term.

The fundamental idea of the Erlangen Programme is usually taken to be that a geometry consists of a space and a group acting on that space. From this point of view it is often contrasted with, for example, the many Riemannian geometries that admit only a trivial group of isometries. In this spirit is then observed that a lot of the work of Élie Cartan and Hermann Weyl can be seen, and was presented, as an attempt to extend the Erlangen Programme to this more general setting. Klein's aims were rather different. There were two: the unification of all geometry within projective geometry and the wish to extend the scope of geometrical thinking.

Klein's main example is that of projective geometry and the way the metrical geometries, most importantly non-Euclidean geometry, fit in as special cases obtained by passing to a subgroup of the projective group. Klein was explicitly concerned to unify geometry by connecting all the examples he knew to projective geometry, which can be regarded as the mother geometry. Curiously, at this stage he was unaware of Möbius's affine geometry, which would have fitted very well into his presentation.

The remarks about analysis situs or topology and the group of all point transformations of a space fit in must be regarded as typical of these grand manifestos. They are strikingly poorly defined, but that was all that could reasonably be said given the mathematics of the day, and quite without any hint of a potential pay-off.

What Klein obviously saw much more clearly was how, on his definition of geometry, geometrical ways of thinking could be brought to bear on topics hitherto regarded as exclusively algebraic. This was to be a leitmotif for Klein: all his life he advocated casting one's ideas in geometrical form as a test of one's understanding and as a way to find new results and better proofs of old ones. Writing in 1923 about his hopes on taking up a professorship in Leipzig in 1880, he said (Klein 1923c, 20):

> I did not conceive of the word geometry one-sidedly as the subject of objects in space, but rather as a way of thinking that can be applied with profit in all domains of mathematics.

His own work frequently demonstrated this point of view, never more so that his work on subgroups of the modular group [19] and in his book on the icosahedron [20].

3 From von Staudt to Klein

The circumstances surrounding the production of the Erlangen Programme are well known and can be described briefly. Felix Klein had gone to the University of Bonn in 1865 at the age of sixteen and a half intending to study physics under Plücker, but he found that Plücker had switched back to the study of geometry, and so he began to work on line geometry in 1871. When Plücker died unexpectedly the next year, Klein was persuaded by Clebsch that only he knew enough to see the second volume of Plücker's study of line geometry into print, and in this way Klein was catapulted into the ranks of research mathematicians without, as he later noted, ever having taken a course in the integral calculus. He was also enabled by Clebsch, he believed, to apply for and obtain a Professorship at the University of Erlangen where von Staudt had worked all his life, and he went there in 1872 – he found it backward and dormant. He gave an inaugural address on his arrival there, which, as Rowe has pointed out, was on mathematics education and is not the Erlangen Programme.[3] The famous programme was distributed as a pamphlet on the day, mailed to a number of universities in Europe, and very likely sat forgotten on library shelves for some years thereafter.

Klein was not, after all, very well-known. There was doubtless a ripple of awareness that somebody had become a Professor at the strikingly young age of 23, but it was in a small, sleepy university. He had understandably very few publications to his name, including some joint papers with his older friend the Norwegian mathematician Sophus Lie, and he had made a brief visit to Paris in 1870 to see Camille Jordan

[3] See [28] and [29].

that had been cut short by the outbreak of the Franco–Prussian war. But the other trip outside Bonn had been a failure: a painful sojourn to Berlin where he had not impressed Weierstrass and Kummer and had in fact laid the foundations of a lifelong hostility between himself and the elite of the older generation of German mathematicians.

His time in Berlin was worthwhile not only because that was where he had met Lie for the first time, and made common cause with him over geometry, which they felt was neglected in Berlin, but because there he had become friends with Otto Stolz. Stolz was seven years older than Klein, and was able to fill in some of the gaps in the younger man's severely compressed education. In particular, he told him of the work of Bolyai and Lobachevsky, and of the work of von Staudt. Klein already knew of Cayley's opinion that "descriptive geometry is all geometry" – that is to say, that projective geometry is all geometry, and he formed the idea, against the opposition of Weierstrass, that non-Euclidean geometry had to fit in there too.

The two papers that Klein wrote on the "so-called non-Euclidean geometry" of 1871 and 1873 in fact do a better job than the Erlangen Programme at explaining how the new, metrical geometry fits into projective geometry, and they remain the clearest example of what he had in mind. These papers are described in full detail in the paper "On Klein's so-called non-Euclidean geometry" by Norbert A'Campo and Athanase Papadopoulos in this volume [1], but it will be helpful to comment on them briefly here. The paper [16] describes non-Euclidean geometry as a space which is mapped to itself by a group. The space is the interior of a non-degenerate conic in the plane, and the group is the group of all projective transformations of the plane that map this space to itself. There is an evident notion of a straight line in this space, namely that part of a projective line that lies in the space, and Klein showed how to equip the space with a metric that made the transformations of this geometry into isometries. In this way, non-Euclidean geometry was exhibited as a geometry on a subset of projective space with a transformation group that is a subgroup of the group of projective. In short: non-Euclidean geometry is a sub-geometry of projective geometry – and that is the theme of the Erlangen Programme. His paper [18] extends these arguments to describe non-Euclidean geometry in any number of dimensions.

Put this way, the full novelty of the Erlangen Programme may not be apparent. The idea that a geometry is a space with a group acting on it and that the geometric properties of the space are those that are invariant under the action of the group was not exactly new. Many geometers would have agreed that they studied Euclidean or projective space, and non-Euclidean geometry was accepted by mathematicians as legitimate – but not by philosophers, which is why Klein called it the 'so-called' non-Euclidean geometry. Mathematicians would also have agreed that one can transform figures in various ways, and indeed that there are different sorts of transformations for different purposes. Plane geometry, for example, admits Euclidean isometries, similarities, and affine transformations, as well as projective ones.

Klein did not propose a wholesale admission of new spaces, indeed he was exclusively concerned with spaces that are subspaces of projective space picked out by suitable subgroups of the projective group. But the clarity with which he asserted that geometry – any geometry – simply *is* a space and a group action on that space was new.

Klein had picked up the group concept either from his own reading or from talking to Sophus Lie. Clebsch had also underlined its importance, and Jordan thanked Clebsch for helping him with finding the groups of various geometric figures in his *Traité* [15] of 1870. And of course Klein had gone to Paris with a view to learning more about group theory from Jordan. Klein took the group concept to mean what we would call a transformation group rather than an abstract group, and indeed it was not even that, because he saw no need in 1872 to specify the existence of inverses for every element.

The concept of a transformation group gave him the first clear way to resolve the well-known nagging doubt about the definition of cross-ratio in projective geometry. Recall that the usual definition of the cross-ratio of four points A, B, C, D on a line is $AB.CD/AD.CB$. But projective transformations do not preserve length, so how can it be claimed that the cross-ratio is the quotient of two products of lengths? A similar problem arises if one defines the cross-ratio of four lines through a point in terms of the sines of the angles between the lines.

Klein's answer was clear. Consider briefly how lengths are introduced in Euclidean geometry. We start with some primitive concepts, such as that of the straight line. We then consider only those transformations that map line segments to line segments and indeed cannot map a line segment onto a proper subset of itself. We now have a suitable candidate to be the unit of length: an arbitrary but fixed line segment AB. Familiar elementary constructions now allow us to replicate the segment indefinitely along the line ℓ that it defines, and to divide it into any number of equal parts, so we are now able to measure lengths along the line that are rational multiples of the unit length. We appeal to the continuity of distance to obtain arbitrary distances along the line ℓ, and then we can measure the length of any line segment in the plane by moving it so that one end point coincides with A and the segment now lies along ℓ. We can also move the unit segment so that one of its end points is in an arbitrary position in the plane, and so that it points in any direction, so we can, for example, choose the perpendicular to ℓ through A and a direction on that line. We are now also in a position to define a rectangular coordinate grid on the plane. Experts will recognize the various axioms that have to be satisfied before this process can be carried out rigorously ([10] is the best modern guide) but this brief account exhibits the salient features we want.

We must have some primitive figures that are mapped to others of the same kind by all the transformations we consider, and in the case of Euclidean geometry this includes the point pair and the idea that a point pair defines a line segment and a line. We require that the transformations are transitive on points and lines, but not point pairs. Klein's insight was that an exactly similar argument involving quadruples of collinear points and transformations that can map any triple of collinear points to any triple of collinear points but is not transitive on quadruples will suffice to define the cross-ratio of a quadruple of points.

Klein learned this insight from von Staudt.

4 Von Staudt

Georg Karl Christian von Staudt is often called the Euclid of synthetic projective geometry, and like his distinguished predecessor he is hard to read. He did at least provide a short Foreword to his *Geometrie der Lage* [31] in which he observed that it had become appropriate to distinguish between the geometry of position and metric geometry on the grounds that in the former there are theorems that do not involve quantities but only ratios. He had therefore sought to give the geometry of position independent foundations and to give an account of those properties of second-order curves and surfaces that can be treated in that spirit. He noted that although every geometry book must start from general considerations most got down too soon to details about the congruence and similarity of triangles and did not deal with several other ideas with the corresponding generality.[4]

He began his own account with the statement that geometry is about an unbounded space, which he took to be three-dimensional. A body is a bounded region of space; bodies are divided by surfaces, surfaces by curves, and lines by points, and the point is indivisible. Every surface has two sides, as does every line in a surface and every point in a line. Bodies, surfaces, lines and points are called geometric figures. Three points A, B, C on a closed curve will be traversed either in the order A, B, C or A, C, B. Three points on a curve that does not intersect itself form a pair of outer points separated by one in the middle, and they divide the line into four segments. And so he went on, making statements that we can mostly rescue and some that we must regard as axioms.

The primitive elements in this geometry are the point and the straight line. A straight line, he said, may be called a ray (Strahl) and is divided by any point on it into two half rays. The useful concepts are that of a pencil of rays (Strahlbündel), which is made up of all the rays through a point, and a pencil of half rays (all the half lines emanating from a point). But von Staudt went on to define many other figures, including conic sections, in ways that did not, even covertly, rely on the concept of distance. He proved a number of fundamental theorems, for example that if a straight line passes through two points of a plane then it lies entirely in the plane, and that if a plane meets a line not lying in the plane in a point then it meets it only in that point, which can be called the trace of the line in the plane or of the plane in the line. Properties of planes and pencils of rays followed, for example, that a pencil of rays contains infinitely many planes, and every plane that has a ray in common with a pencil of rays defines a plane pencil.

In §5 von Staudt introduced infinitely distant elements. Von Staudt said two straight lines that lie in the same plane and do not meet are mutually parallel, and he claimed that through any point not lying on a given line there is a parallel line. Given three planes that meet in three lines, the lines either meet in a point or are mutually parallel. Two planes that do not meet are said to be parallel. Von Staudt noted that in many cases it was possible to define a point given a pencil of lines, and so by extension one could speak of an infinitely distant point as the common point

[4] I give a minimalist account of von Staudt's two books here, which were much more rich and general than I shall describe. For a much fuller account, see [25].

of a pencil of parallel lines. In this way every line is endowed with a unique point at infinity, in each plane there is a line of point at infinity and the collection of all points at infinity may be said to form a plane at infinity.

Two basic figures are said to be related if their elements are in a one-to-one correspondence, and the simplest way in which a straight line s and a plane pencil whose axis does not cut the given straight line can be related is if one considers the straight line as a cut of the plane pencil. Two pencils of rays can then be related if they are cuts of the same plane pencil. Such a relation von Staudt called a perspective relation, and he extended it to included pencils of parallel rays and pencils of parallel planes. This allowed him to consider that there was only one kind of pencil of rays (or of planes)

In §6 von Staudt turned to a discussion of reciprocity or duality. In space, he said, a point and a plane can be interchanged, and every theorem that does not distinguish between proper and improper elements remains valid when this is done. He listed a number of simple examples of this in a two-column format, including statements about pencils. In §7 came a definition of a perspectivity, and a proof of Desargues' theorem and its dual, which uses only incidence considerations because von Staudt was working in three dimensions.

In §8 the idea of a harmonic figure is introduced, via the construction of the fourth harmonic point to three points on a line. In §9 the idea of a projective transformation is introduced: one figure is the projective transform of another figure if every harmonic figure in the one corresponds to a harmonic figure in the other. Such figures are said to be homologous (von Staudt also introduced the symbol $\bar{\wedge}$ for this relationship). Figures related by a perspectivity are projectively related. Duality is admitted, so four points in a line are projectively related to four concurrent lines through the four points. The so-called fundamental theorem of projective geometry was then proved, which states that a projective transformation of a line to another line is determined when it is known on three points, and a projective transformation of a plane to another plane is determined when it is known on four points (no three of which are collinear).[5] Two figures are said (p. 118) by von Staudt to be involution if any two homologous points P and P_1 are such that the homology exchanges P and P_1. We would say that an involution is a transformation of period 2.

Any reading of von Staudt's *Geometrie der Lage* would show that its author aimed at two things: generality (his arguments apply without change to a variety of figures) and independence of metrical or other extraneous considerations. Nowhere is this latter consideration more apparent than in the way he consigned his remarks about distance to an appendix of the book. This opens with the remark (p. 203):

> Just as affinity is a special case of a linear transformation, so is similarity a special case of affinity and congruence a special case of similarity. In similar systems two homologous angles, and in congruent systems two homologous are equal to each other. Two projectively related lines are similar when the infinitely distant points

[5]Here we should note that von Staudt's argument was insufficient, and the role of continuity needed to be further elucidated, as it was by Darboux and Klein in the early 1870s. See [7] and, in more detail, [30].

on each line correspond. If in addition any finite segment of one line is equal to its corresponding segment on the other, then the figures are congruent.

There then followed a detailed account of how the familiar metrical theory of conic sections can be introduced into projective geometry by astute use of the concept of a midpoint and of involutions.

The same attention to generality and autonomy characterize the second of von Staudt's accounts of projective geometry, the *Beiträge zur Geometrie der Lage* [32], published in three parts in 1856 and 1857. This is notable for the introduction of imaginary elements – not, it must be said, in a way that yields complex coordinates[6] – and of the calculus of throws (Würfe). This enabled von Staudt to associate numerical values to such projective figures as four points on a line or four lines through a point, and ultimately to introduce coordinates into projective geometry.[7]

A configuration of four collinear points von Staudt called a throw. He distinguished three degenerate cases: $ABCA$ and $BAAC$ he assigned the symbol 0 to; to $ABCB$ and $BABC$ he assigned the symbol 1; and to $ABCC$ and $CCAB$ he assigned the symbol ∞. At this point in the account these symbols are meaningless. He then developed a theory covering the addition and multiplication of throws, and then defined a way of associating a real number to a throw. In keeping with the philosophy he had followed in the *Geometrie der Lage*, von Staudt put these numerical considerations in an appendix at the end of Book II of the *Beiträge zur Geometrie der Lage*.

The numerical value of a throw depends on an arbitrary constant assigned to an arbitrary segment. Once this choice has been made, the value of a throw is the same for all throws homologous to the given one. Moreover, the value map is a homomorphism (in more modern terms): the value of a sum of two throws is the sum of the values of the throws, and the value of a product of two throws is the product of the values of the throws. Von Staudt then showed how to introduce metrical coordinates into projective space, in such a way that specific pairs of points are a distance 0, 1, or ∞ apart. Something like this had also been outlined by Möbius in terms of what are today called Möbius nets, but von Staudt's treatment was more general and more rigorous. Even so, it was not sufficiently rigorous, as Lüroth was to show in 1875. A fair assessment of von Staudt's presentation of projective geometry would be that it is incomplete, and much is needed to be done here and there to make it a satisfactory account of real projective geometry from an axiomatic point of view, but that it is one of those accounts of a subject good enough to require only repair, not replacement.

What Klein saw in von Staudt's presentation of geometry, allowing for the fact that in several places it needed to be improved, was an autonomous account of projective geometry. It was defined with all the rigor of any other branch of mathematics at the time, and it did not rely on Euclidean geometry in any way. The fundamental operations were certain transformations: involutions and homologies. And quite clearly, metrical, Euclidean geometry was derived as a special case. It remained for

[6]For an account of the fascinating ways in which complex algebraic quantities were interpreted, often in terms of the existence of fixed-point free involutions, see [3].

[7]For a recent account, see the treatment in [25].

him to fit non-Euclidean geometry into this framework, and here he was lucky: non-Euclidean geometry fits in more easily that Euclidean geometry. His account could be easier than von Staudt's treatment of Euclidean geometry, and he could achieve something new in the study of an exciting new geometry.

5 The influence of the Erlangen Programme

An extensive historical literature on the Erlangen Programme culminated in two contrasting papers some twenty years ago. Birkhoff and Bennett [2] who gave a strongly positive assessment of its influence, and Hawkins [11],[8] who, as they noted "challenged these assessments", pointing out that from 1872 to 1890 the E. P. had a very limited circulation; that it was "Lie, not Klein" who developed the theory of continuous groups; that "there is no evidence ... that Poincaré ever studied the Programm;" that Killing's classification of Lie algebras (later "perfected by Cartan") bears little relation to the E. P.; and that Study, "the foremost contributor to ... geometry in the sense of the Erlanger Programm, ... had a strained and distant relationship with Klein."

Birkhoff and Bennett's argument came in two parts. First, they traced the influences on Klein, dwelling on the contributions of Plücker and Clebsch who were decisive in promoting Klein's interest in geometry and noting without further comment the work of von Staudt. They note the intimate connections between Klein and Lie, and they hint at ways in which Klein could have picked up an appreciation of the idea of a group. And then they make the grand claim that

> in 1872, the E. P. was 20 years ahead of its time; it would take at least that long for the new perspective of Klein and Lie to gain general acceptance.

The second part of their paper is an attempt to assess its influence. Since there is no doubt that the Erlangen Programme owed a lot to discussions with Lie, it is right to say that it presents the perspective of Klein and Lie. It is much harder to sort out the influence of the Programme over the subsequent 20 years, and ascribe this part to Klein and that to Lie. The problem is, of course, the steady, and ultimately monumental, build-up of Lie's ideas, and their reworking by Killing, Cartan, and others, and the extent to which this has anything to do with the ideas in the E. P. Birkhoff and Bennett made the astute observation that the message of the E. P. is the way groups act on spaces – it outlines, as they say, a global approach to geometry – whereas Lie's work is overwhelmingly a local theory. So they turn to look at Klein's later work to see how he drew his own lessons, and they observe that Klein was to write extensively on discontinuous groups in the setting of automorphic functions and non-Euclidean geometry (the famous collaboration and competition with Poincaré) and, after his nervous breakdown, on the icosahedron.

---

[8]One can add Hawkins' monumental book [12] on Lie's theory.

Birkhoff and Bennett then note that in the 1880s and still more in the 1890s Klein rose to become the major figure in mathematics in Germany. He positioned himself well to capitalize on the decline of the centre in Berlin, and on reaching Göttingen he made a succession of brilliant hirings that quickly created the leading place for mathematics in the world. They suggest that this amplified his influence: the E. P. was, as it were, the seed of much of Klein's later work, it and the later work inspired others, and the more Klein's stock rose the more attractive his ideas came to be. As they note, their view is in line with the opinions of Max Noether [26] in his obituary of Lie, and Klein's obituarist and successor Richard Courant in [4].

The problem with this account is its naivety. There is no doubt that Klein had a major influence on mathematics, in particular by being a huge force in Göttingen. He actively promoted the geometrical point of view on many topics, and can be regarded as more of a geometer than an analyst or an algebraist. But a closer look at his career shows how much the Göttingen tradition, which he did so much to create in his historical writings, has colored how he is regarded today.

Klein himself said that his research career was brought to a halt by his nervous breakdown and that he never recovered completely: "the centre of my productive thought was, so to speak, destroyed" (see Klein 1923a, 585). The book on the Icosahedron [20] was part of his recovery programme, and thereafter he preferred to collaborate with younger men who, for their part, were happy to take up Klein's good ideas and make them work. But it is not clear, for example, if the four volumes he was to write with Fricke on automorphic functions were very influential, and they are not mentioned by Birkhoff and Bennett. More significantly, in the second half of the 1870s Klein had set himself the task of understanding and advancing the work of Riemann. This was not at all a body of ideas informed by group-theoretic ideas, nor was it projective in spirit. Riemann's ideas about geometry are metrical, deeply tied to topics in complex function theory, and topological. And what is quite clear is that by 1880, when Poincaré emerged on the scene to threaten Klein's rise to the position as the leading mathematician of his generation, Klein still had not fully appreciated what Riemann had been saying.[9]

Klein's major papers in the years 1878–1881 are best exemplified by his paper [19], in which he discussed the geometry associated with the group $PSL(2;7)$. He did find a Riemann surface associated to it, and in true Riemannian spirit he found a way to express it as an algebraic curve, but his take on it was then informed by his appreciation of the fundamental status of projective geometry. In Klein's view, the curve was a branched covering of the Riemann sphere. Poincaré, on the other hand, had little interest in projective geometry, and emphasized the differential geometric side of the story; in his view the curve was a quotient of the non-Euclidean disc. Klein moved fast. He came to the idea of the uniformization theorem before Poincaré, helped no doubt by his better understanding of the Riemannian moduli of an algebraic curve, but with every step he took he was moving away from the stand-

[9] In fact, it is hard to say what was Klein's greatest contribution to mathematics, because nothing he wrote has the status of a masterpiece, a work that changed the subject and did not merely add to it. This may be one reason why the influence of the Erlangen Programme is over-estimated, because it is asked to stand in for the great work on groups and geometry that Klein never wrote.

point he had adopted in the E. P. Equally, Poincaré's way of approaching this topic makes it very clear that he had not been influenced by the E. P. If he had not picked up the importance of groups in his education at the École Polytechnique, or from his own reading, then the most likely influence is Helmholtz's writings on geometry, which stressed the possibility that non-Euclidean geometry could not only be formally the correct description of physical space, but a geometry we could live and experience very much as we (think we) live and experience Euclidean geometry.[10]

It does not seem to me that Birkhoff and Bennett did enough to dislodge Hawkins' key point, (p. 463), that the view that it can be meaningfully regarded as one of the most significant and influential documents in the history of mathematics circa 1872 to 1922 is overly simplistic and essentially unhistorical. Since mathematicians tend to respect priority, the Erlanger Programm was pinpointed as the source of the group theoretic study of geometry, but it should be clear from what has been presented here that such a view, if interpreted as a historical statement of influence, ignores the contributions of Lie and his school. That would be a serious mistake since, as we have seen, no significant development of the ideas of the Erlanger Programm occurred without the involvement of the ideas and results of Lie and his school.

As Hawkins went on to document at length in his book [12], it was Lie's appreciation of the importance of thinking about groups that resulted in a body of work that transformed the whole domain of mathematics, particularly but not only the subjects of differential equations and geometry. Hawkins rightly added that ([11], 463), it would, of course, be equally simplistic and unhistorical to deny to the Erlanger Programm any significance or influence. To mention just one example, the E. P. was much appreciated by Corrado Segre and in due course by his student Fano (see ([12], 251)). But there can be little doubt that the reputation of the E. P., and its existence in at least six translations into other languages, owes more to the importance of Klein in the 1890s than to the impact of the E. P. itself.

In his criticisms of Birkhoff and Bennett, Hawkins often referred to the idea that mathematics is a kind of collage, and ideas occur and are taken up in contexts. This is a much more productive way to view the Erlangen Programme. Throughout the 19th century mathematicians had been finding groups and group actions in different domains in mathematics. The earliest significant occurrence may be in Gauss's *Disquisitiones arithmeticae* [6] in his work on the composition of forms. A powerful expression of the idea came up in Galois's memoirs, although they took a long time to be understood, but quite certainly they were understood and promoted by Jordan in his memoir on groups of motions [13], his paper on Galois theory [14], and his major book the *Traité des Substitutions et des Équations Algébriques* [15]. The use of transformations in various branches of geometry was ubiquitous, the systematic application of transformations (in the form of changes of variable) in the study of differential equations did not begin with Lie, although he took it to new depths of insight and effect. Poincaré too found many applications for the idea of a transformation group (see [8].) We need not look to the 20th century to see that the idea

[10] Influences on the young Poincaré are oddly hard to determine, but the picture he presented throughout his life of geometry emphasized the metrical and topological aspects and its experiential side; axiomatic and projective geometry were never his taste. For a further discussion, see [9].

of a group acting on a space would be a very powerful one. Klein's Erlangen Programme may not have started this runaway chain of ideas, it may not have been the most important single factor in promoting it, and it may even have missed several of its most eloquent examples, but it brought ideas together eloquently and it has validly proved to be a focus around which to organize and extend a profound idea.

Bibliography

[1] N. A'Campo and A. Papadopoulos, On Klein's *Non-Euclidean geometry*. In *Sophus Lie and Felix Klein: The Erlangen program and its impact in mathematics and physics* (ed. L. Ji and A. Papadopoulos), European Mathematical Society, Zürich, 2014.

[2] G. Birkhoff and M. K. Bennett, Felix Klein and his "Erlanger Programm". In *History and philosophy of modern mathematics*, Minnesota Studies in the Philosophy of Science, XI, Univ. Minnesota Press, Minneapolis, 1988, 145–176.

[3] J. L. Coolidge, *The geometry of the complex domain*. Clarendon Press, Oxford, 1924.

[4] R. Courant, Felix Klein. *Jahresbericht D. M. V* 34 (1926), 197–213.

[5] É. Galois, *The mathematical writings of Évariste Galois*, P. M. Neumann (ed.) European Mathematical Society, Heritage of European Mathematics, 2011.

[6] C. F. Gauss, *Disquisitiones arithmeticae*, 1801, English trl. W. C. Waterhouse, A. A. Clarke, Springer 1986.

[7] J. J. Gray, *Worlds out of Nothing; a course on the history of geometry in the 19th century*. 2nd revised edition, Springer, Berlin, 2011.

[8] J. J. Gray, Poincaré and the idea of a group. *Nieuw Archief voor Wiskunde* 13 (2012), 178–186.

[9] J. J. Gray, *Henri Poincaré: a scientific biography*. Princeton University Press, Princeton, 2012.

[10] R. Hartshorne, *Geometry: Euclid and beyond*. Springer, New York, 2000.

[11] T. Hawkins, The Erlanger Programm of Felix Klein: reflections on its place in the history of mathematics. *Historia Mathematica* 11 (1984), 442–470.

[12] T. Hawkins, *Emergence of the theory of Lie groups. An essay in the history of mathematics 1869–1926*. Sources and Studies in the History of Mathematics and Physical Sciences, Springer, 2000.

[13] C. Jordan, Mémoire sur les groupes des mouvements, *Annali di matematiche pura e applicata* 2 (1868), 168–215, 320–345, in *Œuvres* 4, 231–302.

[14] C. Jordan, Commentaire sur Galois. *Mathematische Annalen* 1 (1869) 142–160, in *Œuvres* 1, 211–230.

[15] C. Jordan, *Traité des Substitutions et des Équations Algébriques*, Gauthier-Villars, Paris, 1870, rep. Gabay 1989.

[16] C. F. Klein, Ueber die sogenannte Nicht-Euklidische Geometrie. *Mathematische Annalen* 4 (1871), 573–625. In *Gesammelte Mathematische Abhandlungen* I, (no. XVI) 254–305.

[17] C. F. Klein, Vergleichende Betrachtungen über neuere geometrische Forschungen, Programm zum Eintritt in die philosophische Facultät und den Senat der Universität zu Erlangen, Deichert, Erlangen, 1872. In *Gesammelte Mathematische Abhandlungen* I (no. XXVII) 460–497.

[18] C. F. Klein, Ueber die sogenannte Nicht-Euklidische Geometrie. (Zweiter Aufsatz), *Mathematische Annalen* 6 (1873), 112–145. In *Gesammelte Mathematische Abhandlungen* I (no. XVIII) 311–343.

[19] C. F. Klein, Ueber die Transformation siebenter Ordnung der elliptischen Functionen, *Mathematische Annalen* 14 (1879), 428–471. In *Gesammelte Mathematische Abhandlungen* III (no. LXXXIV) 90–134.

[20] C. F. Klein, Vorlesungen über das Ikosaeder und die Auflösung der Gleichungen vom fünften Grade, Teubner, Leipzig, 1884. English trl. *Lectures on the Icosahedron* G. G. Morrice 1888, Dover reprint 1956.

[21] C. F. Klein, A comparative review of recent researches in geometry. *Bulletin of the New York Mathematical Society* 2 (1893), 215–249. English translation by M. W. Haskell of [17].

[22] C. F. Klein, *Gesammelte Mathematische Abhandlungen* III. Springer, Berlin, 1923.

[23] C. F. Klein, Göttingen Professoren Lebensbilder von eigener hand, 4. Felix Klein. *Göttingen Mittheilungen* 5 (1823).

[24] J. Lüroth, Das Imaginäre in der Geometrie und das Rechnen mit Würfen. Darstellung und Erweiterung der v. Staudt'schen Theorie. *Mathematische Annalen* 8 (1875), 145–214.

[25] P. Nabonnand, La théorie des "Würfe" de von Staudt – une irruption de l'algèbre dans la géométrie pure. *Archive for History of Exact Sciences* 62, 201–242 (2008).

[26] M. Noether, Sophus Lie. *Mathematische Annalen* 53 (1900), 1–41.

[27] G. B. F. , Riemann, Ueber die Hypothesen, welche der Geometrie zu Grunde liegen, *Göttingen Abhandlungen* 13 (1867), 1–20. In *Gesammelte Mathematische Abhandlungen* 304–319, 3rd edition, Springer 1990.

[28] D. E. Rowe, A Forgotten Chapter in the History of Felix Klein's Erlanger Programm. *Historia Mathematica* 10 (1983), 448–457.

[29] D. E. Rowe, Felix Klein's Erlanger Antrittsrede. *Historia Mathematica* 12 (1984), 123–141.

[30] J.-D. Voelke, Le théorème fondamental de la géométrie projective: évolution de sa preuve entre 1847 et 1900. *Archive for History of Exact Sciences* 62 (2008), 243–296.

[31] G. K. C. von Staudt, *Geometrie der Lage*. Bauer und Rasper, Nürnberg, 1847.

[32] G. K. C. von Staudt, *Beiträge zur Geometrie der Lage*. Bauer und Rasper, Nürnberg, 1856/57.

The photo shows the house of Paul Gordan in Göttingen (now Goethestrasse 4, in 1872 Hauptstrasse 204), whom Lie and Klein visited together on September 29, 1872. (Photo: C. Meusburger.) In Klein's private notes, he refers to this visit as follows: "1872: Sept 29, after a visit to Gordan together with Lie. Hauptstrasse 204." One line below, we can read: "26.10. Care of the University Library. Lie (who initiated his program on transformation groups at that time) departed. Erlangen program finished."

The house in Göttingen where Emmy Noether was born (Hauptstrasse 23). (Photo: C. Meusburger.)

Chapter 4

Klein's "Erlanger Programm": do traces of it exist in physical theories?

Hubert Goenner

Contents

1 Introduction

Felix Klein's "Erlanger Programm" of 1872 aimed at characterizing geometries by the invariants of simple linear transformation groups.[1] It was reformulated by Klein in this way: "Given a manifold[ness] and a group of transformations of the same; to develop the theory of invariants relating to that group"[2] ([24]; [25], p. 28). As if he had anticipated later discussions about his program, a slightly different formulation immediately preceding this is: "Given a manifold[ness] and a group of transformations of the same; to investigate the configurations belonging to the manifoldness with regard to such properties as are not altered by the transformations of the group."[3] A wide interpretation of a later time by a mathematician is: "According to F. Klein's viewpoint thus geometrical quantities like distance, angle, etc. are not the fundamental quantities of geometry, but the fundamental object of geometry is the transformation group

[1]We have noticed the difficult relationship of Sophus Lie with regard to F. Klein or W. Killing in connection with priority issues ([52], pp. 365–375). Klein acknowledged S. Lie as "the godfather of my Erlanger Programm" ([25], p. 201). For the historical background on Lie groups, S. Lie and F. Klein cf. [45], [18].

[2]"Es ist eine Mannigfaltigkeit und in derselben eine Transformationsgruppe gegeben. Man soll [...] die Theorie der Beziehungen, welche *relativ* zur *Gruppe* invariant sind, untersuchen." – The translation given is by M. W. Haskell and authorized by Klein; cf. New York Math. Soc. 2, 215–249 (1892/93).

[3]"Es ist eine Mannigfaltigkeit und in derselben eine Transformationsgruppe gegeben; man soll die der Mannigfaltigkeit angehörigen Gebilde hinsichtlich solcher Eigenschaften untersuchen, die durch die Transformationen der Gruppe nicht geändert werden." ([26], pp. 34–35) – In a later annotation reproduced in [27], he denied as too narrow an interpretation of his formulation strictly in the sense of looking only at algebraic invariants.

as a symmetry group; from it, the geometrical quantities only follow" ([49], p. 39). On the other hand, by a physicist, Klein's program is incorrectly given the expression "[...] each geometry is associated with a group of transformations, and hence there are as many geometries as groups of transformations" ([13], p. 2). The two quotations show a vagueness in the interpretation of F. Klein's "Erlanger Programm" by different readers. This may be due to the development of the concepts involved, i.e., "transformation group" and "geometry" during the past century. Klein himself had absorbed Lie's theory of transformation groups (Lie/Engel 1888–1893) when he finally published his Erlanger Program two decades after its formulation. Originally, he had had in mind linear transformations, not the infinitesimal transformations Lie considered.

F. Klein's point of view became acknowledged in theoretical physics at the time special relativity was geometrized by H. Minkowski. Suddenly, the Lorentz (Poincaré) group played the role Klein had intended for such a group in a new geometry, i.e., in space-time. The invariants became physical observables. But, as will be argued in the following, this already seems to have been the culmination of a successful application to physical theories of his program. What has had a lasting influence on physical theories, is the concept of *symmetry* as expressed by (Lie-)transformation groups and the associated algebras with all their consequences. This holds particularly with regard to conservation laws.[4] The reason is that in physical theories, *fields* defined on the geometry are dominant, not geometry itself. Also, for many physical theories a geometry fundamental to them either does not exist or is insignificant. A case in sight is the theory of the fractional quantum Hall effect from which quasi-particles named "anyons" emerge. The related group is the braid group describing topological transformations [17]. What often prevails are geometrical models like the real line for the temperature scale, or Hilbert space, an infinite-dimensional linear vector space, housing the states of quantum mechanical systems. In place of geometries, differential geometrical "structures" are introduced. An example would be field reparametrization for scalar fields in space-time. The fields can be interpreted as local coordinates on a smooth manifold. In the kinetic term of the Lagrangian, a metric becomes visible, which shows the correct transformation law under diffeomorphisms. The direct application of F. Klein's classification program seems possible only in a few selected physical theories. The program could be replaced by a scheme classifying the dynamics of physical systems with regard to symmetry groups (algebras).

The following discussion centers around finite-dimensional continuous groups. Infinite-dimensional groups will be barely touched. (Cf. section 8.) Also, the important application to discrete groups in solid state and atomic physics (e.g., molecular vibration spectra) and, particularly, in crystallography are not dealt with.[5] For the considerations to follow here, the question need not be posed whether a reformulation of Klein's classifying idea appropriate to modern mathematics is meaningful.[6]

---

[4]Important developments following the Noether theorems have been described by Y. Kosmann-Schwarzbach in her book about invariance and conservation laws [31].

[5]A survey of the groups is given in [30]. For finite groups cf. also chapters 1 and 2 of [32]. For the history of the interaction of mathematics and crystallography cf. the book by E. Scholz [46].

[6]Some material in this respect may be found in P. Cartier's essay on the evolution of the concepts of space and symmetry [9].

2 Electrodynamics and Special Relativity

It is interesting that F. Klein admitted that he had overlooked the Galilei-group when writing up his "Erlanger Programm": "Only the emergence of the Lorentz group has led mathematicians to a more correct appreciation of the Galilei–Newton group" ([25], p. 56). It turned out later that the Galilean "time plus space" of this group is more complicated than Minkowski's space-time [36], [35].

What also had not been seen by F. Klein but, more than thirty years after the pronounciation of the "Erlanger Programm", by the mathematicians E. Cunningham and H. Bateman, was that the Maxwell equations in vacuum admit the 15-parameter conformal group as an invariance group [3], ([54], p. 409–436, here p. 423). However, this is a very specific case; if the electromagnetic field is coupled to matter, this group is no longer admitted, in general.

Special relativity, and with it Minkowski space, are thought to form a framework for all physical theories not involving gravitation. Hence, a branch of physics like relativistic quantum field theory in both its classical and quantized versions is included in this application of the "Erlanger Programm."[7] In the beginning of string theory (Veneziano model), the string world sheet was likewise formulated in Minkowski space or in a Lorentz space of higher dimension.

We need not say much more concerning special relativity, but only recall Minkowski's enthusiasm about his new find:

> For the glory of mathematicians, to the infinite astonishment of remaining humanity, it would become obvious that mathematicians, purely in their fantasy, have created a vast area to which one day perfect real existence would be granted – without this ever having been intended by these indeed ideal chaps. (quoted from ([25], p. 77).[8]

3 General Relativity

The description of the gravitational field by a Lorentz-metric, in Einstein's general relativity, was predestined to allow application of Klein's program. The exact solutions of Einstein's field equations obtained at first like the Schwarzschild- and de Sitter solutions as well as the Einstein cosmos, defined geometries allowing 4- and 6-parameter Lie transformation groups as invariance groups. Most of the exact solutions could be found just *because* some invariance group had been assumed in the first place. Later, also algebraical properties of the metrics were taken to alleviate the solution of the non-linear differential equations. In the decades since, it has become

[7] We recall that, on the strictest mathematical level, an unambiguous union of quantum mechanics and special relativity has not yet been achieved. Note also that algebraic quantum field theory does not need full Minkowski space, but can get along with the weaker light-cone structure supplemented by the causality principle.

[8] "Es würde zum Ruhme der Mathematik, zum grenzenlosen Erstaunen der übrigen Menschheit offenbar werden, dass die Mathematiker rein in ihrer Phantasie ein großes Gebiet geschaffen haben, dem, ohne dass es je in der Absicht dieser so idealen Gesellen gelegen hätte, eines Tages die vollendete reale Existenz zukommen sollte."

clear, that the *generic* solution of Einstein's field equations does *not* allow an invariance group – except for the diffeomorphisms Diff(M) of space-time M. As every physical theory can be brought into a diffeomorphism-invariant form, eventually with the help of new geometrical objects, the role of this group is quite different from the one F. Klein had in mind.[9] He was well aware of the changed situation and saved his program by reverting to *infinitesimal* point transformations. He expressed his regret for having neglected, at the time of the formulation of his "Erlanger Programm", Riemann's Habilitationsschrift of 1854 [44], and papers by Christoffel and Lipschitz.[10] In fact, the same situation as encountered in general relativity holds already in Riemannian geometry: generically, no nontrivial Lie transformation group exists. Veblen had this in mind when he remarked:

> With the advent of Relativity we became conscious that a space need not be looked at only as a 'locus in which', but that it may have a structure, a field-theory of its own. This brought to attention precisely those Riemannian geometries about which the Erlanger Programm said nothing, namely those whose group is the identity. [...] ([53], p. 181–182; quoted also by E. T. Bell [4], p. 443).[11]

That general relativity allows only the identity as a Lie transformation group (in the sense of an isometry) to me is very much to the point. Perhaps, the situation is characterized best by H. Weyl's distinction between *geometrical automorphisms* and *physical automorphisms* ([47], p. 17). For general relativity, this amounts to Diff(M) on the one hand, and to the unit element on the other. Notwithstanding the useful identities following from E. Noether's second theorem, all erudite discussions about the physical meaning of Diff(M) seem to be adornments for the fact that scalars are its most general invariants possible on space-time. Usually, physical observables are transforming covariantly; they need not be invariants. While the space-time metric is both an intrinsically geometric quantity and a dynamical physical field, it is not a representation of a finite-dimensional Lie transformation group: F. Klein's program just does not apply. If Einstein's endeavour at a unified field theory built on a more general geometry had been successful, the geometrical quantities adjoined to physical fields would not have been covariants with regard to a transformation group in Klein's understanding.

But F. Klein insisted on having strongly emphasized in his program: "that a point transformation $x_i = \phi(y_1 ... y_n)$ for an infinitely small part of space always has the

[9] Since E. Kretschmann's papers of 1915 and 1917 [33], [34], there has been an extended discussion about an eventual physical content of the diffeomorphism group in general relativity; cf. [40], [41]. It suffers from Einstein's identification of coordinate systems and physical reference systems with the latter being represented by tetrads (frames). These can be adapted to matter variables.

[10] For the contributions of Lipschitz to the geometrization of analytical mechanics cf. ([37], pp. 29–31).

[11] The original quote from Veblen continues with "In such spaces there is essentially only one figure, namely the space structure as a whole. It became clear that in some respects the point of view of Riemann was more fundamental than that of Klein."

character of a linear transformation [...]"[12] ([25], p. 108). A symmetry in general relativity is defined as an isometry through Killing's equations for the infinitesimal generators of a Lie-algebra. Thus, in fact, F. Klein's original program is restricted to apply to the tangent space of the Riemannian (Lorentz) manifold. This is how É. Cartan saw it: a manifold as the envelope of its tangent spaces; from this angle, he developed his theory of groups as subgroups of GL(n,R) with the help of the concept of G-structure.[13] Cartan's method for "constructing finitely and globally inhomogeneous spaces from infinitesimal homogeneous ones" is yet considered by E. Scholz as "a reconciliation of the Erlangen program(me) and Riemann's differential geometry on an even higher level than Weyl had perceived"([47], p. 27).[14]

An extension of general relativity and its dynamics to a Lorentz-space with one time and four space dimensions was achieved by the original Kaluza–Klein theory. Its dimensional reduction to space-time led to general relativity and Maxwell's theory refurbished by a scalar field. Since then, this has been generalized in higher dimensions to a system consisting of Einstein's and Yang–Mills' equations [23], and also by including supersymmetry. An enlargement of general relativity allowing for supersymmetry is formed by supergravity theories. They contain a (hypothetical) graviton as bosonic particle with highest spin 2 and its fermionic partner of spin 3/2, the (hypothetical) gravitino; cf. also Section 6.

4 Phase space

A case F. Klein apparently left aside, is phase space parametrized by generalized coordinates q_i and generalized momenta p_i of particles. This space plays a fundamental role in statistical mechanics, not through its geometry and a possibly associated transformation group, but because of the well-known statistical ensembles built on its decomposition into cells of volume h^3 for each particle, with h being Planck's constant. For the exchange of indistinguishable particles with spin, an important role is played by the permutation group: only totally symmetric or totally anti-symmetric states are permitted. In 2-dimensional space, a statistics ranging continuously between Bose–Einstein and Fermi–Dirac is possible.

The transformation group to consider would be the abelian group of contact transformations (cf. [20]):

$$q'_i = f_i(q_i, p_j)\,,\quad p'_j = g_j(q_i, p_j)\,, \tag{4.1}$$

[12]"dass eine Punkttransformation [...] für eine unendlich kleine Partie des Raumes immer den Charakter einer linearen Transformation hat."

[13]H. Weyl with his concept of purely infinitesimal geometry in which a subgroup $G \subset SL(n, R)$ (generalized "rotations") acts on every tangent space of the manifold, separately, took a similar position ([48], p. 24).

[14]For a Lie group $G \subset L$, the homogeneous space corresponds to $\mathfrak{l}\,\mathfrak{g}^{-1} \cong T_pM$, where $\mathfrak{l}$ and $\mathfrak{g}$ are the respective Lie algebras.

which however is of little importance in statistical mechanics.[15] In some physics textbooks, no difference is made between contact and canonical transformations, cf. e.g., [10]. In others, the concept of phase space is limited to the cotangent bundle of a manifold with a canonical symplectic structure ([1], p. 341). An important subgroup of canonical transformations is given by all those transformations which keep Hamilton's equations invariant for *any* Hamiltonian.[16] Note that for the derivation of the Liouville equation neither a Hamiltonian nor canonical transformations are needed. In this situation, symplectic geometry can serve as a model space with among others, the symplectic groups $SP(n, R)$ acting on it as transformation groups. Invariance of the symplectic form $\Sigma_i^n(dq_i \wedge dp_i)$ implies the reduction of contact to canonical transformations. Symplectic space then might be viewed in the spirit of F. Klein's program. He does not say this but, in connection with the importance of canonical transformations to "astronomy and mathematical physics", he speaks of "quasi-geometries in a R_{2n} as they were developed by Boltzmann and Poincaré [...]" ([29], p. 203).

In analytical mechanics, Hamiltonian systems with conserved energy are studied and thus time-translation invariance is assumed. Unfortunately, in many systems, e.g., those named "dynamical systems", energy conservation does not hold. For them, attractors can be interpreted as geometrical models for the "local asymptotic behavior" of such a system while bifurcation forms a "geometric model for the controlled change of one system into another" ([2], p. XI). Attractors can display symmetries, e.g., discrete planar symmetries [8], etc.

In statistical thermodynamics, there exist phase transitions between thermodynamic phases of materials accompanied by "symmetry breaking." As an example, take the (2nd order) transition from the paramagnetic phase of a particle-lattice, where parallel and anti-parallel spins compensate each other to the ferromagnetic phase with parallel spins. In the paramagnetic state, the full rotation group is a continuous symmetry. In the ferromagnetic state below the Curie-temperature, due to the fixed orientation of the magnetization, the rotational symmetry should be hidden: only axial symmetry around the direction of magnetization should show up. However, in the Heisenberg model (spin 1/2) the dynamics of the system is rotationally invariant also below the Curie point. The state of lowest energy (ground state) is degenerate. The symmetry does not annihilate the ground state. By picking a definite direction, the system spontaneously breaks the symmetry with regard to the full rotation group. When a continuous symmetry is spontaneously broken, massless particles appear called Goldstone(-Nambu) bosons. They correspond to the remaining symmetry. Thus, while the dynamics of a system placed into a fixed external geometry can be invariant under a transformation group, in the lowest energy state the symmetry may be reduced. This situation seems far away from F. Klein's ideas about the classification of geometries by groups.

---

[15] It is only loosely connected with Lie's geometric contact transformation which transforms plane surface elements into each other. Manifolds in contact, i.e., with a common (tangential) surface element remain in contact after the transformation. A class of linear differential equations is left invariant; cf. [28], pp. 19–20.

[16] For contact transformations with higher derivatives cf. [59].

5 Gauge theories

Hermann Weyl's positive thoughts about Klein's program were expressed in a language colored by the political events in Germany at the time:

> The dictatorial regime of the projective idea in geometry was first broken by the German astronomer and geometer Möbius, but the classical document of the democratic platform in geometry, establishing the group of transformations as the ruling principle in any kind of geometry, and yielding equal rights of independent consideration to each and every such group, is F. Klein's 'Erlanger Programm'. (Quoted from Birkhoff & Bennet [6].)

Whether he remembered this program when doing a very important step for physics is not known: H. Weyl opened the road to *gauge theory*. He associated the electromagnetic 4-potential with a connection, at first unsuccessfully by coupling the gravitational and electrodynamic fields (local scale invariance). A decade later, by coupling the electromagnetic field to matter via Dirac's wave function; for the latter he expressly invented 2-spinors. The corresponding gauge groups were R and $U(1)$, respectively. This development and the further path to Yang–Mills theory for non-abelian gauge groups has been discussed in detail by L. O'Raifeartaigh and N. Straumann[17] [42], [43]. Weyl had been convinced about an intimate connection of his gauge theory and general relativity: "Since gauge invariance involves an arbitrary function λ it has the character of 'general' relativity and can naturally only be understood in that context" ([57], translation taken from [43]). But he had not yet taken note of manifolds with a special mathematical structure introduced since 1929, i.e., *fibre bundles*. Fibre bundles are local products of a base manifold (e.g., space-time), and a group. The action of the group creates a fibre (manifold) in each point of the base. Parallel transport in base space corresponds to a connection defined in a section of the bundle. By gauge transformations, a fibre is mapped into itself. In physics, the transformation group may be a group of "external" symmetries like the Poincaré group or of "internal" symmmetries like a Yang–Mills (gauge) group. A well known example is the frame bundle of a vector bundle with structure group GL(n;R). It contains all ordered frames of the vector space (tangent space) affixed to each point of the base manifold. Globally, base and fibres may be twisted like the Möbius band is in comparison with a cylindrical strip.[18] In 1929, Weyl had not been able to see the gauge potential as a connection in a principal fibre bundle. Until this was recognized, two to three decades had to pass.

Comparing the geometry of principal fibre bundles with Riemannian (Lorentzian) geometry, F. Klein's program would be realized in the sense that a group has been built right into the definition of the bundle. On the other hand, the program is limited

[17]The original Yang–Mills gauge theory corresponded to SU(2)-isospin symmetry of the strong interaction.

[18]Since the introduction of fibre spaces by H. Seifert in 1932, at least five definitions of fibre bundles were advanced by different researchers and research groups [38]. The first textbook was written by Steenrod [51].

because the group can be any group. In order to distinguish bundles, different groups have to be selected in order to built, e.g., $SU(2)-$, $SU(3)$-bundles, etc.[19] This is a classification of bundle geometry in a similar sense as isometries distinguish different Lorentz-geometries. To classify different types of bundles is another story.

Moreover, in gauge theories, the relation between observables and gauge invariants is not as strong as one might have wished it to be. E.g., in gauge field theory for non-abelian gauge groups, the gauge-field strength (internal curvature) does not commute with the generators of the group: it is not an immediate observable. Only gauge-invariant polynomials in the fields or, in the quantized theory, gauge-invariant operators, are observables. In contrast, the energy-momentum tensor is gauge-invariant also for non-abelian gauge groups.

In terms of the symmetry[20], gauge invariance is spontaneously broken, both in the case of electroweak and strong interactions.

General relativity with its metric structure is not a typical gauge theory: any external transformation group would not only act in the fibre but also in the tangent space of space-time as well. Thus, an additional structure is required: a soldering form gluing the tangent spaces to the fibres [12]. Many gauge theories for the gravitational field were constructed depending on the group chosen: translation-, Lorentz-, Poincaré, conformal group etc.[21] We will come back to a Poincaré gauge theory falling outside of this Lie-group approach in Section 7.

6 Supersymmetry

Another area in physics which could be investigated as a possible application of Klein's program is supersymmetry-transformations and supermanifolds. Supersymmetry expressed by super-Lie-groups is a symmetry relating the Hilbert spaces of particles (objects) obeying Bose- or Fermi-statistics (with integer or half-integer spin-values, respectively). In quantum mechanics, anti-commuting supersymmetry operators exist mapping the two Hilbert spaces into each other. They commute with the Hamiltonian. If the vacuum state (state of minimal energy) is annihilated by the supersymmetry operators, the 1-particle states form a representation of supersymmetry and the total Hilbert space contains bosons and fermions of *equal* mass.[22] As this is in contradiction with what has been found, empirically, supersymmetry must be broken (spontaneously) in nature.

[19] Elementary particles are classified with regard to local gauge transformations $SU(3)_c \times SU(2)_L \times U(1)_Y$. The index c refers to color-charge, Y to weak hypercharge, and L to weak isospin. For a review of the application of gauge theory to the standard model cf. [55].

[20] $SU(2) \times U(1)$ symmetry of electroweak interactions; approximate flavour $SU(3)$-symmetry of strong interactions.

[21] A recent reader about gauge theories of gravitation is [7].

[22] For the geometry of supersymmetric quantum mechanics cf. e.g., [58]. There, supersymmetric quantum field theory is formulated on certain infinite-dimensional Riemannian manifolds.

For an exact supersymmetry, the corresponding geometry would be supermanifolds, defined as manifolds over superpoints, i.e., points with both commuting coordinates as in a manifold with n space dimensions, and anti-commuting "coordinates" forming a Grassmann-algebra $\zeta, \bar{\zeta}$ ("even" and "odd" elements) [11].[23] As a generalization of Minkowski space, the coset space Poincaré/Lorentz in which the super-Poincaré group acts, is called superspace ([14], Chapter 6), [50], p. 107). Superspace is a space with 8 "coordinates" $z^A = (x^k, \theta^\mu, \bar{\theta}_{\dot{\mu}})$, where x^k are the usual real space-time coordinates plus 4 real (anti-commuting) "fermionic" coordinates from a Weyl-spinor θ^μ and its conjugate $\bar{\theta}_{\dot{\mu}}$).

A super-Lie group G is a Lie group with two further properties: 1) it is a supermanifold the points of which are the group elements of G ; 2) the multiplicative map F: $G \to G \times G$ is differentiable ([11], p. 123).[24] All classical Lie groups have extensions to super-Lie groups. Most important for quantum field theory is the super-Poincaré group and its various associated super-Lie-algebras. The super-Poincaré algebras contain both Lie-brackets and anti-commuting (Poisson) brackets. A superparticle (supermultiplet) corresponds to a reducible representation of the Poincaré algebra.

The *geometry of supermanifolds* seems to play only a minor role in physics. An example for its use would be what has been called the gauging of supergroups [39]. Local super-Lie algebras are important because their representations constitute superfields by which the dynamics of globally or locally supersymmetric physical theories like *supergravity* are built.[25] Supergravity containing no particle of spin larger than 2 can be formulated in Lorentz-spaces up to maximal dimension 11. In space-time, at least 7 supergravities can be formulated. Yet, a geometrical construct like a supermetric is of no physical importance.

This all too brief description is intended to convey the idea that, in physics, the role of supersymmetry primarily is not that of a transformation group in a supermanifold but of a group restricting the dynamics of interacting fields. By calling for invariants with regard to supersymmetry, the choice of the dynamics (interaction terms in the Lagrangian) is narrowed considerably. The supersymmetric diffeomorphism group can be used to formulate supersymmetric theories in terms of differential forms on superspace: "superforms" ([56], Chapter XII). Possibly, B. Julia envisioned the many occuring supersymmetry groups when drawing his illustration for supergravities "A theoretical cathedral" and attaching to the x-axis the maxim: GEOMETRY $\simeq$ GROUP THEORY ([21], p. 357). When the view is narrowed to F. Klein's "Erlanger Programm" as is done here, then the conclusion still is that the program cannot fare better in supergravity than in general relativity.

---

[23] As a supermanifold is not only formed from the usual points with commuting coordinates, another definition has been used: It is a topological space with a sheaf of superalgebras ($\mathbb{Z}_2$-graded commutative algebras).

[24] According to ([11], p. 173–174) *conventional* super-Lie groups and *unconventional* super-Lie groups unrelated to graded algebras must be distinguished.

[25] Superfields can be defined as functions on superspace developed into power series in the nilpotent Grassmann-variables in superspace; the power series break off after the term $\theta\theta\bar{\theta}\bar{\theta}\ a(x)$. Local supersymmetric theories are theories invariant under *supergauge*-transformations.

7 Enlarged Lie algebras

We now come back to space-time and to a generalization of (Lie)-transformation groups acting on it. As insinuated before, for the classification of structures in physical theories the attention should lie rather on the algebras associated with the groups; geometrical considerations intimately related to groups are of little concern. Lie algebras have been generalized in a number of ways. One new concept is "soft", "open" or "nonlinear" Lie algebras, in which the structure constants are replaced by structure functions depending on the generators themselves. They can also be interpreted as infinite-dimensional Lie algebras ([15], pp. 60–61). An example from physics are local supersymmetry transformations (defined to include diffeomorphisms, local Lorentz and local supersymmetry transformations) which form an algebra with structure *functions*. They depend on the symmetry generators themselves ([39], p. 140). Another generalization is "local Lie algebras" which arise as the Lie algebras of certain infinite-dimensional Lie groups. The structure of the Lie algebra is given by:

$$[f_1, f_2] = \Sigma^{n}_{i,j,k} c^{ij}_{k} x^k \partial_i f_1 \partial_j f_2 \,,$$

where f_1, f_2 are smooth functions on a smooth manifold, ∂_k the partial derivatives with respect to local coordinates on M, and c^{ij}_{k} the structure constants of an n-dimensional Lie algebra (cf. [5], section 7). This seems to be a rather special kind of algebra.

Recently, a further enlargment has been suggested called "extended Lie algebras" and in which the structure constants are replaced by functions of the space-time coordinates. In the associated groups, the former Lie group parameters are substituted by arbitrary functions [16]. The Lie algebra elements form an "involutive distribution", a smooth distribution V on a smooth manifold M. The Lie brackets constitute the composition law; the injection $V \hookrightarrow TM$ functions as the anchor map. Thus, this is a simple example for a *tangent Lie algebroid*. In addition to the examples from physics given in [16], the Poincaré gauge theory of F.-W. Hehl et al. seems to correspond to the definition of an extended Lie algebra. In this theory, the difference with the Lie algebra of the Poincaré group is that the structure functions now contain the frame-metric and the gauge fields, i.e., curvature as rotational and torsion as translational gauge field, all dependent on the space-time coordinates [19].

8 Conclusions

In the course of ranging among physical theories with an eye on F. Klein's "Erlanger Programm", we noticed that the focus had to be redirected from groups and geometry to algebras and the dynamics of fields. In particular, with regard to infinite-dimensional groups, the discussion within physical theories of Klein's program would have been easier had it been formulated in terms of algebras. Then, also Virasoro- and Kac–Moody algebras, appearing among others in conformal (quantum) field theory and in string theory could have been included in the discussion.[26] Hopf-algebras occuring in *non-commutative geometry* could have formed another example. With the mentioned change in focus included, the application of Klein's program to physical theories is far more specific than a loosely defined methodological doctrine like the "geometrization of physics" (cf. [22]). While both general relativity and gauge theory can be considered as geometrized, they only partially answer F. Klein's "Erlanger Programm." In physical theories, the momentousness of Lie's theory of transformation groups easily surpasses Klein's classification scheme.

Acknowledgements. For the invitation to contribute to this volume and for his helpful comments and advice I thank A. Papadopoulos, Strasbourg. Inspiring discussions with my colleague Detlef Buchholz, Göttingen, are gratefully acknowledged. My sincere thanks go to Profs. Friedrich Hehl, Köln and Erhard Scholz, Wuppertal whose comments contributed much to the improvement of a first version.

Bibliography

[1] R. Abraham and J. E. Marsden, *Foundations of mechanics*. 2nd revised ed., The Benjamin/Cummings Publ. Co. Inc., Reading 1978.

[2] R. H. Abraham and Christopher D. Shaw, *Dynamics: The geometry of behavior*. Addison-Wesley Publ. Co., Redwood City/ Menlo Park / Reading 1992.

[3] H. Bateman, The Conformal Transformations of a Space of Four Dimensions and their Applications to Geometrical Optics. *Proc. London Math. Soc.* 7 (1908), 70–89.

[4] E. T. Bell, *The development of mathematics*. 2nd ed., McGraw-Hill, New York (1945).

[5] F. A. Berezin, Some remarks about the associative envelope of a Lie algebra. *Functional Analalysis and its Applications* 1, No. 2 (1967), 91–102 .

[6] G. Birkhoff and M. K. Bennett, Felix Klein and his Erlanger Programm. In *History and Philosophy of Modern Mathematics*. Eds. W. Aspray and P. Kitcher, 145–176, University of Minesota Press, Minneapolis 1988.

[7] M. Blagojevic and F. W. Hehl, eds., *Gauge theories of gravitation. A reader with commentaries*. Imperial College Press, London 2013.

[8] N. C. Carter, R. L. Eagles, et al., Chaotic Attractors with Discrete Planar Symmetries. *Chaos, Solitons & Fractals* 9 (1998), 2031–2054.

[9] P. Cartier, A mad day's work: from Grothendieck to Connes and Kontsevich. The evolution of concepts of space and symmetry. *Bull. Amer. Math. Soc.*, 38, No. 4 (2001), 389–408.

[10] H. C. Corben and Ph. Stehle, *Classical Mechanics*. John Wiley and Sons, Inc., New York 1957.

[26] The Virasoro group is an infinite dimensional group related to conformal (quantum) field theory in 2 dimensions. It is defined as $Diff(S^1)$ where S^1 is the unit circle, and its geometry the infinite dimensional complex manifold $Diff(S^1)/S^1$ [60]. In string theory, the Virasoro-*algebra* appears. It is a central extension of a Witt-algebra providing unitary representations. The Witt-algebra is the Lie-algebra of smooth vector fields on S^1.

[11] B. DeWitt, *Supermanifolds*. University Press, Cambridge (GB) 1984.

[12] Ch. Ehresmann, Les connexions infinitésimales dans un espace fibré différentiel. *Colloque de Topologie* (1950), 29–55, Bruxelles.

[13] M. A. Fernandez Sanjuan, Group Contraction and the Nine Cayley–Klein Geometries. *Intern. J. Theor. Phys.*, 23, No. 1 (1984), 1–13.

[14] P. G. O. Freund, *Introduction to supersymmetry*. University Press, Cambridge (GB) 1986.

[15] J. Fuchs and Ch. Schweigert, *Symmetries, Lie algebras and Representations*. University Press, Cambridge (GB) 1997.

[16] H. F. M. Goenner, Weak Lie symmetry and extended Lie algebra. *Journal of Mathematical Physics* 54 041701 (2013); arXiv 1212.1590.

[17] Y. Hatsugai, M. Kohmoto and Y-S Wu, Braid group and anyons on a cylinder. *Phys. Review B* 43, No. 4 (1991), 2661–2677.

[18] T. Hawkins, Line Geometry, Differential Equations and the Birth of Lie's Theory of Groups. In *The History of Modern Mathematics*, vol. I: Ideas and Their Reception. Eds. D. E. Rowe, J. McCleary, pp. 275–327, Academic Press, Boston 1989.

[19] F. W. Hehl, P. von der Heyde, G. D. Kerlick, and J. M. Nester, General relativity with spin and torsion: Foundations and prospects. *Review of Modern Physics* 48 (1976), 393–416.

[20] E. Hölder, Die infinitesimalen Berührungstransformationen der Variationsrechnung. *Jahresber. Deutsche Math.-Vereinigung* 49 (1939), 162–178.

[21] B. Julia, Kac–Moody symmetry of gravitation and supergravity theories. In *Applications of group theory in physics and mathematical physics.* M. Flato, P. Sally and G. Zuckermann, eds., 355–373, American Mathematical Society, Providence 1985.

[22] M. W. Kalinowski, The program of geometrization of physics: some philosophical remarks. *Synthese* 77, No. 2 (1988), 129–138.

[23] R. Kerner, Generalization of the Kaluza–Klein theory for an arbitrary non-abelian gauge group. *Annales de l'Institut Henri Poincaré* 9 (1968), 143–152.

[24] F. Klein, Vergleichende Betrachtungen über neuere geometrische Forschungen. Reprinted in *Math. Ann.* 43, (1893), 63–100.

[25] F. Klein, *Vorlesungen über die Entwicklung der Mathematik im 19. Jahrhundert*, Teil 2, Springer Verlag, Berlin 1927.

[26] F. Klein, *Das Erlanger Programm*, Ostwalds Klassiker 253, Ed. H. Wußing, Akademische Verlagsgesellschaft Geest and Portig, Leipzig 1974.

[27] F. Klein, *Gesammelte Mathematische Abhandlungen*, Band 1, 460–497, Springer-Verlag, Berlin/Heidelberg/New York 1921.

[28] F. Klein, *Lecture Notes in Mathematics. The Evanston Colloquium.* A. Ziwet & H. S. White eds., MacMillan and Co., New York and London 1894.

[29] F. Klein, *Vorlesungen über die Entwicklung der Mathematik im 19. Jahrhundert.* Teil 1, Springer, Berlin 1926.

[30] M. Klemm, *Symmetrien von Ornamenten und Kristallen.* Springer-Verlag, Berlin/Heidelberg/New York 1982.

[31] Y. Kosmann-Schwarzbach, *The Noether theorems. Invariance and conservation laws in the twentieth century.* Sources and Studies in the History of Mathematics and Physical Sciences, Springer-Verlag, New York/Dordrecht/Heidelberg 2002.

[32] Y. Kosmann-Schwarzbach, *Groups and Symmetries. From Finite Groups to Lie groups.* Universitext, Springer, New York/Dordrecht/Heidelberg 2010.

[33] E. Kretschmann, Über die prinzipielle Bestimmbarkeit der berechtigten Bezugssysteme beliebiger Relativitätstheorien (I), (II)". *Annalen der Physik* 48 (1915), 907–942, 943–982.

[34] E. Kretschmann, Über den physikalischen Sinn der Relativitätspostulate. A. Einsteins neue und seine ursprüngliche Relativitätstheorie. *Annalen der Physik* 53 (1917), 575–614.

[35] H. P. Künzle, Galilei and Lorentz structures on space-time: comparison of the corresponding geometry and physics. *Annales de l'Institut Henri Poincaré A* 17, No. 4 (1972), 337–362.

[36] J.-M. Levy-Leblond, Galilei Group and Nonrelativistic Quantum Mechanics. *Journal of Mathematical Physics* 4, No. 6 (1963), 776–789.

[37] J. Lützen, Geometrising configurations. Heinrich Hertz and his mathematical precursors. In *The symbolic universe*. Ed. J. Gray, 25–46, University Press, Oxford 1999.

[38] J. McCleary, A History of Manifolds and Fibre Spaces: Tortoises and Hares, *The History of the Mathematics of the 20th Century*, January 30-February 5, Mathematisches Forschungsinstitut, Oberwolfach, 2000.

[39] P. van Nieuwenhuizen, Gauging of groups and supergroups. In *Applications of group theory in physics and mathematical physics*, M. Flato, P. Sally & G. Zuckermann, eds., 139–162, American Mathematical Society, Providence 1985.

[40] J. D. Norton, The physical content of general covariance. In *Studies in the history of general relativity – Einstein Studies* 3 (1992), Editors J. Eisenstaedt and A. J. Kox, 281–315.

[41] J. D. Norton, Geometries in collision: Einstein, Klein and Riemann. In *The symbolic universe*, Ed. J. Gray, 128–144, University Press, Oxford 1999.

[42] L. O'Raifeartaigh, *The Dawning of Gauge Theory*. University Press, Princeton 1997.

[43] L. O'Raifeartaigh and N. Straumann, Gauge theory: Historical origins and some modern developments. *Reviews of Modern Physics* 72, No. 1 (2000), 1–23.

[44] B. Riemann, Ueber die Hypothesen, welche der Geometrie zu Grunde liegen. *Abhandlungen der Königlichen Gesellschaft der Wissenschaften zu Göttingen.* 13, (1867).

[45] D. E. Rowe, The Early Geometrical Works of Sophus Lie and Felix Klein. In *The History of Modern Mathematics*, vol. I: Ideas and Their Reception, Eds. D. E. Rowe, J. McCleary, pp. 209–273, Academic Press, Boston 1989.

[46] E. Scholz, *Symmetrie – Gruppe – Dualität. Zur Beziehung zwischen theoretischer Mathematik und Anwendungen in Kristallographie und Baustatik des 19. Jahrhunderts*, Science Networks – Historical Studies, vol. 1, Birkhäuser, Basel/Boston/Stuttgart; Deutscher Verlag der Wissenschaften, Berlin (1989).

[47] E. Scholz, The problem of space in the light of relativity: the views of H. Weyl and É. Cartan, http://arxiv.org/abs/1310.7334, to appear in L. Bioesmat-Martagon (Ed.), Eléments d'une biographie de l'espace mathématique, Presses Universitaires de Lorraine, Nancy 2014.

[48] E. Scholz, H. Weyl's and É. Cartan's proposals for infinitesimal geometry in the early 1920s. *Newsletter European Mathematical Society* no. 84 (2012), June 2012, 22–30.

[49] M. Schottenloher, *Geometrie und Symmetrie in der Physik.* Vieweg, Braunschweig/Wiesbaden 1995.

[50] M. F. Sohnius, Introducing supersymmetry. *Physics Reports* 128, Nos. 2 and 3 (1985), 39–204.

[51] N. Steenrod, *The topology of fibre bundles*. Princeton University Press, Princeton 1951.

[52] A. Stubhaug, *The mathematician Sophus Lie. It was the audacity of my thinking*. Springer, Berlin/Heidelberg 2002.

[53] O. Veblen, Differential invariants and geometry. *Atti del Congr. Internat. dei Matematici, Bologna 1928* 1 (1928), 181–189.

[54] A. Warwick, *Masters of Theory: Cambridge and the Rise of Mathematical Physics*. University of Chicago Press, Chicago 2003.

[55] S. Weinberg, The making of the Standard Model. In *Fifty years of Yang–Mills theory*. Ed. G. 't Hooft, 99–117, World Scientific, New Jersey/London/Singapore 2005.

[56] J. Wess and Jonathan Bagger, *Supersymmetry and supergravity*. University Press, Princeton 1983.

[57] H. Weyl, Elektron und Gravitation. I. *Zeitschrift für Physik* 56 (1929), 330–352.

[58] E. Witten, Supersymmetry and Morse theory. *Journal of Differential Geometry* 17 (1982), 661–692.

[59] Y. Yaremko, Contact transformations in classical mechanics. *J. of Nonlinear Mathematical Physics* 4, No.1-2 (1997), 117–123.

[60] B. Zumino, The geometry of the Virasoro group for physicists. In *Particle Physics. Cargèse 1987*, Eds. M. Lévy et al., 81–98, Plenum Press. New York and London 1988.

Chapter 5

On Klein's *So-called Non-Euclidean geometry*

Norbert A'Campo and Athanase Papadopoulos

Contents

1 Introduction

Klein's model of hyperbolic space is well known to geometers. The underlying set is, in the planar case, the interior of an ellipse, and in the three-dimensional case, the interior of an ellipsoid. The hyperbolic geodesics are represented by Euclidean straight lines and the distance between two distinct points is defined as a constant times the logarithm of the cross ratio of the quadruple formed by this pair of points together with the two intersection points of the Euclidean straight line that joins them with the boundary of the ellipsoid, taken in the natural order (see the left hand side of Figure 5.5 below). Klein's formula is the first explicit formula for the distance function in hyperbolic geometry.

A much less known fact is that Klein, besides giving a formula for the distance function in hyperbolic geometry, gave formulae for the distance in spherical and in Euclidean geometry using the cross ratio, taking instead of the ellipse (or ellipsoid) other kinds of conics. In the case of Euclidean geometry, the conic is degenerate.[1] In this way, the formulae that define the three geometries of constant curvature are of the same type, and the constructions of the three geometries are hereby done in a unified

[1] Recall that a conic is the intersection of a cone in 3-space and a plane. Degenerate cases occur, where the intersection is a single point (the vertex of the cone), or a straight line ("counted twice"), or two intersecting straight lines.

way in the realm of projective geometry. This is the main theme of the two papers *On the so-called non-Euclidean geometry, I* and *II* ([33] [35]) of Klein.

In this paper, we review and comment on these two papers.

Klein's construction was motivated by an idea of Cayley. We shall recall and explain Cayley's idea and its developments by Klein. Let us note right away that although Klein borrowed this idea from Cayley, he followed, in using it, a different path. He even interpreted Calyey's idea in a manner so different from what the latter had in mind that Cayley misunderstood what Klein was aiming for, and thought that the latter was mistaken.[2] In fact, Cayley's interest was not particularly in non-Euclidean geometry; it was rather to show the supremacy of projective geometry over Euclidean geometry, by producing the Euclidean metric by purely projective methods. We shall comment on this fact in §8 below.

Klein wrote his two papers in 1871 and 1872, just before he wrote his *Erlangen program* manifesto [34],[3] the famous text in which he proposes a unification of all geometries based on the idea that a geometry should be thought of as a transformation group rather than a space. Although this point of view is familiar to us, and seems natural today, this was not so for mathematicians even by the end of the nineteenth century.[4] Without entering into the details of this philosophical question, let us recall that from the times of Euclid and until the raise of projective geometry, mathematicians were reluctant to the use of transformations – which, classically, carried the name *motion*[5] – as elements in the proof of a geometrical proposition.[6]

Klein's two papers [33] and [35] are actually referred to in the *Erlangen program* text. We shall quote below some of Klein's statements in this program that are very similar to statements that are made in the two papers with which we are concerned.

Another major element in the Erlangen program is the question of finding a *classification* of the various existing geometries using the setting of projective geometry and of the projective transformation groups. Klein's paper [33] constitutes a leading writing on that subject, and it puts at the forefront of geometry both notions of transformation groups and of projective geometry. At the same time, this paper constitutes an important piece of work in the domain of non-Euclidean geometry. Historically, it is probably the most important one, after the writings of the three founders of hyper-

---

[2] Cayley did not understand Klein's claim that the cross ratio is independent of the Euclidean underlying geometry, and therefore he disagreed with Klein's assertion that his construction of the three geometries of constant curvature was based only on projective notions; cf. §5.

[3] The first paper carries the date ("Handed in") "Düsseldorf, 19. 8. 1871" and the second paper: "Göttingen, 8. 6. 1872". Thus the second paper was finished four months before the Erlangen Program, which carries the date "October 1872".

[4] In the introduction of his *Lezioni di geometria proiettiva* (1898) [20], Enriques writes that geometry studies the notion of space and the relation between its elements (points, curves, surfaces, lines, planes, etc.) This was still not the transformation group point of view.

[5] The word *motion*, denoting a rigid transformation, was used by Peano. Hilbert used the word *congruence*.

[6] It is considered that the Arabic mathematician Abu-'Ali Ibn al-Haytham, (d. after 1040), in his book titled *On the known*, developed the first geometrical Euclidean system in which the notion of motion is a primitive notion (see [31] p. 446). Several centuries later, Pasch, Veronese and Hilbert came up with the same idea. In his book *La science et l'hypothèse* (1902), Poincaré discusses the importance of the notion of motion (see [51] p. 60).

bolic geometry (Lobachevsky, Bolyai and Gauss) and after Beltrami's paper [4] on which we shall also comment below.

In the introduction of the first paper [33], Klein states that among all the works that were done in the preceding fifty years in the field of geometry, the development of projective geometry occupies the first place.[7] In fact, the use of the notion of *transformation* and of *projective invariant* by geometers like Poncelet[8] had prepared the ground for Klein's general idea that a geometry is a transformation group. The fact that the three constant curvature geometries (hyperbolic, Euclidean and spherical) can be developed in the realm of projective geometry is expressed by the fact that the transformation groups of these geometries are subgroups of the transformation group of projective transformations. Klein digs further this idea, namely, he gives explicit constructions of distances and of measures for angles in the three geometries[9] using the notion of "projective measure" which was introduced by Cayley about twelve years before him.

Before Klein, Cayley gave a construction of the Euclidean plane, equipped with its metric, as a subset of projective space, using projective notions. This result is rather surprising because *a priori* projective geometry is wider than Euclidean geometry insofar as the latter considers lines, projections and other notions of Euclidean geometry but without any notion of measurement of angles or of distances between points. Introducing distances between points or angle measurement and making the transformation group of Euclidean geometry a subgroup of the projective transformation group amounts to considering Euclidean geometry as a particular case of projective geometry; this was the idea of Cayley and, before him, related ideas were emitted by Laguerre, Chasles and possibly others. We emphasize the fact that Cayley did not use the notion of cross ratio in his definition of the distance and that Klein's definitions of both measures (distances and angles) are based on the cross ratio were new. It was also Klein's contribution that the two non-Euclidean geometries are also special cases of a projective geometry.

Klein starts his paper [33] by referring to the work of Cayley, from which, he says, "one may construct a projective measure on ordinary space using a second degree surface as the so-called fundamental surface." This sentence needs a little explanation. "Ordinary space" is three-dimensional projective space. A "measure" is a way of measuring distances between points as well as angles between lines (in dimension two) or between planes (in dimension three). A measure is said to be "projective" if its definition is based on projective notions and if it is invariant under the projective transformations that preserve a so-called fundamental surface. Finally, the "funda-

[7] A similar statement is made in the introduction of the *Erlangen program* [34]: "Among the advances of the last fifty years in the field of geometry, the development of *projective* geometry occupies the first place."

[8] Poncelet, for instance, made heavy use of projective transformations in order to reduce proofs of general projective geometry statements to proofs in special cases which are simpler to handle.

[9] The reader will easily see that it is a natural idea to define the notion of "angle" at any point in the plane by using a conic (say a circle) at infinity, by taking the distance between two rays starting at a point as the length of the arc of the ellipse at infinity that the two rays contain. However, this is not the definition used by Klein. His definition of angle uses the cross ratio, like for distances between points, and this makes the result projectively invariant.

mental surface" is a quadric, that is, a second-degree surface, which is chosen as a "surface at infinity" in projective space. Cayley called such a surface the *absolute*. In the plane, the quadric is replaced by a conic called the *absolute conic*.

Let us say things more precisely. A fixed quadric is chosen. To define the distance between two points, consider the line that joins them; it intersects the quadric in two points (which may be real – distinct or coincident – or imaginary). The distance between the two points is then taken to be, up to a constant factor, the logarithm of the cross ratio of the quadruple formed by these points together with the intersection points of the line with the quadric taken in a natural order. The cross ratio (or its logarithm) could be imaginary, and the mutiplicative constant is chosen so that the result is real. We shall come back to this definition in §7 of this paper. In any case, the projective measure depends on the choice of a fundamental surface and the definitions of measures on lines and on planes use constructions which are dual to each other. Thus, Klein's construction is based on the fact that two points in the real projective space define a real line, which is also contained in a complex line (its complexification), after considering the real projective space as sitting in the complex projective space. Likewise, a quadric in the real projective space is the intersection with that space of a unique quadric in the complex projective space. If the real line intersects the real quadric in two points, then these two points are real, and in this case the cross ratio is real. In the general case, the complex line intersects the quadric in two points, which may be real or complex conjugate or coincident, and the cross ratio is a complex number. The multiplicative constant in front of the logarithm in the definition of the distance makes all distances real.

In some cases, there is a restriction on the quadric. Namely, in the case where it is defined by the quadratic form $x^2 + y^2 - z^2$, the quadric has an interior and an exterior,[10] and one takes as underlying space the *interior* of the quadric.

Let us recall that in the projective plane, there are only two kinds of non-degenerate conics, viz. the *real conics*, which in homogeneous coordinates can be written as $x^2+y^2-z^2 = 0$, and the *imaginary conics*, which can be written as $x^2+y^2+z^2 = 0$. There is also a degenerate case where the conic is reduced to two coincident lines, which can be written in homogeneous coordinates as $z^2 = 0$, or also $x^2 + y^2 = 0$. (Notice that this is considered as a degenerate because the differential of the implicit equation is zero.) In the way Cayley uses it, a degenerate conic can be thought of as the two points on the circle at infinity whose homogeneous coordinates are $(1, i, 0)$ and $(1, -i, 0)$.

In working with a conic at infinity, Cayley gave a general formula for distances that does not distinguish between the cases where the conic is real or imaginary, but he noted that Euclidean geometry is obtained in the case where the absolute degenerates into a pair of points. Klein made a clear distinction between the cases of a real and an imaginary conic and he obtained the three geometries:

[10] A point is in the interior of the quadric if there is no real tangent line from that point to the quadric (that is, a line that intersects it in exactly one point). Note that this notion applies only to real quadrics, since in the case of a complex quadric, from any point in the plane one can find a line which is tangent to the quadric. This can be expressed by the fact that a certain quadratic equation has a unique solution.

- The elliptic, in the case where the absolute is imaginary. (Notice that in this case, all the real directions are points in the projective space, since none of them intersects the imaginary conic.) The fundamental conic in this case can be taken to be the imaginary circle whose equation is $\sum_i x_i^2 = 0$ (it has no real solutions).
- The hyperbolic, in the case where the absolute is real. In this case, the conic has a well-defined "interior" and "exterior", and the hyperbolic plane corresponds to the interior of the conic.
- The parabolic, in the case where the absolute degenerates into two imaginary points. This is a limiting case of the preceding ones, and it corresponds to Euclidean geometry.

In this way, the three geometries become a particular case of projective geometry in the sense that the transformation group of each geometry is a subgroup of the projective transformation group, namely the group of transformations that fix the given conic.

2 Projective geometry

As we already noted several times, in Klein's program, projective geometry acts as a unifying setting for many geometries. In fact, several theorems in Euclidean geometry (the theorems of Pappus, Pascal, Desargues, Menelaus, Ceva, etc.) find their natural explanation in the setting of projective geometry. Let us say a few words of introduction to this geometry, since it will be important in what follows.

For a beginner, projective geometry is, compared to the Euclidean, a mysterious geometry. There are several reasons for that. First, the non-necessity of any notion of distance or of length may be misleading (what do we measure in this geo-metry?) Secondly, the Euclidean coordinates are replaced by the less intuitive (although more symmetric) "homogeneous coordinates". Thirdly, in this geometry, lines intersect "at infinity". There are points at infinity, there are "imaginary points", and there is an overwhelming presence of the cross ratio, which, although a beautiful object, is not easy to handle. We can also add the fact that the projective plane is non-orientable and is therefore more difficult to visualize than the Euclidean. One more complication is due to the fact that several among the founders of the subject had their particular point of view, and they had different opinions of what the fundamental notions should be.

Using modern notation, the ambient space for this geometry is the n-dimensional[11] projective space $\mathbb{RP}^n$. This is the quotient of Euclidean space $\mathbb{R}^{n+1} \setminus \{0\}$ by the equivalence relation which identifies a point x with any other point λx for $\lambda \in \mathbb{R}^*$. The projective transformations of $\mathbb{RP}^n$ are quotients of the linear transformations of $\mathbb{R}^{n+1}$. The linear transformations map lines, planes, etc. in $\mathbb{R}^{n+1}$ to lines, planes,

[11] Of course, in the early theory, only dimensions two and three were considered.

etc. in $\mathbb{R}^{n+1}$; therefore, they map points, lines, etc. in $\mathbb{RP}^n$ to points, lines, etc. in $\mathbb{RP}^n$. The incidence properties – intersections of lines, of planes, alignment of points, etc. – are preserved by the projective transformations. These transformations form a group called the projective linear group, denoted by $\mathrm{PGL}(n, \mathbb{R})$. There is no metric on $\mathbb{RP}^n$ which is invariant by the action of this group, since this action is transitive on pairs of distinct points. As we already noted, Cayley observed that if we fix an appropriate quadric in $\mathbb{RP}^n$, which he called the *absolute*, we can recover the group of Euclidean geometry by restricting $\mathrm{PGL}(n, \mathbb{R})$ to the group of projective transformations that preserve this quadric. In this way, Euclidean space sits as the complement of the quadric, which becomes the *quadric at infinity*. Klein states in his *Erlangen program* [34]:

> Although it seemed at first sight as if the so-called metrical relations were not accessible to this treatment, as they do not remain unchanged by projection, we have nevertheless learned recently to regard them also from the projective point of view, so that the projective method now embraces the whole of geometry.

Cayley's paper [12] on this subject was published in 1859 and it is abundantly cited by Klein in the two papers which are our main object of interest here.

There is a situation which is familiar to any student in geometry, which is in the same spirit as Cayley's remark. If we fix a hyperplane H in the projective plane $\mathbb{RP}^n$, then the subgroup of the group of projective transformations of $\mathbb{RP}^n$ that preserve H is the affine group. Affine space is the complement of that hyperplane acted upon by the affine group. It is in this sense that "affine geometry is part of projective geometry". In the projective space, at infinity of the affine plane stands a hyperplane. From Klein's point of view, affine geometry is determined by (and in fact it is identified with) the group of affine transformations, and this group is a subgroup of the group of projective transformations. Likewise, Euclidean and hyperbolic geometries are all part of affine geometry (and, by extension, of projective geometry). Furthermore, we have models of the spherical, Euclidean and hyperbolic spaces that sit in affine space, each of them with its metric which is relative to a "conic at infinity". One consequence is that all the theorems of projective geometry hold in these three classical geometries of constant curvature. Klein insists on this fact, when he declares that projective geometry is "independent of the parallel postulate" (see §8 below).

In projective geometry, one studies properties of figures and of maps arising from projections ("shadows") and sections, or, rather, properties that are preserved by such maps. For instance, in projective geometry, a circle is equivalent to an ellipse (or to any other conic), since these objects can be obtained from each other by projection.

Mathematical results where projective geometry notions are used, including duality theory, are contained in the works of Menelaus (1st–2nd c. A. D.), Ptolemy and in the later works of their Arabic commentators; see [52] and [53]. The Renaissance artists used heavily projective geometry; see Figures 5.1 and 5.2 for some examples. A good instance of how projective geometry may be useful in perspective drawing is provided by Desargues' Theorem, which is one of the central theorems of projective geometry and which we recall now.

Consider in the projective plane two triangles abc and ABC. We say that they are *in axial perspectivity* if the three intersection points of lines $ab \cap AB, ac \cap AC, bc \cap$

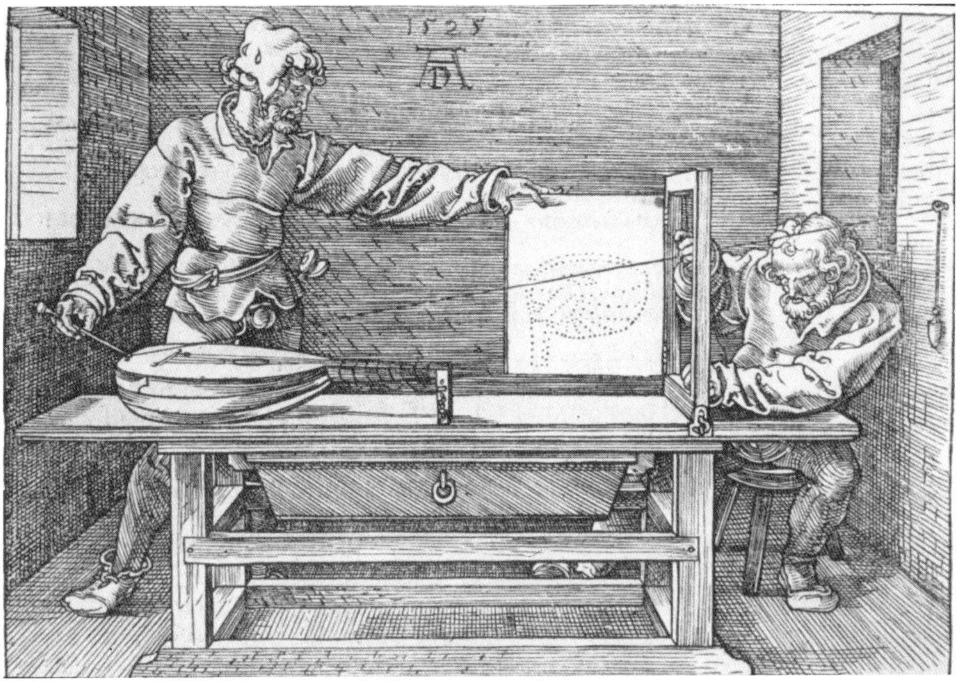

Figure 5.1. *Man drawing a lute*. Woodcut by Albrecht Dürer (1471–1528), from his *Instructions for measuring with compass and ruler*, Book 4, Nürnberg 1525. A drawing device involving a stretched string is used to provide the image under a perspective map. Projective properties of figures are preserved under perspective drawing. The subject of projective geometry was motivated by the art of perspective drawing.

BC are on a common line. We say that the three triangles are *in central perspectivity* if the three lines Aa, Bb, Cc meet in a common point. Desargues' theorem says that for any two triangles, being in axial perspectivity is equivalent to being in central perspectivity.

The theorem is quoted in several treatises of perspective drawing.[12]

Ideas and constructions of projective geometry were extensively used by Renaissance artists like Leon Battista Alberti (1404–1472), Leonardo da Vinci (1452–1519) and Albrecht Dürer whom we already mentioned. All these artists used for instance the principle saying that any set of parallel lines in the space represented by the drawing which are not parallel to the plane of the picture must converge to a common point, called the *vanishing point*.[13] Of course, everything started with the Greeks,

[12]Desargues' theorem was published for the first time by A. Bosse, in his *Manière universelle de M. Desargues pour manier la perspective par petit pied comme le géométral* (Paris, 1648, p. 340). Bosse's memoir is reproduced in Desargues' *Œuvres* (ed. N. G. Poudra, Paris 1884, p. 413–415). Desargues' proof of the theorem uses Menelaus' Theorem. Von Staudt, in his *Geometrie der Lage*, (Nuremberg, 1847) gave a proof of Desargues' theorem that uses only projective geometry notions.

[13]The English term *vanishing point* was introduced by Brook in the treatise [58] that he wrote in 1719, which is also the first book written in English on the art of perspective. The italian expression *punto di fuga* was already

Figure 5.2. *Artist drawing a nude*, woodcut by Albrecht Dürer, from *The Art of Measuring*. Simmern: Hieronymus Rodler 1531.

and two important works on perspective by Euclid (fl. 3rd c. B.C.) and by Heliodorus of Larissa (3d c. A.D.) were translated into Latin in 1573 by the Renaissance mathematician, astronomer and monk Egnazio Danti[14] (1536–1586) [16]. About a quarter of a century later, the italian mathematician and astronomer Guidobaldo del Monte (1545–1607) wrote a treatise in six books (*Perspectivae libri*, published in Pisa in 1600) in which he led the mathematical foundations of perspective drawing that included the vanishing point principle. In this treatise, the author often refers to Euclid's *Elements*. The architect and famous scenographer[15] Nicola Sabbatini (1574–1654) made extensive use of Guidobaldo's theoretical work. Guidobaldo del Monte's book is regarded as a mathematical work on projective geometry. Perspective drawing involves some of the basic operations of projective geometry (projections and sections), and it highlights the non-metrical aspect of that geometry.

It is usually considered that the modern theory of projective geometry started with J.-V. Poncelet (1788–1867), in particular with his two papers *Essai sur les propriétés projectives des sections coniques* (presented at the French Academy of Sciences in 1820) and *Traité des propriétés projectives des figures* (1822). Poncelet tried to eliminate the use of coordinates and to replace them by synthetic reasonings. He made heavy use of duality (also called polarity) theory with respect to a conic. This is based on the simple observation that in the projective plane any two points define a line and any two lines define a point. Using this fact, certain statements in projective geometry remain true if we exchange the words "line" and "point". Using this theory, Brianchon deduced the theorem which bears his name, on the diagonals of a hexagon circumscribed to a conic, from Pascal's theorem on the intersection of pairs of opposite sides of a hexagon inscribed in a conic. Another well-known exam-

used by Alberti and the other italian Renaissance artists. Independently of this aspect of terminology, one should notice that a psychological effort is required when considering parallel lines as lines meeting at a common point.

[14]Danti was, during more than 30 years, professor of mathematics at the University of Bologna, which was one of the most famous universities in Europe (and the oldest one). At his death, Giovanni Antonio Magini, who held a geocentric vision of the world, was chosen as his successor, instead of Galileo.

[15]Designer of theatrical scenery

ple is Menelaus' Theorem which is transformed under duality into Ceva's Theorem. Duality in projective geometry is at the basis of other duality theories in mathematics, for instance in linear algebra, between a finite-dimensional vector space and the vector space of linear forms. To Poncelet is also attributed the so-called *principle of continuity* which roughly says that the projective properties of a figure are preserved when the figure attains a limiting position. This remarkable principle allows one to treat general cases by reducing them to a particular one. For instance, in projective geometry, one can reduce the study of ellipses to the one of circles. One can also reduce the study of general quadrilaterals to that of parallelograms. It is also by the continuity principle that Poncelet could assert that points or lines, which disappear at infinity, become imaginary and can therefore be recovered, and one can make appropriate statements about them. Poncelet is also the first who considered that in the (projective) plane, the points at infinity constitute a line.

Among the other founders of projective geometry, we mention J. Brianchon (1783–1864), A. F. Möbius (1790–1868), M. Chasles (1793–1880), K. G. K. von Staudt (1798–1867) and J. Steiner (1796–1863). We shall refer to some of them in the text below. The Göttingen lecture notes of Klein [36] (1889–90) contain notes on the history of projective geometry (p. 61 & ff.). We also refer the reader to the survey [21] by Enriques, in which the works of Cayley and of Klein are analyzed. A concise modern historical introduction to projective geometry is contained in Gray's *Worlds out of nothing* [26].

Perhaps Poncelet's major contribution, besides the systematic use of polarity theory, was to build a projective geometry which is free from the analytic setting of his immediate predecessors and of the cross ratio (which he called the *anharmonic ratio*, and so does Klein in the papers under consideration). Chasles, in his 1837 *Aperçu historique sur l'origine et le développement des méthodes en géométrie*, highlights the role of transformations in geometry, in particular in projective geometry, making a clear distinction between the metric and the projective (which he calls "descriptive") properties of figures. Von Staudt insisted on the axiomatic point of view, and he also tried to build projective geometry independently from the notions of length and angle. One should also mention the work of E. Laguerre (1843–1886), who was a student of Chasles and who, before Cayley, tried to develop the notions of Euclidean angle and distance relatively to a conic in the plane, cf. [41] p. 66. Laguerre gave a formula for angle measure that involves the cross ratio. It is important nevertheless to note that Laguerre, unlike Klein, did not consider this as a possible *definition* of angle. Laguerre's formula originates in the following problem that he solves: Given a system of angles $A, B, C, \dots$ of a certain figure F in a plane, satisfying an equation

$$F(A, B, C, \dots),$$

find a relation satisfied by the image angles $A', B,' C', \dots$ when the figure F is transformed by a projective transformation (which Laguerre calls a homography). The solution that Laguerre gives is that $A', B,' C', \dots$ satisfy the relation

$$F\left(\frac{\log a}{2\sqrt{-1}}, \frac{\log b}{2\sqrt{-1}}, \frac{\log c}{2\sqrt{-1}}, \dots\right),$$

where $a, b, c, \ldots$ is the cross ratio of the quadruple of lines made by two sides of the angles $A, B, C, \ldots$ together the lines $AP, AQ, BP, BQ, CP, CQ, \ldots$ which are the images of the lines made by $A, B, C, \ldots$ and the two cyclic points of the plane of F. Laguerre notes that Chasles, in his *Traité de géométrie supérieure* [14], p. 446, gave a solution to this problem in the case where the angles $A, B, C, \ldots$ share the same vertex or when they are equal. It seems that neither Cayley nor Klein, at the beginning of their work on this subject, were aware of Laguerre's work. Klein mentioned Laguerre's work in his later 1889–90 lecture notes ([36], p. 47 and 61). In the *Gesammelte Mathematische Abhandlungen* ([38], vol. 1, p. 242) Klein declares that at the time he wrote his paper [33], he was not aware of Laguerre's ideas. This work of Laguerre is also mentioned in [17] and [55]. See also [39] for notes of Klein on Laguerre's work, and [61] for some comments on Laguerre's formulae.

For Klein, the subject of projective geometry includes both a synthetic and an analytic aspect.[16] In his historical remarks contained in his lecture notes ([36] p. 61), he makes a distinction between the French and the German school of projective geometry and he notes that in the beginning of the 1850s, the French school had a serious advance over the German one, the latter still distinguishing between the projective and the metric properties.[17]

It is also fair to recall that this nineteenth-century activity on projective geometry was preceded by works of the Greeks, in particular by the work of Apollonius on the Conics (where the notion of a polar line with respect to a conic appears for the first time), by works of Pappus, and by the much later works of several French mathematicians, including G. Desargues (1591–1661), who tried to give a firm mathematical foundation to the perspective theory that was used on a heuristic basis by painters and architects, and then, B. Pascal (1623–1662), who was influenced by Desargues, and G. Monge (1746–1818). Both Desargues and Pascal applied the ideas of projective geometry in the study of conics, considering the work of Apollonius from a new perspective. Monge is considered as the father of descriptive geometry (1795).

One important factor that emanates from all this is that the ellipse, parabola and hyperbola can be considered as one and the same object, but seen from different point of views. This is of paramount importance in the work of Klein in which we are concerned.

[16]We can quote here Klein, from his *Erlangen program* [34]: "The distinction between modern synthetic and modern analytic geometry must no longer be regarded as essential, inasmuch as both subject-matter and methods of reasoning have gradually taken a similar form in both. We choose therefore in the text as a common designation of them both the term *projective geometry*."

[17]In his *Erlangen program* [34], Klein writes: "Metrical properties are to be considered as projective relations to a fundamental configuration, the circle at infinity" and he adds in a note: "This view is to be regarded as one of the most brilliant achievements of [the French school]; for it is precisely what provides a sound foundation for that distinction between properties of position and metrical properties, which furnishes a most desirable starting-point for projective geometry." Regarding Klein's comments on the difference between the French and the German schools, one may remember the context of that time, namely the French–German war (July 19, 1870 – January 29, 1871), opposing the French Second Empire to the Prussian Kingdom and its allies; France suffered a crushing defeat and lost the Alsace-Moselle, which became the German Reichsland Elsaß-Lothringen. In this and in other of Klein's historical notes, mathematics outweighed German nationalism.

3 Non-Euclidean geometry

In this section, we recall a few facts on the birth and the reception of non-Euclidean geometry, and on its relation with Klein's work on projective geometry.

Nikolai Ivanovich Lobachevsky was the first to publish a treatise on hyperbolic geometry, namely, his *Elements of Geometry* [44] (1829). The two other founders of the subject are János Bolyai and Carl Friedrich Gauss. For more than 50 years, their works remained unknown to the mathematical community. Lobachevsky's work was acknowledged as being sound only ten years after his death (1856), when Gauss's correspondence was published.[18]

This geometry first attracted the attention of Cayley and then Beltrami. Beltrami started by publishing two articles on the subject, *Saggio di Interpretazione della geometria non-Euclidea* [4] (1868) which concerns the two-dimensional case and *Teoria fondamentale degli spazii di curvatura costante* [6] (1868–69) which concerns the three-dimensional case.[19] We shall elaborate on these works below.

Lobachevsky, Bolyai and Gauss developed the hyperbolic geometry system starting from the axiomatic point of view. This consists in drawing conclusions from the axioms of Euclidean geometry with the parallel axiom replaced by its negation. No Euclidean model is involved in this approach (and no Euclidean model existed at that time). Beltrami, in his paper [4], was the first to establish the relation between hyperbolic geometry and negative curvature. None of the three founders of hyperbolic geometry used the notion of curvature. It is true that at that time, curvature in the Riemannian geometry sense was not yet discovered, but Gauss had already introduced

---

[18] In a letter he wrote to his friend the astronomer H. C. Schumacher, dated 28 November 1846, Gauss expresses his praise for Lobachevsky's work, cf. [24] p. 231–240, and it was after the publication of this letter that Lobachevsky's works attracted the attention of the mathematical community. Lobachevsky was never aware of that letter. Gauss's correspondence was published during the few years that followed Gauss's death, between 1860 and 1865.

[19] Eugenio Beltrami (1835–1900) was born in a family of artists. He spent his childhood in a period of political turbulence: the Italian revolutions, the independence war, and eventually the unification of Italy. He studied mathematics in Pavia between 1853 and 1856, where he followed the courses of Francesco Brioschi, but due to lack of money or may be for other reasons (Loria reports that Beltrami was expelled from the university because he was accused of promoting disorders against the rector [47]), he interrupted his studies and took the job of secretary of the director of the railway company in Verona. The first mathematical paper of Beltrami was published in 1862. In the same year he got a position at the University of Bologna. He later moved between several universities, partly because of the changing political situation in Italy, and he spent his last years at the university of Rome. A stay in Pisa, from 1863 to 1866, was probably decisive for his mathematical future research; he met there Betti and Riemann (who was in Italy for health reasons). Two of the most influential papers of Beltrami are quoted in the present survey, [4] (1868) and [6] (1869). They were written during his second stay in Bologna where he was appointed on the chair of rational mechanics. His name is attached to the Beltrami equation which is at the basis of the theory of quasiconformal mappings, and to the Laplace–Beltrami operator. Besides mathematics, Beltrami cultivated physics, in particular thermodynamics, fluid dynamics, electricity and magnetism. He translated into Italian the work of Gauss on conformal representations. He contributed to the history of mathematics by publishing a paper on the work of Gerolamo Saccheri (1667–1733) on the problem of parallels (*Un precursore italiano di Legendre e di Lobatschewski*, 1889), comparing this work to the works of Borelli, Clavius, Wallis, Lobachevsky and Bolyai on the same subject, and highlighting the results on non-Euclidean geometry that are inherent in that work. Besides mathematics, Beltrami cultivated music, and also politics. In 1899, he became (like his former teacher Brioschi) senator of the Kingdom of Italy.

the notion of surface curvature and he had showed that it can be defined independently of any embedding in an ambient Euclidean space. Gauss did not make explicit the relation between curvature and the geometry of the hyperbolic plane.

The definition of spherical geometry as a system which can also be defined using the notion of constant positive curvature is due to Riemann. It is also a geometrical system which is at the same level as Euclidean geometry, where the "lines" are the great circles of the sphere but where there are no disjoint lines. It is also good to recall that, unlike hyperbolic geometry, the geometrical system of the sphere cannot be obtained from the Euclidean one by modifying only one axiom, since not only the Euclidean parallel axiom is not valid on the sphere, but other axioms as well, e.g. the one saying that lines can be extended indefinitely.[20] Riemann established the bases of spherical geometry in his famous habilitation lecture *Über die Hypothesen, welche der Geometrie zu Grunde liegen* (On the hypotheses which lie at the foundations of geometry) (1854) [54]. It is generally accepted that the three geometries – Euclidean, hyperbolic and spherical – appear clearly for the same time at the same level, as "the" three geometries of constant curvature in the paper [33] of Klein. However, one can mention a letter from Hoüel to De Tilly, dated April 12, 1872, in which he writes[21] (see [29]):

> The idea of the three geometries is not due to Klein: it goes back to Lejeune–Dirichlet, who has thoroughly meditated upon this subject, but who, unfortunately, did not leave us anything written.

[20]The intuition that there are exactly *three* geometries, which correspond to the fact that the angle sum in triangles is less than, equal, or greater than two right angles (these are the hyperbolic, Euclidean and spherical geometry respectively) can be traced back to older works. In the memoir *Theorie der Parallellinien* [42] of Johann Heinrich Lambert, (1728–1777) written in 1766, that is, more than 100 years before Klein wrote his memoir [33], the author, attempting a proof of Euclid's parallel postulate, developed a detailed analysis of geometries that are based on three assumptions, concerning a class of quadrilaterals, which are now called *Lambert* or *Ibn al-Haytham-Lambert quadrilaterals*. These are quadrilaterals having three right angles, and the assumptions Lambert made are that the fourth angle is either acute, right or obtuse. These hypothesis lead respectively to hyperbolic, Euclidean and spherical geometry. One must add that Lambert was not the first to make such a study of these quadrilaterals. Saccheri and, before him, Abū 'Alī Ibn al-Haytham and 'Umar al-Khayyām (1048–1131) made similar studies. Of course, in all these works, the existence of hyperbolic geometry was purely hypothetical. The approaches of these authors consisted in assuming that such a geometry exists and in trying to deduce a contradiction. We refer the interested reader to the recent edition of Lambert's work [48], with a French translation and mathematical comments.

[21]A few words are needed on Hoüel and de Tilly, two major major figures in the history of non-Euclidean geometry but whose names remain rather unknown to most geometers. Guillaume-Jules Hoüel (1823–1886) taught at the University of Bordeaux. He wrote geometric treatises giving a modern view on Euclid's *Elements*. He was working on the impossibility of proving the parallel postulate when, in 1866, he came across the writings of Lobachevsky, and became convinced of their correctness. In the same year, he translated into French Lobachevsky's *Geometrische Untersuchungen zur Theorie der Parallellinien* together with excerpts from the correspondence between Gauss and Schumacher on non-Euclidean geometry, and he published them in the *Mémoires de la Société des Sciences physiques et naturelles de Bordeaux*, a journal of which he was the editor. Hoüel also translated into French and published in French and Italian journals works by several other authors on non-Euclidean geometry, including Bolyai, Beltrami, Helmholtz, Riemann and Battaglini. Barbarin, in his book *La Géométrie non Euclidienne* ([3] p. 12) writes that Hoüel, "who had an amazing working force, did not hesitate to learn all the European languages in order to make available to his contemporaries the most remarkable mathematical works." Hoüel also solicited for his journal several papers on hyperbolic geometry, after

In Klein's paper [33], while the three geometries are placed at the same level of importance, Euclidean geometry acts as a *transitional* geometry between the other two. Klein writes about this:

> Straight lines have no points at infinity, and indeed one cannot draw any parallel at all to a given line through a point outside it.
>
> A geometry based on these ideas could be placed alongside ordinary Euclidean geometry like the above-mentioned geometry of Gauss, Lobachevsky and Bolyai. While the latter gives each line two points at infinity, the former gives none at all (i.e. it gives two imaginary points). Between the two, Euclidean geometry stands as a transitional case; it gives each line two coincident points at infinity.

We develop this idea of "transitional geometry" in our paper [1] in this volume.

Today, people are so much used to these ideas that it is hard for them to appreciate their novelty for that epoch and their importance. Let us recall in this respect that Klein's paper came out only three years after Beltrami published his two famous papers in which he confirmed that Lobachevsky's researches on hyperbolic geometry were sound.

As we shall see later in this paper, to prove that Cayley's constructions lead to the non-Euclidean geometries, Klein argued in a synthetic way, at the level of the axioms, showing that the characteristics of the Lobachevsky and of the spherical geometries are satisfied in the geometry defined by this distance function. But Klein also described the differential-geometric aspects, introducing a notion of curvature which he showed is equivalent to Gauss's surface curvature.

In the rest of this paper, we shall present the basic ideas contained in Klein's two papers, making connections with other ideas and works on the same subject.

4 Preliminary remarks on Klein's papers

In this section, we start by summarizing the important ideas contained in Klein's two papers. We then discuss the reception of these ideas by Klein's contemporaries and by other mathematicians. We then make some remarks on the names *hyperbolic, elliptic* and *parabolic* geometries that were used by Klein.

---

the French Academy of Sciences, in the 1870s, decided to refuse to consider papers on that subject. We refer the interested reader to the article by Barbarin [2] and the forthcoming edition of the correspondence between Hoüel and de Tilly [29]. Beltrami had a great respect for Hoüel, and there is a very interesting correspondence between the two men, see [8]. It appears from these letters that Beltrami's famous *Saggio di Interpretazione della geometria non-Euclidea* [4] arose from ideas that he got after reading Lobachevsky's *Geometrische Untersuchungen zur Theorie der Parallellinien* in the French translation by Hoüel, see [8] p. 9. For a detailed survey on the influence of Hoüel's work see [10].

Joseph-Marie de Tilly (1837–1906) was a member of the Royal Belgian Academy of Sciences, and he was also an officer in the Belgian army, teaching mathematics at the Military School. In the 1860s, de Tilly, who was not aware of the work of Lobachevsky, developed independently a geometry in which Euclid's parallel postulate does not hold. One of his achievements is the introduction of the notion of distance as a primary notion in the three geometries: hyperbolic, Euclidean and spherical. He developed an axiomatic approach to these geometries based on metric notions, and he highlighted some particular metric relations between finite sets of points; see for instance his *Essai sur les Principes Fondamentaux de la Géométrie et de la Mécanique* [59] and his *Essai de Géométrie analytique générale* [60].

Klein's major contributions in these two papers include the following:

1. An explanation of the notion of Cayley measure and its representation, and its inclusion in two important settings: transformation groups and curvature.
2. A realization of Lobachevsky's geometry as a metric space (and not only as a system of axioms).
3. The construction of a new model of Lobachevsky's geometry, by taking, in Cayley's construction, the "absolute" to be an arbitrary real second-degree curve in the projective plane and showing that the interior of that curve, equipped with some adequate structure, is a model of Lobachevsky's geometry. Although the idea for the construction originates in Cayley's work (Cayley gave a formula for a distance function without realizing that the resulting metric space is the Lobachevsky space), and although the construction of such a model for the hyperbolic plane (but without the distance function) had been made three years earlier by Beltrami in this paper [4] (1868) in which he realized that the Euclidean segments of the disk are models for the geodesics of hyperbolic space, Klein gave the first explicit distance function for hyperbolic geometry. At the same time, he made the first link between hyperbolic geometry and projective geometry.
4. A unified setting for Euclidean, hyperbolic and spherical geometries, as these three geometries can be considered as projective geometries. Although it is well known that Klein gave a formula for the hyperbolic metric using cross ratio, it is rather unknown to modern geometers that Klein also gave in the same way formulae for the elliptic and for the Euclidean distance functions using the cross ratio.

   Cayley expressed the advantage of Klein's distance formula in his comments on his paper [12] contained in his *Collected mathematical papers* edition ([13] Vol. II, p. 604):

> In his first paper, Klein substitutes, for my $\cos^{-1}$ expression for the distance between two points,[22] a logarithmic one; viz. in linear geometry if the two fixed points are A, B then the assumed definition for the distance of any two points P, Q is
> $$\text{dist.}\,(PQ) = s \log \frac{AP.BQ}{AQ.BP};$$
> this is a great improvement, for we at once see that the fundamental relation, dist. (PQ) + dist. (QR) = dist. (PR), is satisfied.

[22] The formulae to which Cayley refers are contained in his paper [12] p. 584–585. After giving these formulae in the case where the absolute is a general conic, he writes:

> The general formulae suffer no *essential* modifications, but they are greatly simplified by taking for the point-equation of the absolute
> $$x^2 + y^2 + z^2 = 0,$$

We note that Cayley did not introduce the cross ratio in his definition of distances, but he showed that his formulae are invariant by the action of the projective geometry transformations.

5. Likewise, Klein gave a formula for the dihedral angle between two planes as a cross ratio between four planes, the additional two planes being the tangent planes to the fixed conic passing through the intersection of the first two planes.
6. The conclusion that each of the three geometries is consistent if projective geometry is consistent.
7. The idea of a *transitional geometry*. This is a geometrical system in which one can transit continuously from spherical to hyperbolic geometry, passing through Euclidean geometry.
8. The introduction of the names hyperbolic, parabolic and elliptic for the Lobachevsky, Euclidean and spherical geometries respectively, thus making the relation with other settings where the three names were already used. We shall discuss this at the end of this section.

We shall elaborate on all these items below.

Klein's papers are sometimes difficult to read and they were received by the mathematical community in diverse manners. Let us quote, for example, Darboux, from his obituary concerning Henri Poincaré [18]:

> Mr. Felix Klein is the one who removed these very serious objections [concerning non-Euclidean geometry] by showing in a beautiful memoir that a geometry invented by the famous Cayley and in which a conic called the absolute provides the elements of all measures and enables, in particular, to define the distance between two points, gives the most perfect and adequate representation of non-Euclidean geometry.

---

or, what is the same, for the line-equation

$$\xi^2 + \eta^2 + \zeta^2 = 0.$$

In fact, we then have for the expression of the distance of the points $(x, y, z), (x', y', z')$,

$$\cos^{-1} \frac{xx' + yy' + zz'}{\sqrt{x^2 + y^2 + z^2}\sqrt{x'^2 + y'^2 + z'^2}};$$

for that of the lines $(\xi, \eta, \zeta), (\xi', \eta', \zeta')$,

$$\cos^{-1} \frac{\xi\xi + \eta\eta' + \zeta\zeta'}{\sqrt{\xi^2 + \eta^2 + \zeta^2}\sqrt{\xi'^2 + \eta'^2 + \zeta'^2}};$$

and that for the point (x, y, z) and the line (ξ', η', ζ'),

$$\cos^{-1} \frac{\xi' x + \eta' y + \zeta' z}{\sqrt{x^2 + y^2 + z^2}\sqrt{\xi'^2 + \eta'^2 + \zeta'^2}}.$$

The reader will notice the analogy between these formulae and the familiar formula for distance in spherical geometry (the "angular distance"), which, Klein also establishes in his paper [33]; see Formulae (5.7) and (5.8) in the present paper.

On the other hand, Genocchi[23] wrote, regarding the same matter ([25] p. 385):

> From the geometric point of view, the spirit may be shocked by certain definitions adopted by Mr. Klein: the notions of distance and angle, which are so simple, are replaced by complicated definitions [...] The statements are extravagant.

Hans Freudenthal (1905–1990), talking about Klein's analysis of the work of von Staudt on the so-called "fundamental theorem of projective geometry" in which Klein discusses some continuity issues that were missing in von Staudt's arguments,[24] writes [23]:

> [In 1873], logical analysis was not the strong point of Klein, and what he wrote on that question in the years that followed is as much confusing as possible.

It is true that Klein was much more interested in ideas (and he had profound ideas) than in writing up rigorous proofs.

We end this section with two remarks. The first one concerns the title of the two papers [33] and [35], and the second concerns the adjectives "hyperbolic", elliptic" and "parabolic".

Klein's title, *On the so-called Non-Euclidean geometry*, may be considered as having a negative connotation, and indeed it does. This is also the title of a note (Note No. 5) at the end of his *Erlangen program* text. In that note, Klein writes:

> We associate to the name Non-Euclidean geometry a crowd of ideas that have nothing mathematical, which are accepted on the one hand with as much enthusiasm that they provoke aversion on the other hand, ideas in which, in any case, our exclusively mathematical notions have nothing to do.

However, Klein, in his later writings, used extensively the term "non-Euclidean geometry", without the adjective "so-called".

Now about the names of the three geometries.

Klein coined the expressions "elliptic", "hyperbolic" and "parabolic" geometry as alternative names for spherical, Lobachevsky and Euclidean geometry respectively. In his *Über die sogenannte Nicht-Euklidische Geometrie* [33], he writes (Stillwell's translation p. 72):

[23] Angelo Genocchi (1817–1889) was an Italian mathematician who made major contributions in number theory, integration and the theory of elliptic functions. Like Cayley, he worked for several years as a lawyer, and he taught law at the University of Piacenza, but at the same time he continued cultivating mathematics with passion. In 1859, he was appointed professor of mathematics at the University of Torino, and he remained there until 1886. During the academic year 1881–82, Guiseppe Peano served as his assistant, and he subsequently helped him with his teaching, when Genocchi became disabled after an accident. Genocchi's treatise *Calcolo differenziale e principii di calcolo integrale con aggiunte del Dr. Giuseppe Peano*, written in 1864, was famous in the Italian universities.

[24] This continuity issue is mentioned in Chapter 2 of this volume [27], and its is discussed in detail in [62].

> Following the usual terminology in modern geometry, these three geometries will be called *hyperbolic, elliptic or parabolic* in what follows, according as the points at infinity of a line are real, imaginary or coincident.

The expression "following the usual terminology" may be related to the fact that in the projective plane, a hyperbola meets the line at infinity in two points, a parabola meets it is one point and an ellipse does not meet it at all. In a note in [57] (Note 33), Stillwell recalls that before Klein, the points on a differentiable surface were called hyperbolic, elliptic or parabolic when the principal tangents at these points are respectively real, imaginary or coincident. He also recalls that Steiner used these names for certain surface involutions, an involution being called hyperbolic, elliptic or parabolic when the double points arising under it are respectively real, imaginary or coincident. The name *non-Euclidean geometry* is due to Gauss.[25]

Soon after Klein introduced this terminology, Paul du Bois-Reymond (1831–1889) introduced (in 1889) the classification of second-order differential operators into "elliptic", "hyperbolic" and "parabolic".

Let us note finally that it is easy to be confused concerning the order and the content of the two papers of Klein, if one looks at the French versions. The papers appeared in 1871 and 1873 respectively, under the titles *Über die sogenannte Nicht-Euklidische Geometrie* and *Über die sogenannte Nicht-Euklidische Geometrie (Zweiter Aufsatz)*. In 1871, and before the first paper [33] was published, a short version, presented by Clebsch,[26], appeared in the *Nachrichten von der Kgl. Gesellschaft der Wissenschaften zu Göttingen*, under the title *F. Klein, Über die sogenannte Nicht-Euklidische Geometrie. Vorgelegt von A. Clebsch.* The same year, a translation of this short paper appeared under the title *Sur la géométrie dite non euclidienne, de Félix Klein*, in the *Bulletin de sciences mathématiques et astronomiques*, translated by Hoüel. A translation by Laugel of the first paper (1871) appeared much later in the *Mémoires de la Faculté des Sciences de Toulouse* under the title *Sur la Géométrie dite non euclidienne, par Mr. Félix Klein*, in 1898. In the volume [57] (1996) which contains translations by Stillwell of some of the most important sources on non-Euclidean geometry, only the first paper [33] by Klein is included, under the title *On the so-called non-Euclidean geometry*. It is followed by a short excerpt (6 lines) of the second paper.

---

[25] Gauss used it in his correspondence with Schumacher.

[26] Alfred Clebsch (1833–1872) was a young professor at Göttingen, who was responsible for Klein's first invitation at that university, in 1871. He was well aware of Cayley's work on invariant theory, and he transmitted it to Klein. Klein stayed in Göttingen a few months, and then moved to Erlangen, where he was appointed professor, again upon the recommendation of Clebsch. He came back to the University of Göttingen in 1886, and he stayed there until his retirement in 1913. Clebsch was also the founder of the *Mathematische Annalen*, of which Klein became later the main editor. See also [27].

5 The work of Cayley

In this section, we comment on the idea of Cayley[27] which acted as a motivation for Klein's work.

Let us quote again Klein, from the introduction to his paper [33]:

> *It is our purpose to present the mathematical results of these works [of Gauss, Lobachevsky and Bolyai], insofar as they relate to the theory of parallels, in a new and intuitive way, and to provide a clear general understanding.*
>
> The route to this goal is through projective geometry. By the results of Cayley, one may construct a projective measure on ordinary space using an arbitrary second degree surface as the so-called fundamental surface. Depending on the type of the second degree surface used, this measure will be a model for the various theories of parallels in the above-mentioned works. But it is not just a model for them; as we shall see, it precisely captures their inner nature.

The paper to which Klein refers is Cayley's *Sixth Memoir upon Quantics*[28] [12] which appeared in 1859. In this paper, Cayley asserts that descriptive geometry (which is the name he used for projective geometry) "is all geometry", an idea which was taken up by Klein later on.[29] In particular, Cayley considered that projective geometry includes metrical geometry (which is the name he used for Euclidean geometry) as a special case. In Cayley's words: "A chief object of the present memoir

---

[27] Arthur Cayley (1821–1895) was born in the family of an English merchant who was settled in Saint-Petersburg. The family returned to England when the young Arthur was eight. Cayley is one of the first discoverers of geometries in dimensions greater than three. To him is attributed the introduction of the term "n-dimensional space" and the invention of matrices, which lead to examples of n^2-dimensional spaces (cf. Cayley's *Memoir on the theory of matrices*, Phil. Trans. of the Royal Society of London, 1858). Cayley is also one of the main inventors of the theory of invariants. These include invariants of algebraic forms (the determinant being an example), and algebraic invariants of geometric structures and the relations they satisfy ("syzygies"). Cayley studied mathematics and law. As a student in mathematics, he was very talented and he wrote several papers during his undergraduate studies, three of which were published in the *Cambridge Mathematical Journal.* The subject included determinants, which became later one of his favorite topics. After completing a four-year position at Cambridge university, during which he wrote 28 papers for the Cambridge journal, Cayley did not succeed in getting a job in academics. He worked as a lawyer during 14 years, but he remained active in mathematics; he wrote during these years about 250 mathematical papers. In 1863, he was appointed professor of mathematics at Cambridge. His list of papers includes about 900 entries, on all fields of mathematics of his epoch. The first definition of an abstract group is attributed to him, cf. [11]. Cayley proved that every finite group G is isomorphic to a subgroup of a symmetric group on G. His name is attached to the famous *Cayley graph* of a finitely generated group, an object which is at the basis of modern geometric group theory. Cayley was the first mathematician who wrote on the work on Lobachevsky (cf. Cayley's *Note on Lobachevsky's imaginary geometry*, *Philosophical Magazine*, 1865, p. 231–233), but he failed to realize the importance of these ideas. Referring to Lobachevsky's trigonometric formulae, Cayley writes: "I do not understand, but it would be very interesting to find a *real* geometrical interpretation of the last mentioned system of equations." In his review of Cayley's *Collected Mathematical Papers* edition in 13 volumes, G. B. Halsted writes: "'Cayley not only made additions to every important subject of pure mathematics, but whole new subjects, now of the most importance, owe their existence to him. It is said that he is actually now the author most frequently quoted in the living world of mathematicians" [28]. We refer the reader to the biography by Crilly [15] which is regrettably short of mathematical detail, but otherwise very informative and accurate.

[28] In Cayley's terminology, a quantic is a homogeneous polynomial.

[29] In fact, the statement is also true if we interpret it in the following sense (which, however, is not what Cayley meant): Most of the geometers at the time Cayley made that statement worked on projective geometry.

is the establishment, upon purely descriptive principles, of the notion of distance." At first sight, there is something paradoxical in this statement, because length is not a projective notion. In fact, in his foundational work on descriptive geometry, and in particular in his famous 1822 *Traité* [49], Poncelet had already stressed on the distinction between the metrical properties (namely, those that involve distance and angle), which are not preserved by projective transformations, and the projective (which he calls "descriptive") properties, which are precisely the properties preserved by projective transformations, e.g. alignment of points, intersections of lines, etc. Thus, in principle, there are no distances, no circles and no angles in projective geometry. Cayley, followed by Klein, was able to define such notions using the concepts of projective geometry by fixing a quadric in projective space, in such a way that these properties are invariant under the projective transformations that fix the quadric.

The cross ratio of four points is a projective invariant, and in some sense it is a complete projective invariant, since a transformation of projective space which preserves the cross ratio of quadruples of aligned points is a projective transformation. Therefore, it is natural to try to define distances and angles using the cross ratio. This is what Klein did. Likewise, it was an intriguing question, addressed by Klein, to try to express the concept of parallelism in Euclidean and in hyperbolic geometry using projective notions, although parallelism is a priori not part of projective geometry.

Cayley defined a geometry which is non-Euclidean, but did not realize that it coincides with the Lobachevsky geometry. Let us quote Cayley's paper (the conclusion):

> I have, in all that has preceded, given the analytical theory of distance along with the geometrical theory, as well for the purpose of illustration, as because it is important to have an analytical expression of a distance in terms of the coordinates; but I consider the geometrical theory as perfectly complete in itself: the general result is as follows; viz. assuming in the plane (or space of geometry of two dimensions) a conic termed the *absolute*, we may by means of this conic, by descriptive constructions, divide any line or range of points whatever, and any point or pencil of lines whatever, into an infinite series of infinitesimal elements, which are (as a definition of distance) assumed to be equal; the number of elements between any two points of the range or two lines of the pencil, measures the distance between the two points or lines; and by means of the pencil, measures the distance between the two points or lines; and by means of the quadrant, as a distance which exists as well with respect to lines as points, we are enabled to compare the distance of two lines with that of two points; and the distance of a point and a line may be represented indifferently as the distance of two points, or as the distance of two lines.
>
> In ordinary spherical geometry, the theory undergoes no modification whatever; the absolute is an actual conic, the intersection of the sphere with the concentric evanescent sphere.
>
> In ordinary plane geometry, the absolute degenerates into a pair of points, viz. the points of intersection of the line at infinity with any evanescent circle, or what is the same thing, the absolute is the two circular points at infinity. The general theory is consequently modified, viz. there is not, as regards points, a distance such as the quadrant, and the distance of two lines cannot be in any way compared with the distance of two points; the distance of a point from a line can be only represented as a distance of two points.

> I remark in conclusion that, *in my point of view*, the more systematic course in the present introductory memoir on the geometrical part of the subject of quantics, would have been to ignore altogether the notions of distance and metrical geometry; for the theory in effect is, that the metrical properties of a figure are not the properties of the figure considered *per se* apart from everything else, but its properties when considered in connexion with another figure, viz. the conic termed the absolute. The original figure might comprise a conic; for instance, we might consider the properties of the figure formed by two or more conics, and we are then in the region of pure descriptive geometry by fixing upon a conic of the figure as a standard of reference and calling it the absolute. Metrical geometry is thus a part of descriptive geometry, and descriptive is *all* geometry and reciprocally; and if this can be admitted, there is no ground for the consideration in an introductory memoir, of the special subject of metrical geometry; but as the notions of distance and of metrical geometry could not, without explanation, be thus ignored, it was necessary to refer to them in order to show that they are thus included in descriptive geometry.

In his *Lectures on the development of mathematics in the XIXth century* [40] (1926–1927), Klein recounts how he came across Cayley's ideas (p. 151):

> In 1869, I had read Cayley's theory in the version of Fiedler[30] of Salmon's *Conics*. Then, I heard for the first time the names of Bolyai and Lobatscheffski, from Stolz,[31] in the winter of 1869/70, in Berlin. From these indications I had understood very little things, but I immediately got the idea that both things should be related. In February 1870, I gave a talk at Weierstrass's seminar on the Cayley metric.[32] In my conclusion, I asked whether there was a correspondence with Lobatscheffski. The answer I got was that these were two very different ways of thinking, and that for what concerns the foundations of geometry, one should start by considering the straight line as the shortest distance between two points. I was daunted by this negative attitude and this made me put aside the insight which I had. [...]
>
> In the summer of 1871, I came back to Göttingen with Stolz. [...] He was above all a logician, and during my endless debates with him, the idea that the non-Euclidean geometries were part of Cayley's projective geometry became very clear to me. I imposed it on my friend after a stubborn resistance. I formulated this idea in a short note that appeared in the Göttingen Nachrichten, and then in a first memoir, which appeared in Volume 4 of the *Annalen*.

A couple of pages later, Klein, talking about his second paper [35], writes (see also [7]):

> I investigated in that paper the foundations of von Staudt's [geometric] system, and had a first contact with modern axiomatics. [...] However, even this extended presentation did not lead to a general clarification. [...] Cayley himself mistrusted my reasoning, believing that a "vicious circle" was buried in it.

---

[30]Wilhelm Fiedler (1832–1911)

[31]Otto Stolz (1842–1905) was a young mathematician at the time when Klein met him. He obtained his habilitation in Vienna in 1867 and, starting from 1869, he studied in Berlin under Weierstrass, Kummer and Kronecker. He attended Klein's lecture in 1871 and he remained in contact with him. He became later a successful textbook writer.

[32]In 1870, Weierstrass started at the university of Berlin a seminar on non-Euclidean geometry.

Cayley was more interested in the foundational aspect of projective geometry and his approach was more abstract than that of Klein. In a commentary on his paper [12] in his *Collected mathematical papers* edition [13] (Vol. II, p. 605), he writes:

> As to my memoir, the point of view was that I regarded "coordinates" not as distances or ratios of distances, but as an assumed fundamental notion not requiring or admitting of explanation. It recently occurred to me that they might be regarded as mere numerical values, attached arbitrarily to the point, in such wise that for any given point the ratio $x : y$ has a determinate numerical value, and that to any given numerical value of $x : y$ there corresponds a single point. And I was led to interpret Klein's formulæ in like manner; viz. considering A, B, P, Q as points arbitrarily connected with determinate numerical values a, b, p, q, then the logarithm of the formula would be that of $(a-p)(b-q) \div (a-q)(b-q)$. But Prof. Klein called my attention to a reference (p. 132 of his second paper) to the theory developed in Staudt's Geometrie der Lage, 1847. The logarithm of the formula is $\log(A, B, P, Q)$ and, according to Staudt's theory (A, B, P, Q), the anharmonic ratio of any four points, has independently of any notion of distance the fundamental properties of a numerical magnitude, viz. any two such ratios have a sum and also a product, such sum and product being each of them like a ratio of four points determinable by purely descriptive constructions.

Cayley refers here to von Staudt's notion of a point as a *harmonic conjugate* relatively to three other points, a definition which was also meant to be independent of any notion of distance ([56] p. 43).

Let us end this section by quoting J. E. Littlewood from his Miscellany [43], where he stresses the importance of Cayley's idea:

> The question recently arose in a conversation whether a dissertation of 2 lines could deserve and get a Fellowship. I had answered this for myself long before; in mathematics the answer is yes. Cayley's projective definition of length is a clear case if we may interpret "2 lines" with reasonable latitude. With Picard's Theorem[33] it could be literally 2, one of statement, one of proof. [...] With Cayley the importance of the idea is obvious at first sight."[34]

Finally, we point out to the reader that when Cayley talks about a metric space, he does not necessarily mean a metric space as we intend it today. We recall that the axioms of a distance function were formulated by Maurice Fréchet (1878–1973) in his thesis, defended in 1906. The idea of a "metric" was somehow vague for Cayley and Klein, but it included nonnegativeness and the triangle inequality.

---

[33] Littlewood is talking here about Picard's theorem saying that if $f : \mathbb{C} \to \mathbb{C}$ is an entire and non-constant function, then it is either surjective or it misses only one point.

[34] Littlewood adds: With Picard the situation is clear enough today (innumerable papers have resulted from it). But I can imagine a referee's report: "Exceedingly striking and a most original idea. But, brilliant as it undoubtedly is, it seems more odd than important; an isolated result, unrelated to anything else, and not likely to lead anywhere."

6 Beltrami and the Beltrami–Cayley–Klein model of the hyperbolic plane

Beltrami's discovery of the Euclidean model for hyperbolic space was a major step in the development of hyperbolic geometry. Although Beltrami did not write any major text on the relation between non-Euclidean and projective geometry, he was well aware of the works of Cayley and Klein, and he could have contributed to it. Let us start by quoting Klein on Beltrami's involvement in this subject. This is extracted from the introduction to [33] (Stillwell's translation p. 73):

> Since it will be shown that the general Cayley measure in space of three dimensions covers precisely the hyperbolic, elliptic and parabolic geometries, and thus coincides with the assumption of constant curvature, one is led to the conjecture that the general Cayley measure agrees with the assumption of constant curvature in any number of dimensions. This in fact is the case, though we shall not show it here. It allows one to use formulae, in any spaces of constant curvature, which are presented here assuming two or three dimensions. It includes the facts that, in such spaces, geodesics can be represented by linear equations, like straight lines, and that the elements at infinity form a surface of second degree, etc. These results have already been proved by Beltrami, proceeding from other considerations; in fact, it is only a short step from the formulae of Beltrami to those of Cayley.

In fact, Beltrami, two years before Klein published his first paper [33], wrote the following to Hoüel (letter dated July 29, 1869 [8] p. 96–97).[35]

> The second thing [I will add] will be the most important, if I succeed in giving it a concrete form, because up to now it only exists in my head in the state of a vague conception, although without any doubt it is based on the truth. This is the conjecture of a straight analogy, and may be an identity, between pseudo-spherical geometry[36] and the theory of Mr. Cayley on the *analytical origin of metric ratios*, using the *absolute* conic (or quadric). However, since the theory of invariants plays there a rather significant role and because I lost this a few years ago, I want to do it again after some preliminary studies, before I address this comparison.

Three years later, in a letter to Hoüel, written on July 5, 1872, Beltrami acknowledges the fact that Klein outstripped him ([8] p. 165):

> The principle which has directed my analysis[37] is precisely that which Mr. Klein has just developed in his recent memoir[38] on non-Euclidean geometry, for 2-dimensional spaces. In other words, from the analytic point of view, the geometry of spaces of constant curvature is nothing else than Cayley's doctrine of the absolute. I regret very much to have let Mr. Klein supersede me on that point, on which I had already assembled some material, and it was my mistake of not giving enough weight to

[35] The translation from the French are ours.

[36] This is the term used by Beltrami to denote hyperbolic geometry.

[37] Beltrami refers here to a note [5] which he had just published in *Annali di Matematica*.

[38] Beltrami refers here to Klein's paper [33].

> this matter. Beside, this point of view is not absolutely novel, and it is precisely for that reason that I was not anxious to publish my remark. It is intimately related to an already old relation of Mr. Chasles concerning the angle between two lines regarded as an anharmonic ratio (Geom. sup. art. 181) [14]) and to a theorem of Mr. Laguerre Verlay[39] (Nouv. Ann. 1853, Chasles, Rapport sur les progrès de la géométrie, p. 313). All that Cayley did is to develop an analytic algorithm and, above all, to show that in the general geometry, the theory of rectilinear distances responds exactly to that of angle distances in ordinary geometry. He also showed how and under what circumstances the Euclidean theory of distance differs from the general theory, and how it can be deduced from it by going to the limit.

Finally, we quote a letter that Beltrami wrote in the same year to D'Ovidio[40] (December 25, 1872, cited in [47] p. 422–423).

> When I learned about the theory of Cayley, I realized that his *absolute* was precisely this *limit* locus which I obtained from the equation $w = 0$, or $x = 0$, and I understood that the identity of the results was due to the following circumstance, that is, in (the analytic) projective geometry one only admits *a priori* that the linear equations represent lines of shortest distance, so that in this geometry one studies, without realizing it, spaces of constant curvature. I was wrong in not publishing this observation, which has been made later on by Klein, accompanied by many developments of which, for several of them, I had not thought.

We saw that in the case where the fundamental conic used in Cayley's construction is real, the measure defined on the interior of the conic gives a model of Lobachevsky's geometry. Klein recovered in this way the model which Beltrami had introduced in his paper [4] in which he noticed that the Euclidean straight lines in the unit disc behave like the non-Euclidean geodesics. It was Klein who provided this model with an explicit distance function, namely, the distance defined by the logarithm of the cross ratio, and he also noticed that the circle, in Beltrami's model, can be replaced by an ellipse.

Although this was not his main goal, Klein used this model to discuss the issue of the non-contradiction of hyperbolic geometry. This was also one of Beltrami's achievements in his paper [4].[41] We mention that the non-contradiction issue as well as the relative non-contradiction issue (meaning that if one geometry is contradictory, then the others would also be so) among the three geometries was one of the major

---

[39] Edmond Laguerre-Verlay (or, simply, Laguerre) (1834–1886) studied at the École Polytechnique, and after that he became an officer in the army. In 1883, he was appointed professor at Collège de France and two years later he was elected at the French Academy of Sciences. Laguerre was a specialist of projective geometry and analysis. His name is connected with orthogonal polynomials (the *Laguerre polynomials*). He is the author of 140 papers and his collected works were edited by Hermite, Poincaré and Rouché.

[40] Enrico D'Ovidio (1842–1933) was an Italian geometer who is considered as the founder of the famous Turin geometry school. Like Klein, he worked on the question of deriving the non-Euclidean metric function from concepts of projective geometry, paving the way for subsequent works of Giuseppe Veronese, Corrado Segre and others. D'Ovidio was known for his outstanding teaching, his excellent books, and his care for students. Guiseppe Peano, Corrado Segre, Guido Castelnuovo and Beppo Levi were among his students.

[41] We recall that this model was discovered by Beltrami four years before he discovered his famous pseudo-spherical model.

concerns of Lobachevsky, see e.g. his *Pangeometry* and the comments in the volume [45]. It is also important to recall that while Beltrami's Euclidean model showed that hyperbolic geometry is consistent provided Euclidean geometry is, Klein's work shows that Euclidean, spherical and hyperbolic geometries are consistent provided projective geometry is consistent.

We also mention that in §14 of the paper [33], while he computed the curvature of the metric, Klein obtained the expression, in polar coordinates, of the so-called Poincaré metric of the disk.

In his lecture notes [36] (p. 192), Klein writes the following:[42]

> [...] it is the merit of Beltrami's Saggio, to emphatically have called attention to the fact that the geometry on surfaces of constant negative curvature really corresponds to non-Euclidean hyperbolic geometry.

On p. 240, Klein discusses topology, and says that for 2-dimensional spaces of positive curvature, instead of working like Beltrami on the sphere, where two geodesic lines intersect necessarily intersect in two points, one can work in elliptic space, where geodesics intersect in only one point.[43]

7 The construction of measures

We now return to Klein's papers. The core of the paper [33] starts at §3, where Klein describes the construction of one-dimensional projective measures, that is, measures on lines and on circles. The one-dimensional case is the basic case because higher-dimensional measures are built upon this case.[44] Klein refers to the one-dimensional case as the *first kind*. There are two sorts of *measures* to be constructed: measures on points and measures on angles. The measure function on points satisfies the usual properties of a distance function[45] except that it can take complex values. The measure for angles is, as expected, defined only up to the addition of multiples of 2π, and at each point it consists of a measure on the pencil of lines that pass through that point. It can also take complex values. Klein specifies the following two properties that ought to be satisfied by measures:

1. the measures for points satisfy an additivity property for triples of points which are aligned;
2. the two measures (for points and for lines) satisfy the property that *they are not altered by a motion in space*.

[42] The English translations of our quotes from [36] were made by Hubert Goenner.

[43] It is interesting that in the 1928 edition of Klein's course on non-Euclidean geometry [37], the editor Rosemann removed almost all of Klein's remarks concerning Beltrami's contributions made in his course of 1989/90 [36] while Klein was alive. (This remark was made to the authors by Goenner.)

[44] In fact, Klein considers only spaces of dimensions two and three. At the end of his memoir, he says that the ideas can obvioulsy be generalized to higher dimensions.

[45] It is considered that the first formal statement of the axioms of a distance function as we know them today is due to Fréchet in his thesis [22] (1906), but the nineteenth-century mathematicians already used this notion, and they were aware of geometries defined by distance functions.

Property (a) says that the projective lines are geodesics for these measures. A metric that satisfies this property is called (in modern terms) *projective*. The motions that are considered in Property (b) are the projective transformations that preserve a conic, which is termed the *basic figure*. Klein then addresses the question of the classification of measures, and he notes that this depends on the classification of the transformations of the basic figures, which in turn depends on the number of fixed points of the transformation. Since the search for fixed points of such transformations amounts to the search for solutions of a degree-two equation, the transformations that preserve the basic figure fall into two categories:

1. Those that fix two (real or imaginary) points of the basic figure, and this is the generic case. They are termed measures *of the first kind.*
2. Those that fix one point of the basic figure. They are termed measures *of the second kind.*

Klein describes in detail the construction of measures on lines. The overall construction amounts to a division of the circle (seen as the projective line) into smaller and smaller equal parts, using a projective transformation. Thus, if we set the total length of the circle to be 1, the first step will provide two points at mutual distance $\frac{1}{2}$, the second step will provide three points at mutual distance $\frac{1}{3}$, and so forth. Passing to the limit, we get a measure on the circle which is invariant by the action of the given projective transformation.

More precisely, Klein starts with a transformation of the projective line of the form $z \mapsto \lambda z$, with λ real and positive. The transformation has two fixed points, called *fundamental elements*, the points 0 and ∞. Applying the transformation to a point z_1 on the line, we obtain the sequence of points $z_1, \lambda z_1, \lambda^2 z_1, \lambda^3 z_1, \ldots$. In order to define the measure, Klein divides, for any integer n, the line into n equal parts using the transformation $z' = \lambda^{\frac{1}{n}} z$. The nth root determination is chosen in such a way that $\lambda^{\frac{1}{n}} z$ lies between z and λz. The distance between two successive points is then defined as the $\frac{1}{n}$th of the total length of the line. Iterating this construction, for any two integers α and β, the distance between z_1 and a point of the form $\lambda^{\alpha+\frac{\beta}{n}} z_1$ is set to be the exponent $\alpha + \frac{\beta}{n}$, that is, the logarithm of the quotient $\frac{\lambda^{\alpha+\frac{\beta}{n}} z_1}{z_1}$ divided by $\log \lambda$. By continuity, we can then define the distance between two arbitrary points z and z_1 to be the logarithm of the quotient $\frac{z}{z_1}$ divided by the constant $\log \lambda$. The constant $\frac{1}{\log \lambda}$ is denoted henceforth by c.

Klein shows that the measure defined in this way is additive, that the distance from a point to itself is zero, and that the distance between two points is invariant by any linear transformation that fixes the fundamental elements $z = 0$ and $z = \infty$. He then observes that the quotient $\frac{z}{z'}$ may be interpreted as the cross ratio of the quadruple $0, z, z', \infty$. Thus, the distance between two points z, z' is a constant multiple of the logarithm of the cross-ratio of the quadruple $0, z, z', \infty$. In particular, the distance between the two fundamental elements is infinite.

Figures 5.3 and 5.4 reproduce Klein's constructions of equidistant points in the cases of hyperbolic and spherical geometry.

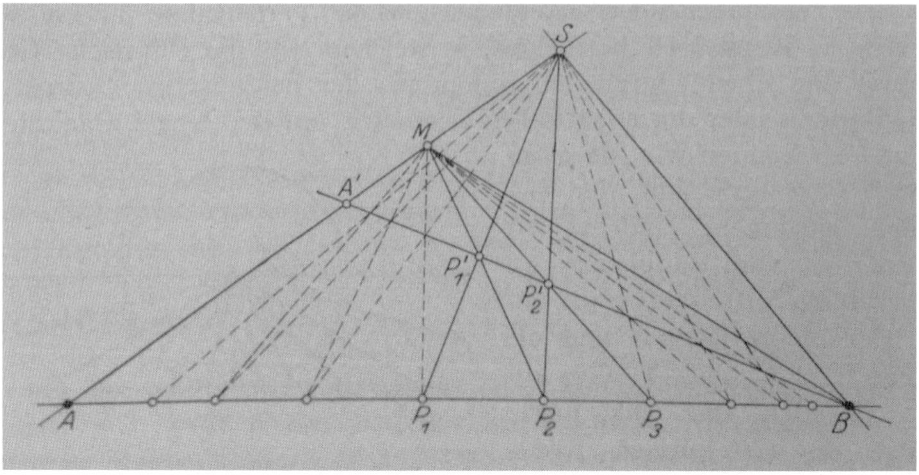

Figure 5.3. The construction of equidistant points on the hyperbolic line, using the invariance of the cross ratio. The drawing is extracted from [37] (p. 172).

In §4 of his paper, Klein extends the distance function $c \log \frac{z}{z'}$ to pairs of points on the complex line joining the points 0 and ∞, after choosing a determination of the complex logarithm. He then gives an expression for a general result where he assumes, instead of the special case where the two fundamental elements are 0 and ∞, that these points are the solutions of a second-degree equation

$$\Omega = az^2 + 2bz + c = 0.$$

For two arbitrary points given in homogeneous coordinates, (x_1, x_2) and (y_1, y_2), setting

$$\Omega_{xx} = ax_1^2 + 2bx_1x_2 + cx_2^2,$$
$$\Omega_{yy} = ay_1^2 + 2by_1y_2 + cy_2^2$$

and

$$\Omega_{xy} = ax_1y_1 + 2b(x_1y_2 + x_2y_1) + cx_2y_2,$$

the distance between the two points is

$$c \log \frac{\Omega_{xy} + \sqrt{\Omega_{xy}^2 - \Omega_{xx}\Omega_{yy}}}{\Omega_{xy} - \sqrt{\Omega_{xy}^2 - \Omega_{xx}\Omega_{yy}}}. \tag{5.1}$$

Later on in Klein's paper, the same formula, with the appropriate definition for the variables, defines a measure between angles between lines in a plane and between planes in three-space.

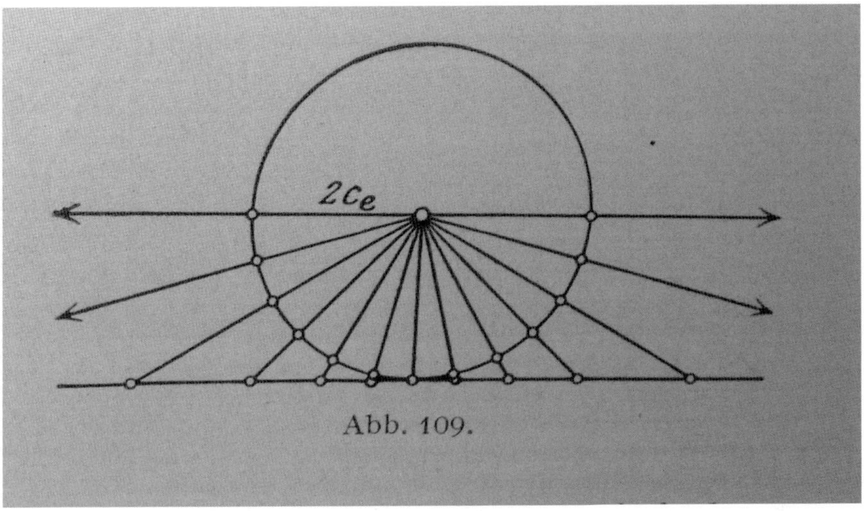

Figure 5.4. The construction of equidistant points in the case of spherical geometry. The drawing is from [37] (p. 171).

In §5, Klein derives further properties of the construction of the measure, distinguishing the cases where the two fundamental points are respectively real distinct, or conjugate imaginary, or coincident.

The first case was already treated; the two fundamental points are infinite distance apart, and they are both considered at infinity. He observes that this occurs in hyperbolic geometry, a geometry where any line has two points at infinity.

The case where the two fundamental points are conjugate imaginary occurs in elliptic geometry where a line has no point at infinity. Klein shows that in this case all the lines are finite and have a common length, whose value depends on the constant c that we started with. The distance 5.1 between two points becomes

$$2ic \arccos \frac{\sqrt{\Omega_{xy}}}{\sqrt{\Omega_{xx}\Omega_{yy}}}. \tag{5.2}$$

Klein notes that a particular case of this formula appears in Cayley's paper, who used only the value $-\frac{i}{2}$ for c and when, consequently, the term in front of arccos is equal to 1.

Klein studies the case where the two fundamental points coincide in §6. This case concerns Euclidean (parabolic) geometry. It is more complicated to handle than the other cases and it needs a special treatment. One complication arises from the fact that in this case Equation (5.2), which has a unique solution, leads to distance zero between the points x and y. The problem is resolved by considering this case as a limit of the case where the equation has two distinct solutions. Klein derives from there the formula for the distance on a line in which there is a unique point at infinity, that is, a unique point which is infinitely far from all the others.

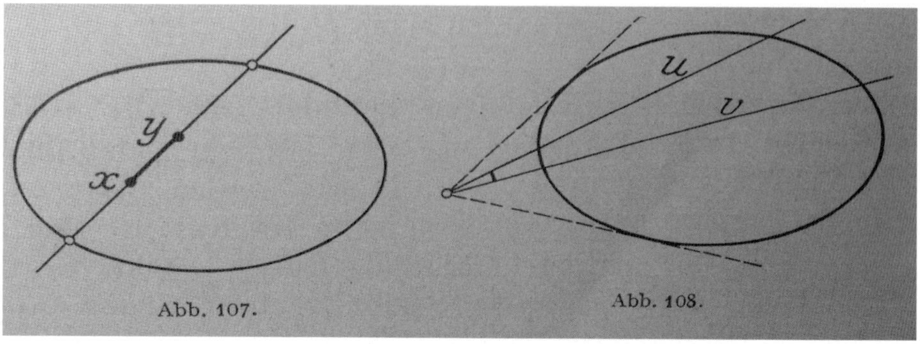

Figure 5.5. The distance between the two points x and y is computed as the cross ratio of the four points (figure to the left hand side). The angle between two lines u and v is computed as the cross ratio of the four lines (figure to the right hand side). The drawing from is from [37] (p. 165).

In §7, Klein introduces a notion of *tangency of measures at an element*. For this, he introduces two measures associated to a basic figure of the first kind, which he calls "general" and "special", and which he terms as "tangential". The overall construction amounts to the definition of infinitesimal geometric data, and it is also used to define a notion of a *curvature of a general measure*. The sign and the value of this curvature depend on some notion of deviation, which he calls "staying behind or running ahead" of the general measure relative to the special measure. He shows that the value of this geometrically defined curvature is constant at every point, and equal to $\frac{1}{4c^2}$, where c is the characteristic constant of the general measure. Using Taylor expansions, Klein shows that the three geometries (elliptic, parabolic and hyperbolic) are tangentially related to each other, which is a way of saying that infinitesimally, hyperbolic and spherical geometry are Euclidean. The value of c is either real or imaginary so that one can get positive or negative curvature.[46]

In §8, Klein outlines the construction of the measure for basic figures of the second kind, that is, measures on planes and measures on pencils of dihedral angles between planes. He uses for this an auxiliary conic. This is the so-called *fundamental conic* (the conic that is called the *absolute* by Cayley). Each projective line intersects this conic in two points (real, imaginary or coincident). The case where the points are real is represented in the left hand side in Figure 5.5. The two points play the role of fundamental points for the determination of the metric on that line, and the problem of finding a measure is reduced to the 1-dimensional case which was treated before. The fundamental conic is the locus of points which are infinitely distant from all others.

Measures on rays in the plane are based on the fact that at each point, there are rays that start at that point and that are tangent to the conic. Again, these rays are solutions of a certain quadratic equation and they may be distinct real, distinct imaginary or coincident. The two tangent rays are taken to be the fundamental rays for the angle determination in the sense that the angle between two arbitrary rays is then taken to be the cross ratio of the quadruple formed by these rays and the fundamental rays.

[46]The result should be real, and for that reason, c has sometimes to be taken imaginary. This is to be compared with the fact that some (real) trigonometric functions can be expressed as functions with imaginary arguments.

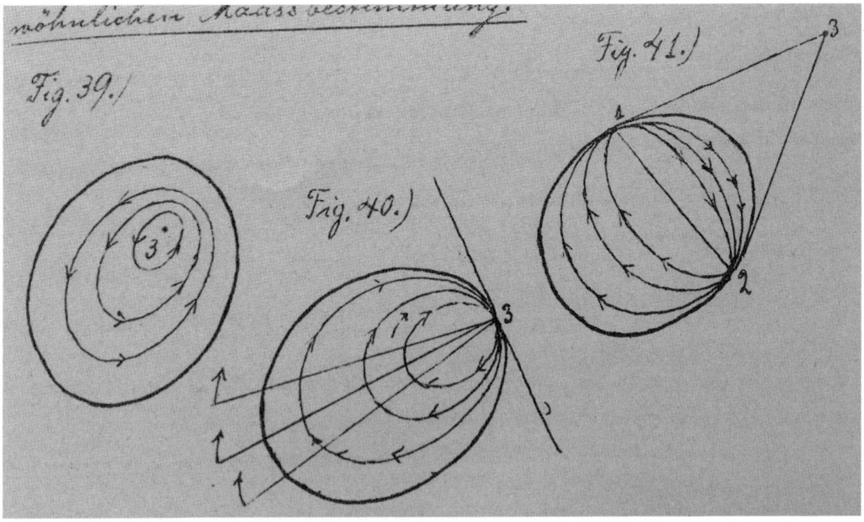

Figure 5.6. The three types of motions of non-Euclidean geometry inside an ellipse. Klein liked to describe them as "rotations", with center either in the space, or at the boundary of the space, or outside the space. In §9 of [33], Klein writes: *Each motion of the plane is a rotation about a point. All other points describe circles with this point as centre* (p. 93 of the English translation). The drawing is from [36] (p. 149).

This is represented in the right hand side of Figure 5.5. The multiplicative constant is not necessarily the same in the formulae giving the measures on lines and on rays.

Klein then determines an analytic expressions for these measures. It turns out that the formulae are the same as those obtained in §4. If the equation of the fundamental conic is

$$\Omega = \sum_{i,j=1}^{3} a_{ij} x_i y_j = 0,$$

then the distance between the two points x and y, in homogeneous coordinates (x_1, x_2, x_3) and (y_1, y_2, y_3), is

$$c \log \frac{\Omega_{xy} + \sqrt{\Omega_{xy}^2 - \Omega_{xx}\Omega_{yy}}}{\Omega_{xy} - \sqrt{\Omega_{xy}^2 - \Omega_{xx}\Omega_{yy}}} \tag{5.3}$$

where Ω_{xx}, Ω_{yy}, etc. are the expressions obtained by substituting in Ω the coordinates (x_1, x_2, x_3) of a point x or (y_1, y_2, y_3) of a point y, etc. Equivalently, we have

$$2ic \arccos \frac{\sqrt{\Omega_{xy}}}{\sqrt{\Omega_{xx}\Omega_{yy}}}. \tag{5.4}$$

That is, one obtains again Formulae (5.1) and (5.2) of §4.

Concerning measures on angles, the equations have a similar form. One takes the equation of the fundamental conic in line coordinates to be

$$\Phi_{u,v} = \sum_{i,j=1}^{3} A^{ij} u_i v_j = 0.$$

The distance between the two points u and v in homogeneous coordinates (u_1, u_2, u_3) and (v_1, v_2, v_3) is then

$$c' \log \frac{\Phi_{uv} + \sqrt{\Phi_{uv}^2 - \Phi_{uu}\Phi_{vv}}}{\Phi_{uv} - \sqrt{\Phi_{uv}^2 - \Phi_{uu}\Phi_{vv}}} \tag{5.5}$$

or, equivalently,

$$2ic' \arccos \frac{\sqrt{\Phi_{uv}}}{\sqrt{\Phi_{uu}\Omega_{vv}}}. \tag{5.6}$$

where Ω_{uu}, Ω_{vv} have the same meaning as before.

The constant c' is in general different from c. In general, the constants are chosen so that the result is real.

The measures on points and on lines are defined by similar formulae. This is a consequence of the fact that they are solutions of second-degree equations, and that the coefficients of the two equations are related to each other by the duality in projective geometry. Duality is discussed in the next section.

§9 concerns the properties of the projective transformations of the plane that preserve a conic. Klein points out that there is a "threefold infinity" of such transformations (in other words, they form a 3-dimensional group), and he starts a classification of such transformations, based on the fact that each transformation fixes two points of the conic and reasoning on the line connecting them, on the tangents at these points, on their point of intersection, and working in the coordinates associated to the triangle formed by the connecting line and the two tangents. The classification involves the distinction between real conics with real points and real conics without real points. The aim of the analysis is to prove that the transformations that map the conic into itself preserve the metric relations between points and between angles. There is also a polar duality determined by the conic. With this duality, a quadruple formed by two points and the intersection of the line that joins them with the conic corresponds to a quadruple formed by two lines and tangents to the conic that pass through the same point. This correspondence preserves cross ratios. The duality is such that the distance between two points is equal to the angle between the dual lines. This is a generalization of the polar duality that occurs in spherical geometry.

After the discussion of projective measures between points in §9, Klein considers in §10 measures for angles in pencils of lines and of planes. In this setting, he uses a *fundamental cone of second degree* instead of the fundamental conic. He also appeals to polarity theory and he makes a relation with the measure obtained in the previous section. The result brought out at the end of the previous section is interpreted here as saying that the angle between two planes is the same as the angle between their normals, and this has again an interpretation in terms of spherical geometry duality.

In §11, Klein develops a model for spherical geometry that arises from his measures associated to conics.

When the fundamental conic is imaginary, setting $c = c_1\sqrt{-1}$ and $c' = c'_1\sqrt{-1}$, the measures for lines and for angles are found to be respectively

$$2c_1 \arccos \frac{xx' + yy' + zz'}{\sqrt{x^2 + y^2 + z^2}\sqrt{x'^2 + y'^2 + z'^2}} \tag{5.7}$$

and

$$2c'_1 \arccos \frac{uu' + vv' + ww'}{\sqrt{u^2 + v^2 + w^2}\sqrt{u'^2 + v'^2 + w'^2}}, \tag{5.8}$$

which are the familiar formulae for angle measure on a sphere. In particular, the distance between any two points is bounded, as expected. In fact, all lines are closed, they have finite length, and these lengths have a common value, $2c_1\pi$, which is (up to a constant multiple) the angle sum of a pencil, which is $2c'_1\pi$. The point measure is completely similar to the angle measure. This again can be explained by the duality between points and lines in spherical geometry. Klein concludes from this fact that "plane trigonometry, under this measure, is the same as spherical trigonometry" and that "the plane measure just described is precisely that for elliptic geometry." By choosing appropriately the constants c_1 and c'_1, the angle sum of any pencil becomes π and the maximal measure between points becomes also π. Klein also deduces that in that geometry, the angle sum of a plane triangle is greater than π, as for spherical triangles, and only equal to π for infinitesimally small triangles.

In §12, Klein describes the construction that leads to hyperbolic geometry. This is the case where the absolute is a real fundamental conic in the plane. In this case, the constant c that appears in the general formula 5.1 for distances is taken to be real. The points in the plane are divided into three classes: the points inside the conic, the points on the conic and the points outside the conic. The points inside the conic are those that admit no real tangent line to the conic. The points on the conic are those that admit one real tangent line. The points outside the conic are those that admit two real tangent lines.

Likewise, the lines in the plane are divided into three classes: the lines that meet the conic in two real points, those that meet the conic in a unique (double) real point and those that do not meet the conic in any real point. Klein claims that this case corresponds to hyperbolic geometry. To support this claim, he writes:

> The geometry based on this measure *corresponds completely with the idea of hyperbolic geometry*, when we set the so far undetermined contant c'_1 equal to $\frac{1}{2}$, making the angle sum of a line pencil equal to π. In order to be convinced of this, we consider a few propositions of hyperbolic geometry in somewhat more detail.

The propositions that Klein considers are the following:

1. Through a point in the plane there are two parallels to a given line, i.e. lines meeting the points at infinity of the given line.
2. The angle between the two parallels to a given line through a given point decreases with the distance of the point from the line, and as the point tends to infinity, this angle tends to 0, i.e. the angle between the two parallels tends to zero.
3. The angle sum of a triangle is less than π. For a triangle with vertices at infinity, the angle sum is zero.
4. Two perpendiculars to the same line do not meet.
5. A circle of infinite radius is not a line.

Klein notes that these properties are satisfied by his geometry. This is not a full proof of the fact that the geometry defined using the distance function he described *is* hyperbolic geometry, but it is a strong indication of this fact. In fact, it is surely possible, but very tedious, to show that all the axioms of hyperbolic geometry are satisfied by his geometry. Klein then adds (Stillwell's translation p. 99):

> Finally, the *trigonometric formulae* for the present measure are obtained immediately from the following considerations. In §11 we have seen that, on the basis of an imaginary conic in the plane and the choice of constants $c = c_1 i, c' = c'_1 i = \frac{\sqrt{-1}}{2}$, the trigonometry of the plane has the same formulae as spherical trigonometry when one replaces the sides by sides divided by $2c_1$. The same still holds on the basis of a real conic. Because the validity of the formulae of spherical trigonometry rests on analytic identities that are independent of the nature of the fundamental conic. The only difference from the earlier case is that $c_1 = \frac{c}{i}$ is now imaginary.
>
> *The trigonometric formulae that hold for our measure result from the formulae of spherical trigonometry by replacing sides by sides divided by* $\frac{c}{i}$.
>
> But this is the same rule one has for the trigonometric formulae of hyperbolic geometry. The constant c is the characteristic constant of hyperbolic geometry. One can say that planimetry, under the assumption of hyperbolic geometry, is the same as geometry of a sphere with the imaginary radius $\frac{c}{i}$.
>
> The preceding immediately gives a model of hyperbolic geometry, in which we take an arbitrary real conic and construct a projective measure on it. Conversely, if the measure given to us is representative of hyperbolic geometry, then the infinitely distant points of the plane form a real conic enclosing us, and the hyperbolic geometry is none other than the projective measure based on this conic.

§13 concerns parabolic geometry. In this case, the fundamental figure at infinity is a degenerate conic. It is reduced to a pair of points, and, in Klein's words, it constitutes a "bridge between a real and an imaginary conic section." The metric obtained is that of Euclidean geometry and the line joining the pair of points at infinity (the degenerate conic) is the familiar line at infinity of projective geometry. In this sense, parabolic geometry is regarded as a transitional geometry, sitting between hyperbolic and elliptic geometry. To understand how this occurs, Klein gives the example of a degeneration of a hyperbola. A hyperbola has a major and a minor axis, which are

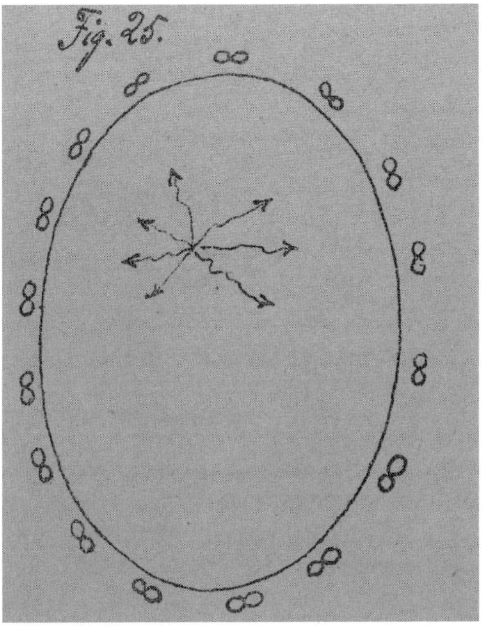

Figure 5.7. The conic at infinity, or the "absolute", in the case of a real conic. The drawing is from [36].

symmetry axes, the major axis being the segment joining the two vertices a and $-a$ (and the length of this axis is therefore equal to $2a$) and the minor axis being perpendicular to the major one, with vertices at points b and $-b$, of length $2b$. The major and minor axes are also the two perpendicular bisectors of the sides of a rectangle whose vertices are on the asymptotes of the hyperbola. In a coordinate system where the two axes are taken as the major and major axes, the equation of the hyperbola is $\frac{x^2}{a^2} - \frac{y^2}{b^2} = 1$. The minor axis is also called the imaginary axis because of the minus sign occurring in this equation.

The degeneration of the hyperbola into two imaginary points is obtained by keeping fixed the imaginary axis and shrinking to zero the major axis. Meanwhile, the two branches of the hyperbola collapse to the line carried by the minor axis, covering it twice. This line represents a degenerate conic, and in fact, as Klein points out, insofar as it is enveloped by lines, it is represented by the two conjugate imaginary points. The associated measure on the plane is called a *special measure*, because it uses a pair of points instead of a fundamental conic. Klein obtains an analytic formula that gives the associated distance between points. Starting with the general expression

$$2ic \arcsin \frac{\sqrt{\Omega_{xy}^2 - \Omega_{xx}\Omega_{yy}}}{\sqrt{\Omega_{xx}\Omega_{yy}}},$$

where $\Omega = 0$ is the equation of the conic and Ω_{xx}, Ω_{xy} and Ω_{yy} are as he defined in §4, and taking limits when the conic degenerates to a pair of points, he deduces that the distance between two points (x, y) and (x', y') is

$$\frac{C}{k^2}\sqrt{(x-x')^2+(y-y')^2}$$

which up to a constant factor is the Euclidean distance in coordinates. He concludes this section with the following:

> We want to stress that with imaginary fundamental points the trigonometric formulae become the relevant formulae of parabolic geometry, so the angle sum of a triangle is exactly π, whereas with a real fundamental conic it is smaller, and with an imaginary conic it is larger.

In §14, Klein considers again the notion of a "measure on a plane which is tangent to a general measure at a point", that is, he considers infinitesimal distances. This leads him again to the definition of a notion of curvature which is equivalent to Gaussian curvature. Klein then uses a duality, where the dual of a point of the given geometry is a "line at infinity" which is the polar dual of the given point with respect to the fundamental conic. He obtains a qualitative definition of curvature which turns out to be equivalent to the Gaussian curvature. This leads to the conclusion that the curvature of a general measure is the same at all points and is equal to $\frac{-1}{4c^2}$, that it is positive if the fundamental conic is imaginary (the case of elliptic geometry) and negative if the fundamental conic is real (the case of hyperbolic geometry). In the transitional case (parabolic geometry), which is a limiting case in which the fundamental conic degenerates into a pair of imaginary points, the curvature is zero. Klein concludes this section with the following statement:

According to whether we adopt the hypothesis of an elliptic, hyperbolic or parabolic geometry, the plane is a surface with constant positive, constant negative or zero curvature.

In §15, Klein talks about a continuous transition from hyperbolic to parabolic and from spherical to parabolic. Let us quote him (Stillwell's translation p. 107):

> If we are actually given parabolic geometry we can immediately constructs a geometry which models hyperbolic geometry by constructing a general measure with real fundamental conic, tangential to the given special measure at a point of our choice. We achieve this by describing a circle of radius $2c$ centred on our point, and using it as the basis for a projective measure with the constant c determining the distance between two points and the constant $c' = \frac{\sqrt{-1}}{2}$ determining the angles between two lines. This general measure approaches the given parabolic measure more closely as c becomes larger, coinciding with it completely when c becomes infinite.
>
> In a similar way we construct a geometry that shows how elliptic geometry can tend toward the parabolic. To do this it suffices to give a pure imaginary value $c_1 i$ to the c we used previously. Then we fix a point at distance $2c_1$ above the given point of contact and take the distance between two points of the plane to be c_1 times the angle the two points subtend at the fixed point. The angle between the two lines

in the plane is just the angle they subtend at the fixed point. The resulting measure approaches more closely to the parabolic measure the greater c_1 is, and it becomes equal to it when c_1 is infinite.

When elliptic or hyperbolic geometry is actually the given geometry one can in this way make a model presenting its relationship with parabolic or the other geometry.

§16 concerns projective measures in space. The same procedure as before is used, with a second-degree fundamental surface in 3-space instead of the fundamental curve in the plane. The case where the fundamental surface is imaginary leads to elliptic geometry. The case where it is *real* and *not ruled*, leads to hyperbolic geometry. The case where the fundamental surface degenerates to a conic section leads to parabolic geometry, and this conic section becomes the *imaginary circle at infinity*. The case where the fundamental surface is *real* and *ruled*, that is, a one-sheeted hyperboloid, is not related to any of the classical geometries, it leads to a geometry which is not locally Euclidean but pseudo-Euclidean.

The title of §17 is "The independence of projective geometry from the theory of parallels." Klein observes that projective geometry insofar as it uses the notions of homogeneous coordinates and the cross ratio, is defined in the setting of parabolic geometry. He notes however that in the same way as one can construct projective geometry starting with parabolic geometry, one can construct it also on the basis of hyperbolic and elliptic geometry. He then notes that projective geometry can be developed without the use of any measure, using the so-called incidence relations, referring to the work of von Staudt.

In the conclusion to the paper (§18), Klein notes that by a consideration of the sphere tangent to the fundamental surface, one is led to only the three geometries considered, elliptic, hyperbolic and the transitional one, that is, the parabolic.

8 Klein's second paper

The second paper ([35] same title, Part II), appeared two years after the first one. It has a more general character, it is in the spirit of his *Erlangen program*, and it is less technical than the first paper. There does not seem to be an available English translation of that paper. In this paper, Klein gives some more details on results he obtained in the first paper. Let us quote Klein from his Göttingen lecture notes of Klein [36] (p. 286–287; Goenner's translation):

> When the accord of Cayley's measure geometry and non-Euclidean geometry was stated *, it became essential to draw conclusions from it. On these conclusions I wish to attach most importance, although they were developed more in detail only in the second paper, when I noticed that the very same [conclusions] did appear to other mathematicians not as self-evident as for myself. [...] (footnote) * Beltrami and Fiedler also had noticed this accord, as they later wrote to me.

We now give a brief summary of the content of the paper [35].

This paper has two parts. In the first part, Klein develops the idea of a transformation group that characterizes a geometry. In the second part, he develops an idea concerning projective geometry which he had also mentioned in the first paper [33], namely, that projective geometry is independent from the Euclidean parallel postulate (and from Euclidean geometry).[47] Klein insisted on this fact, because, as he wrote, some mathematicians thought that there was a vicious circle in his construction of Euclidean geometry from projective geometry, considering that the definition of the cross ratio uses Euclidean geometry, since it involves a compounded ratio between four Euclidean segments. Some also thought that there was a contradiction in Klein's reasoning, since in spherical and hyperbolic geometry Euclid's parallel axiom is not satisfied, so a formula for the metric defining these geometries cannot be based on the distance function of Euclidean geometry where the parallel axiom is satisfied. In fact, as was already recalled above, in his *Geometrie der Lage* [56], von Staudt had already worked out a purely projective notion of the cross ratio, independent of any notion of distance.[48] In his *Lectures on the development of mathematics in the XIXth century* (1926–1927), [40] Klein returns to the history and he writes the following (Vol. 1, p. 153):

> More important is the objection I received from mathematicians. In my paper written in Volume IV of the *Annalen*, I did not expect the logical difficulties that the problem raised, and I had started an innocent use of metric geometry. It is only at the end that I mentioned in a very brief way the independence of projective geometry from any metric, referring to von Staudt. I was accused from everywhere of making circular reasoning. The purely projective definition of von Staudt of the cross ratio as a number was not understood, and people stood firmly on the idea that this number was only given as a cross ratio of four Euclidean numbers.

We also quote Cayley's citation of R. S. Ball [13] (Vol. II, p. 605):

> I may refer also to the memoir, Sir R. S. Ball "On the theory of content," *Trans. R. Irish Acad.* vol. XXIX (1889), pp. 123–182, where the same difficulty is discussed. The opening sentences are – "In that theory [Non-Euclidian geometry] it seems as if we try to replace our ordinary notion of distance between two points by the logarithm of a certain anharmonic ratio. But this ratio itself involves the notion of distance measured in the ordinary way. How then can we supersede the old notion

---

[47]One of the basic features of projective geometry is that in the arguments that involves lines, unlike in Euclidean geometry, one does not have to distinguish between the cases where the lines intersect or are parallel. In projective geometry, any two distinct points define one line, and any two distinct lines intersect in one point. We already mentioned that this principle is at the basis of duality theory in projective geometry.

[48]We can quote here Klein from his *Erlangen program* [34]: "We might here make mention further of the way in which *von Staudt* in his *Geometrie der Lage* (Nürnberg, 1847) develops projective geometry, – i.e., that projective geometry which is based on the group containing all the real projective and dualistic transformations." And in a note, he adds: "The extended horizon, which includes *imaginary* transformations, was first used by *von Staudt* as the basis of his investigation in his later work, *Beiträge zur Geometrie der Lage* (Nürnberg, 1856–60)." For the work of von Staudt and in his influence of Klein, and for short summaries of the *Geometrie der Lage* and the *Beiträge zur Geometrie der Lage*, we refer the reader to the paper [27] in this volume.

of distance by the non-Euclidian notion, inasmuch as the very definition of the latter involves the former?"

From this, let us conclude two different things:

1. There was a great deal of confusion about Klein's ideas, even among the most brilliant mathematicians.
2. The mathematicians were not only interested in formulae, but they were digging in the profound meaning that these formulae express.

We end this paper with a brief summary of the content of [35], since no available translation exists. The reader can compare the content of this paper with the summary of the *Erlangen program* lecture given in Chapter 1 of this volume [27].

The introduction contains historical recollections on the works of Cayley and von Staudt which, according to Klein, did not have yet any applications. Klein then recalls that the "problem of parallels", that is, the problem of deciding whether Euclid's parallel axiom follows or not from the other axioms of Euclidean geometry was settled. He mentions the works of the founders of modern geometry, and he says that each of them brought new mathematical concepts, in particular, new examples of spaces of constant curvature. At the same time, several open questions remain to be solved, and other things need to be made more precise. Klein then mentions the spaces of variable curvature constructed by Riemann. All these works contribute to new points of view on spaces and on mechanics. He also recalls that there is a difference between the metrical and the projective points of view, and he declares that the geometries of constant curvature should be simpler to study.

In Section 1 of the first part of the paper, Klein considers the concept of higher dimensions. He mentions the relation between constant curvature manifolds and projective manifolds.[49] Analytic geometry allows the passage to higher dimensions, working in analogy with the low dimensions that we can visualize. He points out that on a given line, we can consider either the real points or all points. He then recalls the definition of the cross ratio.

Section 2 concerns transformations. Klein explains the notion of composition of transformations, and he considers in particular the case of collineations, forming a group. He then presents the idea of group isomorphism. The reader should recall that these ideas were relatively new at that time.

Section 3 concerns "invariant", or "geometric", properties. A property is geometric if it is independent of the location in space. A figure should be indistinguishable from its symmetric images. The properties that we seek are those that remain invariant by the transformations of the geometry.

In Section 4, Klein develops the idea that the methods of a given geometry are characterized by the corresponding groups. This is again one of the major ideas that he expresses in his *Erlangen program*. He elaborates on the significance of projective

[49] In [34], the term "Mannigfaltigkeiten", which for simplicity we translate by "manifold", is usually translated by "manifoldness". See the comments in [27] in this volume on the meaning of the word manifoldness. We recall that the notion of manifold as we intended today, defined by coordinate charts, was given much later.

geometry, and in particular on the transformations that leave invariant the imaginary circle at infinity. The methods depend on the chosen transformation group.

The discussion is also confirmed by Klein in his lecture notes of 1898/90 ([36], p. 120; Goenner's translation):

> In contradistinction [to Helmholtz], I had the generic thought that, in studying manifolds under the viewpoint of giving them a *geometric* character, one can put ahead any transformation group [...].* Above all it is advisable to chose the collineations (linear transformations) as such a group. [...] This then is the specially so-called *invariant theory*. – (footnote) * Ann. VI, p. 116 et seq., as well as notably the Erlangen program.

Section 5 concerns generalizations to higher-dimensional spaces. The simplest transformation groups are the groups of linear transformations. They give rise to projective geometry. Although there is no distance involved, this is considered as a geometry. Klein introduces the word "invariant theory", where we have no distance involved, but we look for invariant objects. Modern algebra is helpful in that study. In the case where we have a metric, we have an invariant quadratic form. He declares that he will study the case where there is none. He introduces a notion of differential of a map. At the infinitesimal level, the differential behaves like a linear map.

In Section 6, Klein considers spaces of constant nonzero curvature. He refers to Beltrami, who showed that in such a space we can define geodesics by equations that are linear in the appropriate coordinates. He raises the question of understanding the transformations of a manifold of constant curvature in the projective world, and this is done in linearizing them. Indeed, by choosing adequate coordinates, the group of transformations that we attach to a manifold of constant curvature is contained in the group of linear transformations. (*"The transformation group of a constant curvature geometry is reducible to a transformation group which preserves a quadratic form."*) Klein says however that there is a difference between his viewpoint and the one of Beltrami, namely, Klein starts with complex variables and then he restricts to real variables. This gives a uniform approach to several things.

Elliptic space is obtained from the sphere by identifying antipodal points so that there is a unique geodesic connecting two points. In higher dimensions similar objects exist.

Sections 7 and 8 concern the description of constant curvature manifolds in terms of projective notions, and Klein recalls the definition of the distance using the cross ratio.

In Section 9, Klein defines a point at infinity of the space as a representative of a class of geodesic lines.

The subject of the second part of the paper is the fact that, following von Staudt, one can construct projective space independently of the parallel axiom.

In the first section of this part, Klein explains various constructions that are at the basis of projective geometry. He also introduces the betweenness relation. He talks about lines and pencils of planes, of the cross ratio and of the notion of harmonic division, and he states the fact that there is a characterization of 2-dimensional projective geometry. He recalls that von Staudt, in his work, used the parallel axiom, but that without essential changes one can recover the bases of projective geometry without the parallel axiom. He then studies the behavior of lines and planes, and the notion of asymptotic geodesics. This section also contains a detailed discussion of von Staudt's axiomatic approach.

Section 2 concerns the "formulation of a proposition which belongs to the general theory of *Analysis situs*." Klein explains how one can attach coordinates to points.

In Section 3, he returns to the bases of systems of planes and their intersection.

In Section 4, he elaborates on the notion of harmonic element, and on the notion of betweenness among points.

In Section 5, Klein expands on the work of van Staudt on projective transformations.

In Section 6, he talks about pencils of planes and about duality and ideal points; a point at infinity defines as a class of lines which do not intersect.

Section 7 concerns the cross ratio and homogeneous coordinates.

In Section 8, Klein gives an analytical proof of the main theorem of projective geometry. He says that there is a characterization of 2-dimensional projective geometry.

To conclude this section, let us insist on the fact that beyond their immediate goal (which is an important one), the two papers by Klein are full of interesting historical comments and references to works of other mathematicians. They are the expression of the elegant style and the great erudition which characterizes Klein's writings in general. The reader should remember that in 1871, at the time he wrote the paper [33], Klein was only 22 years old.

Let us note as a conclusion that Klein, in this work and is later works, was one of the earliest promoters of hyperbolic geometry, and that he used it extensively in his later work, notably, in the subject which Poincaré called later on *Kleinian groups*, that is, discrete groups of fractional linear transformations acting on hyperbolic space.

9 Poincaré

In his paper [50] (1887), Poincaré describes a construction of a set of geometries, using quadrics in three-space. The theory of associating a geometry to a quadric is of course related to the theory developed by Klein, although the point of view is different. Whereas in Klein's (and Cayley's) construction, the quadric is at infinity, the geometry, in the case developed by Poincaré, *lives* on the quadric.

Let us recall that a *quadric*, also called (by Poincaré) a *quadratic surface*, ("surface quadratique") is a surface in Euclidean three-space which is the zero locus of a degree-two polynomial equation in three variables. There is a projective characterization of quadrics, which is coordinate-free: a quadric is a surface in projective space whose plane sections are all conics (real or imaginary). It follows from this definition that the intersection of any line with a quadric consists of two points, which may be real or imaginary, unless the line belongs to the quadric. Furthermore, the set of all tangents to a quadric from an arbitray point in space is a cone which cuts every plane in a conic, and the set of contact points of this cone with the quadric is also a conic.

There are well-known classification of quadrics; some of them use coordinates and others are coordinate-free. Chapter 1 of the beautiful book of Hilbert and Cohn-Vossen, *Geometry and the imagination* [30], concerns quadrics. The equation of a conic can be put into normal form. Like for conics (which are the one-dimensional analogues of quadrics), there are some nondegenerate cases, and some degenerate cases. Poincaré obtained the two non-Euclidean geometries as geometries living on non-degenerate quadrics, and Euclidean geometry as a geometry living on a degenerate one. This is very close to the ideas of Klein.

There are nine types of quadrics. Six of them are *ruled surfaces* (each point is on at least one straight line contained in the surface); these are the cone, the one-sheeted hyperboloid, the hyperbolic paraboloid and the three kind of cylinders (the elliptic, parabolic and hyperbolic). The three non-ruled quadrics are the ellipsoid, the elliptic paraboloid and the two-sheeted hyperboloid. These three surfaces do not contain any line.

The one-sheeted hyperboloid and the hyperbolic paraboloid are, like the plane, *doubly ruled*, that is, each point is on at least two straight lines.

Three types of nondegenerate quadrics which possess a center: the ellipsoid, the two-sheeted hyperboloid and the one-sheeted hyperboloid.

Poincaré starts with a quadric in $\mathbb{R}^3$, called the *fundamental surface*. On such a surface, he defines the notions of *line*, of *angle* between two lines and of *length* of a segment.

Given a quadric, the locus of midpoints of the system of chords that have a fixed direction is a plane, called a *diametral plane* of the quadric. In the case where the quadric has a center, a diametral plane is a plane passing through the center.

Like in the case of the sphere, with its great circles and its small circles, Poincaré calls a *line* an intersection of a quadric surface with a diametral plane, and a *circle* an intersection of a quadric with an arbitrary plane.

Angles are then defined using the cross ratio. Given two lines l_1 and l_2 passing through a point P, Poincaré considers the quadruple of Euclidean lines formed by the tangents to l_1 and l_2 and the two rectilinear generatrices of the surface that pass through the point P. There are two generatrices at every point of the surface, and they may be real or imaginary. Poincaré defines the *angle* between l_1 and l_2 as the logarithm of the cross ratio of the four Euclidean lines (l_1, l_2 and the two generatrices) in the case where the two generatrices are real (and this occurs of the surface is a one-sheeted hyperboloid), and this logarithm divided by $\sqrt{-1}$ in the case where the generatrices are imaginary.

Now given an arc of a line of the quadric, consider the cross ratio of the quadruple formed by the two extremities of this arc and the two points at infinity of the conic. The *length* of this arc is the logarithm of the cross ratio of this quadruple of points in the case where the conic is a hyperbola, and the logarithm of this cross ratio divided by $\sqrt{-1}$ otherwise.

Poincaré then says that there are relations between lengths and distances defined in this way, and that such relations constitute a set of theorems which are analogous to those of plane geometry. He calls the collection of theorems associated to a given quadric a *quadratic geometry*. There are as many quadratic geometries as there are kinds of second degree surfaces, and Poincaré goes on with a classification of such geometries.

In the case where the fundamental surface is an ellipsoid, the geometry obtained is spherical geometry.

In the case where the fundamental surface is a two-sheeted hyperboloid, the geometry obtained is the Lobachevsky (or hyperbolic) geometry.

In the case where the fundamental surface is an elliptic paraboloid, the geometry obtained is the Euclidean and Poincaré says that this geometry is a limiting geometry of each of the previous two.

There are other geometries, e.g. the one-sheeted hyperboloid and its various degenerate cases. Some of the degenerate geometries give the Euclidean geometry. But the one-sheeted hyperboloid itself gives a geometry which Poincaré highlights, as being a geometry which was not been studied yet, and in which the following three phenomena occur:

1. The distance between two points on the fundamental surface which are on a common rectilinear generatrix is zero.
2. There are two sorts of lines: lines of the first kind, which correspond to the elliptic diametral sections, and lines of the second kind, which correspond to the hyperbolic diametral sections. It is not possible, by a real motion, to bing a line of the first kind onto a line of the second kind.
3. There is no nontrivial real symmetry which sends a line onto itself. (Such a symmetry is possible in Euclidean geometry; it is obtained by a 180° rotation centered at a point on the line.)

This geometry is in fact the one called today the planar *de Sitter geometry*.

Poincaré, in this paper, does not mention Klein, but he thoroughly mentions Lie, and he considers this work as a consequence of Lie's work on groups. In the second part of the paper, titled "Applications of group theory", Poincaré gives a characterization of the transformation group of each of these geometries. This is done in coordinates, at the infinitesimal level, in the tradition of Lie. He considers (p. 215) that "geometry is nothing else than the study of a group", and this brings us again to Klein's *Erlangen program*.

10 Conclusion

Our aim in this chapter was to present to the reader of this book an important piece of work of Felix Klein. We also tried to convey the idea that mathematical ideas may occur at several people at the same time, when time is ready for that. Each of us has a special way of thinking, and it often happens that works on the same problem, if they are not collective, complement each other. We also hope that this chapter will motivate the reader to go through the original sources.

Let us conclude with the following two problems.

1. We already noted that Klein, in his development of the three geometries in his papers [33] and [35], considers Euclidean geometry as a *transitional* geometry. In this way, Euclidean geometry corresponds to a limiting case of the absolute, in which the fundamental conic degenerates to a pair of imaginary points. We developed the notion of transitional geometry in the paper [1], in a way different from Klein's, and we studied in which manner the fundamental notions of geometry (points, lines, distances, angles, etc.) as well as the trigonometric formulae transit from one geometry to another. An interesting problem is to make the same detailed study of transition of these fundamental notions in the context of Klein's description of the geometries.
2. Hilbert developed a generalization of the Klein model of hyperbolic geometry where the underlying set is the interior of an ellipsoid to a geometry (called Hilbert geometry) where the underlying set is an arbitrary open convex set in $\mathbb{R}^n$. The distance between two points x and y in Hilbert's generalization is again the logarithm of the cross ratio of the quadruple consisting of x and y and the two intersection points of the Euclidean line that joins these points with the boundary of the ellipsoid, taken in the natural order. We propose, as a problem, to develop generalizations of the two other geometries defined in the way Klein did it, that are analogous in some way to the generalization of hyperbolic geometry by Hilbert geometry.

Acknowledgements. The authors would like to thank Jeremy Gray and François Laudenbach for remarks on a preliminary version of this paper, and especially Hubert Goenner for his careful reading and for interesting notes. The authors are supported by the French ANR project FINSLER and the GEAR network of the National Science Foundation (GEometric structures And Representation varieties). Part of this work was done during a stay of the two authors at the Erwin Schrödinger Institute (Vienna). The second author was also supported by a Tubitak 2221 grant during a visit to Galatasaray University (Istanbul).

Bibliography

[1] N. A'Campo and A. Papadopoulos, Transitional geometry. In *Sophus Lie and Felix Klein: The Erlangen program and its impact in mathematics and physics*, ed. L. Ji and A. Papadopoulos, European Mathematical Society Publishing House, Zürich, 2015, 217–235.

[2] P. Barbarin, La correspondence entre Hoüel et de Tilly. *Bulletin des Sciences Mathématiques*. Tome L, 1926, 50–61.

[3] P. Barbarin, *La Géométrie non Euclidienne (suivie de Notes Sur la Géométrie non Euclidienne dans ses Rapports avec la Physique Mathématique par A. Buhl)*. Gauthier-Villars, Paris, Third edition 1928. (First edition, 1902.)

[4] E. Beltrami, Saggio di Interpretazione della geometria non-Euclidean. (Essay on the interpretation of non-Euclidean geometry) *Giornale di Matematiche* vol. VI (1868), p. 284–322. Beltrami's *Works*, Vol. I, pp. 374–405. French translation: Essai d'interprétation de la géométrie noneuclidienne, Trad. J. Hoüel, *Annales scientifiques de l'École Normale Supérieure*, Sér. 1, Tome VI (1869), p. 251–288. English translations by J. Stillwell, in [57] 7–34.

[5] E. Beltrami, Osservatione sulla Nota del Prof. L. Schlaefli alla Memoria del sig. Beltrami "Sugli spazii di curvatura costante". *Annali di matematica pura ed applicata* (2) 5 (1871–1873), 194–198; Opere, II, 385–389.

[6] E. Beltrami, Teoria fondamentale degli spazii di curvatura costante. (Fundamental theory of spaces of contant curvature). *Annali di matematica pura et applicata* serie II, Tomo II (1868–69), pp. 232–255. Beltrami's *Works*, Vol. I, pp. 406–429. French translation, by J. Hoüel, *Théorie fondamental des espaces de courbure constante*, *Annales scientifiques de l'École Normale Supérieure*, Sér. 1, Tome VI, (1869), pp. 347–375. English translation by J. Stillwell, in [57] 41–62.

[7] G. Birkhoff and M. K. Bennett, Felix Klein and his "Erlanger Programm". *History and philosophy of modern mathematics*, Minnesota Stud. Philos. Sci. XI (Minneapolis, MN, 1988), 145–176.

[8] L. Boi, L. Giacardi and R. Tazzioli (ed.) *La découverte de la géométrie non euclidienne sur la pseudosphère. Les lettres d'Eugenio Beltrami à Jules Hoüel (1868–1881).* (The discovery of non-Euclidean geometry on the pseudosphere. Eugenio Beltrami's letters to Jules Hoüel (1868–1881).) Introduction, notes and commentary by L. Boi, L. Giacardi and R. Tazzioli. Preface by Ch. Houzel and E. Knobloch. Collection Sciences dans l'Histoire. Paris, Librairie Scientifique et Technique Albert Blanchard (1998).

[9] R. Bonola, *La geometria non-euclidea, Esposizione storico-critica del suo sviluppo.* First edition, Ditta Nicola Zanchinelli editore, Bologna 1906. English translation by H. S. Carslaw: *Non-Euclidean Geometry, A critical and historical study of its development*, 1912. Reprinted by Dover, 1955.

[10] G. Brunel, Notice sur l'influence scientifique de Guillaume-Jules Hoüel. In *Mémoires de la Société des Sciences Physiques de Bordeaux*. Tome III (4) 1–72 (1888).

[11] A. Cayley, On the theory of groups, as depending on the symbolic equation $\theta^n = 1$. *Philosophical Magazine*, 4th series, 7 (42): 40–47, 1854.

[12] A. Cayley, A sixth Memoir upon Quantics. *Philosophical Transactions of the Royal Society of London*, Vol. 149 (1859), 61–90.

[13] *The Collected Mathematical Papers of Arthur Cayley*, 13 volumes. 1889–1897. Andrew Russell Forsyth, Ed. Cambridge U. Press.

[14] M. Chasles, *Traité de géométrie supérieure*. Paris, 1852.

[15] A. J. Crilly, *Arthur Cayley: Mathematician laureate of the Victorian age*. Johns Hopkins University Press, 2006.

[16] E. Danti, La Prospettiva di Euclide: Nella quale si tratta di quelle cose, che per raggi diritti si veggomo, et di quelle, che con raggi reflessi nelli Specchi appariscomo, tradotta dal R. P. M. Egnatio Danti, Con alcune sue annotationi de' luoghi piu importanti ; insieme con La prospettiva di Eliodoro Larisseo, cavata della Libreria Vaticana, e tradotta dal medesimo nuovamente data in luce, Fiorenza, stamperia de' Giunti, 1573.

[17] G. Darboux, *Étude sur le développement des méthodes géométriques*. Congrès des sciences et des arts à Saint-Louis, Gauthier-Villars, Paris, 1904.

[18] G. Darboux, Éloge historique d'Henri Poincaré, lu dans la séance publique annuelle du 15 décembre 1913, Comptes rendus de l'Académie des Sciences, 1913, Œuvres de Poincaré, 2, VII–LXXI.

[19] F. Engel, Nikolaj Iwanowitsch Lobatschefskij, Zwei Geometrische Abhandlungen aus dem russischen übersetzt mit Ammerkungen und mit einer Biographie des Verfassers. German translation of Lobachevsky's *Elements of Geometry* and *New elements of geometry, with a complete theory of parallels*. Leipzig, Teubner, Leipzig 1898.

[20] F. Enriques, *Lezioni di geometria proiettiva*. N. Zanichelli, Bologna, 1898.

[21] F. Enriques, Prinzipien der Geometrie. In *Encyklopädie der mathematischen Wissenschaften mit Einschluss ihre Anwendungen* (Encyclopedia of mathematical sciences including their applications). Ed. F. Klein, W. Meyer and H. Burkhardt, B. G. Teubner Verlag, Leipzig, 1898–1933. Vol. 3, part I, p. 1–129, 1907. French edition, Encyclopédie des sciences mathématiques pures et appliquées, ed. J. Molk, Paris, Gauthier-Villars, 1911–1915 and 1955.

[22] M. Fréchet, Sur quelques points du calcul fonctionnel, Thèse présentée à la Faculté des Sciences de Paris, 1906. Published in *Rendiconti del Circolo Matematico di Palermo*, 22, (1906), Issue 1, 1–72.

[23] H. Freudenthal, *La topologie dans les fondements de la géométrie*. ICM, Amsterdam (1954).

[24] C. F. Gauss, *Collected Works*. Vol. VIII, Königliche Gesellshaft der Wissenschaften, Göttingen, 1900.

[25] A. Genocchi, Intorno ad una dimostrazione di Daviet de Foncenex. *Atti della Reale Accademia delle Scienze di Torino* IV (1869), p. 323–327.

[26] J. J. Gray, *Worlds Out of Nothing. A Course in the History of Geometry in the 19th Century*. Springer Undergraduate Mathematics Series, 2007. Second edition, 2011.

[27] J. J. Gray, Klein and the Erlangen Programme. In *Sophus Lie and Felix Klein: The Erlangen program and its impact in mathematics and physics*, ed. L. Ji and A. Papadopoulos, European Mathematical Society Publishing House, Zürich, 2015, 59–73.

[28] G. B. Halsted, The Collected Mathematical Papers of Arthur Cayley. *The American Mathematical Monthly*, Vol. 6, No. 3 (1899), 59–65.

[29] Ph. Henry et Ph. Nabonnand, Conversations avec Jules Hoüel: Regards sur la géométrie non euclidienne et l'analyse infinitésimale vers 1875. La correspondence active et passive de Jules Hoüel avec Joseph Marie De Tilly (1870–1885) et Gaston Darboux (décembre 1874–avril 1875), to appear, 2014.

[30] D. Hilbert and S. E. Cohn-Vossen, *Geometry and the imagination*. Chelsea (1952). Translated from the German edition titled *Anschauliche Geometrie*. 1932.

[31] Ibn al-Haytham, Les Connus. In *Les mathématiques infinitésimales du IXe au XIe siècle, vol. III: Ibn al-Haytham*. Théorie des coniques, constructions géométriques et géométrie pratique, London, al-Furqān, p. 446.

[32] F. Klein, Über die sogenannte Nicht-Euklidische Geometrie. Vorgelegt von A. Clebsch. Nachricheten von der Kgl. Gesellschaft der Wissenschaften zu Göttingen, Nr. 17, (30 August 1871). French version with the title : F. Klein, Sur la géométrie dite non euclidienne, translated by J. Hoüel, Bull. sci. math. et astr., tome 2 (1871), p. 341–351.

[33] F. Klein, Über die sogenannte Nicht-Euklidische Geometrie. *Mathematische Annalen* IV (1871), 573–625. English translation (On the co-called non-Euclidean geometry) by J. Stillwell in [57]. French translation (Sur la Géométrie dite non euclidienne) by L. Laugel, Mém. de la Fac. des Sciences de Toulouse, t. XI (1898), p. 334–335.

[34] F. Klein, *Vergleichende Betrachtungen über neuere geometrische Forschungen.* Verlag von Andreas Deichert, Erlangen, 1872. Published later in *Mathematische Annalen* 43 (1893), 63–100. Included in [38], Vol. 1, p. 460 ff., Springer-Verlag, Berlin, 1921. English translation by M. W. Haskell, A comparative review of recent researches in geometry. *Bulletin of the New York Mathematical Society* 2 (1892–1893), 215–249.

[35] F. Klein, Über die sogenannte Nicht-Euklidische Geometrie (Zweiter Aufsatz). *Mathematische Annalen* VI (1873), 112–145.

[36] F. Klein, Nicht-Euklidische Geometrie, I. Vorlesung, gehalten während des Wintersemesters 1889–90, Augearbeitet von F. Schilling, Göttingen, 1893.

[37] F. Klein, *Vorlesungen über Nicht-Euklidische Geometrie.* Für Druck neu bearbeitet von W. Rosemann, Berlin, Springer, 1928.

[38] F. Klein, *Gesammelte Mathematische Abhandlungen.* Hrsg. von R. Fricke und A. Ostrowski (von F. Klein mit ergänzenden zusatzen versehen), 3 volumes, Springer Verlag, Berlin, 1921.

[39] F. Klein, *Vorlesungen über höhere Geometrie.* Dritte Auflage, Springer Verlag 1926.

[40] F. Klein, Vorlesungen über die Entwicklung der Mathematik im 19. Jahrhundert (Lectures on the development of mathematics in the XIXth century), 2 volumes, Julius Springer Verlag, Berlin 1926 and 1927, reprinted in 1 volume, Springer-Verlag, Berlin, Heidelberg, New York, 1979. English translation: *Development of Mathematics in the 19th Century.* Ed. R. Herrmann, Brookline, Math. Sci. Press, 1979.

[41] E. Laguerre, Sur la théorie des foyers, *Nouvelles annales de mathématiques*, 12, 1853, p. 57–66.

[42] J. H. Lambert, *Theorie der Parallellinien.* In: P. Stäckel et F. Engel, *Die Theorie der Parallellinien von Euklid bis auf Gauss, eine Urkundensammlung zur Vorgeschichte der nichteuklidischen Geometrie* (The theory of parallels from Euclid to Gauss: A collection of sources for the prehistory of non-Euclidean geometry). B. G. Teubner, Leipzig, 1895. Presentation, French translation and comments in [48].

[43] J. E. Littlewood, *Littlewood's Miscellany.* Ed. Béla Bollobás, Cambridge, Cambridge University Press, 1986.

[44] N. I. Lobachevsky, Elements of geometry (Russian), *Kazansky Vestnik*, Issue 25: Feb. and Mar., 1829, 178–187; Apr., 1829 228–241. Issue 27: Nov. and Dec., 1829, 227–243. Issue 28: Mar. and Apr., 1830, 251–283; Jul. and Aug., 1830, 571–683. German translation in [19].

[45] N. I. Lobachevsky, *Pangeometry.* New edition with an English translation and a commentary by A. Papadopoulos. Heritage of European Mathematics, Vol. 4, European Mathematical Society, Zürich, 2010.

[46] N. I. Lobachevsky, *Geometrische Untersuchungen zur Theorie der Parallellinien* (Geometrical researches on the theory of parallels). Kaiserl. russ. wirkl. Staatsrathe und ord. Prof. der Mathematik bei der Universität Kasan. Berlin 1840. In der Finckeschen Buchhandlung. 61 pages. French translation by J. Hoüel, Etudes géométriques sur la théorie des parallèles, *Mémoires de la Société de sciences physiques et naturelles de Bordeaux.* t. IV, 1866, 83–182, and Paris, Gauthier-Villars, 1866. Reprinted by Hermann, Paris, 1900. English translation by G. B. Halsted, Geometrical researches on the theory of parallels, first published by the University of Texas at Austin, 1891. Reprinted in Bonola [9].

[47] G. Loria, Eugenio Beltrami e le sue opere matematiche. *Bibliotheca mathematica* (1901) No. 2, 392–440.

[48] A. Papadopoulos and G. Théret, *La théorie des lignes parallèles de Johann Heinrich Lambert*. Collection Sciences dans l'Histoire, Librairie Scientifique et Technique Albert Blanchard, Paris, 2014.

[49] J.-V. Poncelet, *Traité des propriétés projectives des figures : ouvrage utile à ceux qui s'occupent des applications de la géométrie descriptive et d'opérations géométriques sur le terrain*. Ed. Bachelier, Paris, 1822.

[50] H. Poincaré, Sur les hypothèses fondamentales de la géométrie. *Bulletin de la Société mathématique de France* 15 (1887) 203–216.

[51] H. Poincaré, *La science et l'hypothèse*. Bibliothèque de philosophie scientifique, Paris, Flammarion, 1902.

[52] R. Rashed and A. Papadopoulos, On Menelaus' *Spherics* III.5 in Arabic mathematics. I: *Ibn 'Irāq, Arabic Science and Philosophy* (Cambridge University Press), vol. 24 (2014), 1–68.

[53] R. Rashed and A. Papadopoulos, Critical edition with English translation and mathematical commentary of Menelaus' *Spherics*, based on the Arabic text of al-Harawī, to appear.

[54] G. F. B. Riemann, *Über die Hypothesen, welche der Geometrie zu Grunde liegen* (On the Hypotheses which lie at the Foundations of Geometry), Inaugural dissertation, Göttingen, 1854, Gesammelte Werke, 1. Aufl. p.254–269.

[55] E. Rouché, Edmond Laguerre, sa vie et ses travaux. *Nouvelles annales de mathématiques* 6, 1887, 105–173.

[56] K. G. C. von Staudt, *Geometrie der Lage*. Nürnberg, Verlag von Bauer und Raspe (Julius Merz), 1847.

[57] J. Stillwell, *Sources of Hyperbolic geometry*. History of Mathematics Vol. 10, AMS-LMS, 1996.

[58] B. Taylor, *New principles of linear perspective, or the art of designing on a plane the representations of all sorts of objects, in a more general and simple method than has been done before*. London, R. Knaplock, 1719.

[59] J.-M. de Tilly, *Essai sur les Principes Fondamentaux de la Géométrie et de la Mécanique*. Mémoires de la Société des Sciences Physiques et Naturelles de Bordeaux, 1880.

[60] J.-M. de Tilly, *Essai de Géométrie analytique générale*. Mémoires de l'Académie Royale de Belgique, 1892.

[61] M. Troyanov, On the origin of Hilbert geometry. In *Handbook of Hilbert geometry*, ed. A. Papadopoulos and M. Troyanov, European Mathematical Society Publishing House, Zürich, 2014.

[62] J.-D. Voelke, Le théorème fondamental de la géométrie projective: évolution de sa preuve entre 1847 et 1900. *Archive for History of Exact Sciences* 62 (2008), 243–296.

Chapter 6

What are symmetries of PDEs and what are PDEs themselves?

Alexandre Vinogradov

Contents

This is neither a research nor a review but some reflections about the general theory of (nonlinear) partial differential equations (N)PDEs and its strange marginal status in the realm of modern mathematical sciences.

Since a long time a "zoological-botanical" approach dominates the study of PDEs, and, especially, of NPDEs. Namely, single equations coming from geometry, physics, mechanics, etc. were "tamed and cultivated" like single animals/plants of a practical or theoretical interest. As a rule, for each of these equations were found some prescriptions for the treatment motivated by some concrete *external*, i.e., physical, etc. reasons, but not based on the knowledge of their *intrinsic* mathematical nature. Mainly, these prescriptions are focused on how to construct the solutions rather than to answer numerous questions concerning global properties of the PDE itself.

Modern genetics explains what are living things, their variety and how to treat them to get the desired result. Obviously, a similar theory is indispensable for PDEs, i.e., a solid, well established *general theory*. The recent spectacular progress in genetics became possible only on the basis of not less spectacular developments in chemistry and physics in the last century. Similarly, the general state of the art in mathematics 50–60 years ago was not sufficiently mature to think about the general theory of PDEs. For instance, the fact that an advanced homological algebra will become an inherent feature of this theory could have been hard to imagine at that time.

Recent developments in the general theory of PDEs are revealing more and more its intimate relations with quantum mechanics, quantum field theory and related areas of contemporary theoretical physics, which, also, could be hardly expected a priori. Even more, now we can be certain that the difficulties and shortcomings of current physical theories are largely due to this historically explainable ignorance.

In this survey, we informally present in a historical perspective problems, ideas and results that had led to the renaissance of a general theory of PDEs after the long dead season that followed the pioneering S. Lie opera. Our guide was a modern interpretation of the Erlangen program in the form of the principle : look for the symmetries and you will find the right way. Also, one of our goals was to show that this theory is not less noble part of pure mathematics than algebraic geometry, which may be viewed as its zero-dimensional subcase. The paradox is that the number of mathematicians who worked on this theory does not exceed the number of those who studied Kummer surfaces.

Warnings. The modern general theory of PDEs is written in a new, not commonly known mathematical language, which was formed in the past 30–40 years and was used by a very narrow circle of experts in this field. This makes it impossible to present this theory to a wide mathematical audience, to which this survey is addressed, in its native language. This is why the author was forced to be sometimes rather generic and to refer to some "common places" instead. His apology is in a maxim attributed to Confucius: "An ordinary man wonders marvelous things, a wise man wonders common places."

Notation. Throughout the paper we use $\Lambda^k(M)$ (resp., $D_k(M)$) for the $C^\infty(M)$-module of k-th order differential forms (resp., k-vector fields) on a smooth manifold M. For the rest the notation is standard.

1 A brief history of nonlinear partial differential equations

Sophus Lie was the pioneer who sought for an order in the primordial chaos reigning in the world of NPDEs at the end of the nineteenth century. The driving force of his approach was the idea to use symmetry considerations in the context of PDEs in the same manner as they were used by E. Galois in the context of algebraic equations. In the initial phase of realization of this program, Lie was guided by the principle "chercher la symétrie" and he discovered that behind numerous particular tricks found by hand in order to solve various concrete differential equations there are groups of transformations preserving these equations, i.e., their symmetries. Then, based on these "experimental data", he developed the machinery of transformation groups, which allows one to systematically compute what is now called *point* or *classical* symmetries of differential equations. Central in Lie theory is the concept of an *infinitesimal transformation* and hence of an *infinitesimal symmetry*. Infinitesimal symmetries of a differential equation or, more generally, of an object in differential geometry, form a Lie algebra. This Lie's invention is among the most important in the history of mathematics.

The computation of classical symmetries of a system of differential equations leads to another *nonlinear* system, which is much more complicate than the original one. Lie resolved this seemingly insuperable difficulty by passing to infinitesimal symmetries. In order to find them one has to solve an overdetermined system of *linear* differential equations, which is a much easier task and it non infrequently allows a complete solution. Moreover, by exponentiating infinitesimal symmetries one can find almost all finite symmetries. A particular case of this mechanism is the famous relation between Lie algebras and Lie groups.

Initial expectations that groups of classical symmetries are analogues of Galois group for PDEs had led to a deep delusion. Indeed, computations show that this group for a generic PDE is trivial. This was one of the reasons why the systematic applications of Lie theory to differential equations was frozen for a long time and the original intimate relations of this theory with differential equations were lost. Only much later, in the period 1960–1970, L. V. Ovsiannikov and his collaborators resumed these relations (see [43, 20]) and now they are extending in various directions.

Contact geometry was another important contribution of S. Lie to the general theory of NPDEs. Namely, he discovered that symmetries of a first order NPDE imposed on one unknown function are *contact transformations*. These transformations not only mix dependent and independent variables but their derivatives as well. For this reason they are much more general than the above-mentioned point transformations, which mix only dependent and independent variables. Moreover, it turned out that contact symmetries are sufficient to build a complete theory for this class of equations, which includes an elegant geometrical method of construction of their solutions. In this sense contact symmetries play the role of a Galois group for this class of equations. On the other hand, the success of contact geometry in the theory of first order NPDEs led to the suspicion that classical symmetries form just a small part of all true symmetries of NPDEs. But the question of what are these symmetries remained unanswered for a long time up to the discovery of *integrable systems* (see below). But some signs of an implicit use of such symmetries in concrete situations can already be found in works of A. V. Bäcklund and E. Noether.

A courageous attempt to build a general theory of PDEs was undertaken by Charles Riquier at the very end of the 19th century. "Courageous" because at that time the only way to deal with general PDEs was to manipulate their coordinate descriptions. His results were then gathered in a handsome book [46] of more than 600 pages presenting, from the modern point of view, the first systematically developed general theory of formal integrability. This book is full of cumbersome computations, and the results obtained are mostly of a descriptive nature and do not reveal structural units of the theory. Nevertheless, it demonstrated that a general theory of PDEs, even at a formal level, is not impossible. Moreover, Riquier showed that the formal theory duly combined with the Cauchy–Kowalewski theorem lead to various existence results in the class of analytic functions such as the famous Cartan–Kähler theorem (see [7, 24]). In its turn Riquier's work motivated Élie Cartan to look for a coordinate-free language for the formal/analytical theory and it led him to the theory of *differential systems* based on the calculus of differential forms (see [7]). Cartan's theory was later developed and extended by E. Kähler [24], P. K. Rashevsky [45], M. Kuranishi [32] and others. The reader will find its latest version in [6].

In the middle of the twentieth century the theory of differential systems circulated in a narrow group of geometers as the most general theory of PDEs. However, this was an exaggeration. For instance, there were no relations between this theory and the theory of linear PDEs, which was in a booming growth at that time. Moreover, this theory did not produce any, worth to be mentioned, application to the study of concrete PDEs. We can say that it is even hardly possible to imagine that the study of the Einstein or Navier–Stokes equations will become easier after being converted into differential systems. Therefore, the apparatus of differential forms did not confirm the expectations to become a natural base language for the general theory of PDEs, but it became one of the basic instruments in modern differential geometry and in many areas of its applications.

The original Riquier approach was improved and developed by M. Janet ([23]). But, unfortunately, his works were for a long period shadowed by the works of É. Cartan. Their vitality was confirmed much later at the beginning of the new era for NPDEs (see [44]). This era implicitly starts with the concept of a jet bundle launched by Ch. Ehresmann (see [12]). Ehresmann himself did not develop applications of jet bundles to PDEs. But, fortunately, this term became a matter of fashion and later was successfully used in various areas of differential geometry. In particular, D. C. Spencer and H. Goldschmidt essentially used jet bundles in their new theory of formal integrability by inventing a new powerful instrument, namely, the Spencer cohomology (see [50, 44]). In this way were discovered the first structural blocks of the general theory of PDEs, and this new theory demonstrated some important advantages in comparison with the theory of differential systems.

The Ehresmann concept of jet bundle is, however, too restrictive to be applied to general NPDEs and for this reason should be extended to that of jet space of submanifolds. Namely, the k-th order jet bundle $J^k(\pi)$ associated with a smooth bundle $\pi : E \to M$ consists of k-th order jets of sections of π, while the k-th order jet space $J^k(E,n)$ consists of k-th order jets of n-dimensional submanifolds of the manifold E. Jet spaces are naturally supplied with a structure, called the *Cartan distribution*, which allows an interpretation of functions defined on them as nonlinear differential operators. Differential equations in the standard but coordinate independent meaning of this term are naturally interpreted as submanifolds of jet spaces. The first systematic study of geometry of jet spaces was done by A. M. Vinogradov and participants of his Moscow seminar in the 1970s (see [60, 70, 30]). Later, on this basis, it was understood that various natural differential operators and constructions that are necessary for the study of a system of PDEs of order k do not live necessarily on the k-th order jet space but involve jet spaces of any order. This is equivalent to saying that a conceptually complete theory of PDEs is possible only on *infinite order* jet spaces. A logical consequence of this fact is that objects of the *category of partial differential equations* are *diffieties*, which duly formalize the vague idea of the "space of all solutions" of a PDE. Diffieties are a kind of infinite-dimensional manifolds, and the specific differential calculus on them, called *secondary calculus*, is a native language to deal with PDEs and especially with NPDEs (see [66, 27, 29]).

Below we shall show how to come to secondary calculus and hence to the *general theory* of (nonlinear) partial differential equations by trying to answer the question "what are symmetries of a PDE". It is worth stressing that Klein's Erlangen program was a good guide in this expedition, which was decisive in finding the right way at some crucial moments.

2 Evolution of the notion of symmetry for differential equations

A retrospective view on how the answer to the question "what are symmetries of a PDE" evolved historically will be instructive for our further discussion. From the very beginning this question was more implicitly than explicitly related with the answer to question "what is a PDE". It seems that the apparent absurdity of this question prevented its exact formulation and hence slowed down the development of the general theory.

Below we shall use the following notation. If $\sigma = (\sigma_1, \dots, \sigma_n)$ is a multiindex, then $|\sigma| = \sigma_1 + \dots + \sigma_n$, and

$$\frac{\partial^{|\sigma|} f(x)}{\partial x^\sigma} = \frac{\partial^s f(x)}{\partial x_1^{\sigma_1} \dots x_n^{\sigma_n}}, \quad |\sigma| = s, \quad f(x) = f(x_1, \dots, x_n).$$

We assume that $\frac{\partial^{|\sigma|} f}{\partial x^\sigma} = f$ if $\sigma = (0, \dots, 0)$.

Standard ("classical") definition. According to the commonly accepted point of view a system of PDEs is a set of expressions

$$F_i(x, u, u_{[1]}, \dots, u_{[k]}) = 0, \quad i = 1, 2, \dots, l, \tag{6.1}$$

where $x = (x_1, \dots, x_n)$ are *independent variables*, $u = (u^1, \dots, u^m)$ *dependent* ones, and $u_{[s]}$ stands for the totality of symbols u^i_σ, $1 \le i \le m$, with $|\sigma| = s$. Further we shall use short "PDE" for "system of PDEs" and, accordingly, write

$$F(x, u, u_{[1]}, \dots, u_{[k]}) = 0 \quad \text{assuming that} \quad F = (F_1, \dots, F_l).$$

Solutions. A system of functions $f_1(x), \dots, f_m(x)$ is a solution of the PDE (6.1) if the substitutions $\frac{\partial^{|\sigma|} f^i}{\partial x^\sigma} \to u^i_\sigma$ transform the expressions (6.1) to functions of x that are identically equal to zero.

This traditional view on PDEs is presented in all, modern and classical, textbooks. For instance, in Wikipedia one may read that a PDE is "an equation that contains unknown multivariable functions and their partial derivatives" or "une équation aux dérivées partielles (EDP) est une équation dont les solutions sont les fonctions inconnues vérifiant certaines conditions concernant leurs dérivés partielles."

Symmetries: the first idea. The "common sense" coherent with this point of view suggests to call a symmetry of PDE (6.1) transformations

$$x_i = \phi_i(\bar{x}_1, \dots, \bar{x}_n),\ i = 1, \dots, n, \quad u^j = \psi^j(\bar{u}^1, \dots, \bar{u}^m),\ j = 1, \dots, m, \tag{6.2}$$

of dependent and independent variables that "preserve the form" of relations (6.1). More exactly, this means that the so-obtained functions $\bar{F}_i = \bar{F}_i(\bar{x}, \bar{u}, \bar{u}_{[1]}, \dots, \bar{u}_{[s]})$s are linear combinations of functions F_is with functions of $x, u, u_{[1]}, \ldots, u_{[k]}$ as coefficients. Here we used the confusing classical notation where (x, u) stands for coordinates of the image of the point $(\bar{x}, \bar{u})$. Also, it is assumed that the transformations of the symbols u^i_σ are naturally induced by those of x and u.

Many fundamental equations in physics and mechanics inherit space-time symmetries, and these are "first idea" symmetries. Very popular in mechanics of continua dimensional analysis is also based on the so-understood concept of symmetry (see [4, 5, 43]).

Example 2.1. The Burgers equation $u_t = u_{xx} + uu_x$ is invariant, i.e., symmetric, with respect to space shifts ($x = \bar{x} + c, t = \bar{t}, u = \bar{u}$), time shifts ($x = \bar{x}, t = \bar{t} + c, u = \bar{u}$) and the passage to another Galilean inertial frame moving with the velocity v ($x = \bar{x} + vt, t = \bar{t}, u = \bar{u}$). This equation possesses also *scale* symmetries: $x = \lambda\bar{x}$, $t = \lambda^2\bar{t}$, $u = \lambda^{-1}\bar{u}$, $\lambda \in \mathbb{R}$.

The above definition of a symmetry is based on the a priori premise that the division of variables into dependent and independent ones is an indispensable part of the definition of a PDE. However, many arguments show that this point of view is too restrictive. In particular, numerous tricks that were found by hands to resolve various concrete PDEs involves transformations which do not respect this division. For instance, transformations

$$x = \frac{\bar{x}}{\tau\bar{t} + 1}, \qquad t = \frac{\bar{t}}{\tau\bar{t} + 1}, \qquad u = \bar{u} + \tau(\bar{t}\bar{u} - \bar{x})$$

depending on a parameter $\tau \in \mathbb{R}$ leave the Burgers equations invariant. They, however, do not respect sovereignty of the dependent variable u.

In this connection, a more obvious and important argument is that

> *what is called functions in the traditional definition of a PDE are not, generally, functions but elements of coordinate-wise descriptions of certain objects, like tensors, submanifolds, etc.*

Indeed, if the dependent variables u are components of a tensor, then a transformation of independent variables induces automatically a transformation of independent ones. So, the division of variables into dependent and independent ones cannot, in principle, be respected in such cases. Moreover, this, as banal as well-known observation, which is nevertheless commonly ignored, poses a question

> *what are "independent variables", i.e., what are mathematical objects that are the subject of PDEs?*

The "obvious" answer that these are "objects that are described coordinate-wisely by means of functions" is purely descriptive and hence not very satisfactory. In fact, this question is neither trivial, nor stupid, and, in particular, its analysis directly leads to the conception of jets (see below).

Symmetries: the second idea. Under the pressure of the above arguments it seems natural to call a symmetry of the PDE (6.1) a transformation of independent and dependent, which respect the status of independent variables only, i.e., a transformation of the form

$$\begin{cases} x_i = & \phi_i(\bar{x}_1, \ldots, \bar{x}_n), \quad i = 1, \ldots, n, \\ u^j = & \psi^j(\bar{x}_1, \ldots, \bar{x}_n, \bar{u}^1, \ldots, \bar{u}^m), \quad j = 1, \ldots, m. \end{cases} \tag{6.3}$$

This idea is consistent with many situations in physics and mechanics where space-time coordinates play the role of independent variables, while "internal" characteristics of the considered continua, fields, etc. refer to dependent ones. Mathematically, these quantities are represented as sections of suitable fiber bundles, and transformations that preserve the bundle structure are exactly of the form (6.3). Gauge transformations in modern physics are of this kind.

On the other hand, since the second half of 18th century, the development of differential geometry put in light various problems related with surfaces and, later, with manifolds and their maps (see [41, 15, 10]) formulated in terms of PDEs. A surface in the 3-dimensional Euclidean space E^3 is not, generally, the graph of a function. So, the phrase that the equation

$$(1 + u_x^2)u_{yy} - 2u_x u_y u_{xy} + (1 + u_y^2)u_{xx} = 0 \tag{6.4}$$

is the equation of minimal surfaces is not, rigorously speaking, true. More exactly, it is true only locally for surfaces of the form $z = u(x, y)$ with (x, y, z) being standard Cartesian coordinates in E^3. So, the question "what is the true (global) equation of minimal surfaces" should be clarified. This question, which was historically ignored, becomes, however, rather relevant if one thinks about global topological properties of minimal surfaces. Also, isometries of E^3 preserve the class of minimal surfaces and hence they must be considered as symmetries of the "true" equation of minimal surfaces in any reasonable sense of this term. But, generally, these transformations do not respect the status of both independent and dependent variables. This and many other similar examples show that the second idea is still too restrictive.

Symmetries: the third idea. The next obvious step is to consider transformations

$$x_i = \phi_i(\bar{x}_1, \ldots, \bar{x}_n, \bar{u}^1, \ldots, \bar{u}^m), \qquad u^j = \psi^j(\bar{x}_1, \ldots, \bar{x}_n, \bar{u}^1, \ldots, \bar{u}^m), \\ i = 1, \ldots, n, \quad j = 1, \ldots, m. \tag{6.5}$$

as eventual symmetries of PDEs. In this form the idea of a symmetry of a PDE was formulated by S. Lie at the end of 19th century and was commonly accepted for a long time up to the discovery of integrable systems at the late 1960s, when some strong doubts about it arose. Symmetries of the form (6.5) are called *point symmetries* in order to distinguish them from *contact symmetries* (see below) and more general ones that recently emerged.

Symmetries: the fourth incomplete idea. But Lie himself created the ground for such doubts by developing the theory of first order PDEs in the form of contact geometry. From this point of view natural candidates for symmetries of such a PDE are contact transformations, which mix independent and dependent variables and their first order derivatives in an almost arbitrary manner. In particular, this means that, generally, transformations of dependent and independent variables involve also first derivatives. i.e.,

$$\begin{cases} x_i = & \phi_i(\bar{x}_1, \ldots, \bar{x}_n, \bar{u}, \bar{u}_{x_1}, \ldots, \bar{u}_{x_m}), \quad i = 1, \ldots, n, \\ u = & \psi^j(\bar{x}_1, \ldots, \bar{x}_n, \bar{u}, \bar{u}_{x_1}, \ldots, \bar{u}_{x_m}). \end{cases} \tag{6.6}$$

Transformations (6.6) are to be completed by transformations of first derivatives

$$x_i = \phi_i(\bar{x}_1, \ldots, \bar{x}_n, \bar{u}, \bar{u}_{x_1}, \ldots, \bar{u}_{x_m}), \quad i = 1, \ldots, n,$$

in a way that respects the "contact condition" $d\bar{u} - \sum_{i=1}^{n} \bar{u}_{x_i} dx_i = 0$.

Contact transformations can be naturally extended to transformations of higher order derivatives and, therefore, considered as candidates for true symmetries of PDEs with one dependent variable. For instance, as such they are very useful in the study of Monge–Ampère equations (see [33]). In other words, the third idea becomes too restrictive, at least, for equations with one dependent variable.

The above discussion leads to a series of questions:

Question 1 What are analogues of contact transformations for PDEs with more than one dependent variable?

Question 2 Are there higher order analogues of contact transformations, i.e., transformations mixing dependent and independent variables with derivatives of order higher than one?

To answer this questions we, first, need to bring the traditional approach to PDEs to a more conceptual form. In particular, a coordinate-free definition of a PDE equivalent to the standard one is needed. This is done in the next section.

3 Jets and PDEs

Various objects (functions, tensors, submanifolds, smooth maps, geometrical structures, etc.) that are subject of PDEs may be interpreted as submanifolds of a suitable manifold. For instance, functions, sections of fiber bundles, in particular, tensors, and smooth maps may be geometrically viewed as the corresponding graphs. So, we assume this unifying point of view and interpret PDEs as *differential restrictions* imposed on submanifolds of a given manifold.

3.1 Jet spaces Thus, objects of our further considerations will be n-dimensional submanifolds of an $(n+m)$-dimensional submanifold E. Let $L \subset E$ be such a submanifold. In order to locally describe it in a local chart $(y_1, \ldots, y_{n+m})$ we must choose n coordinates among these which are independent on L, say, $(y_{i_1}, \ldots, y_{i_n})$ and declare the remaining y_js to be dependent ones. The notation $x_1 = y_{i_1}, \ldots, x_n = y_{i_n}$, $u^1 = y_{j_1} \ldots, u^m = y_{j_m}$ with $\{j_1, \ldots, j_m\} = \{1, \ldots, n+m\} \setminus \{i_1, \ldots, i_n\}$ stresses this artificial division of local coordinates into dependent and independent ones. We shall refer to (x, u) as a *divided chart.* By construction, L is locally described in this divided chart by equations of the form $u^i = f^i(x)$, $i = 1, \ldots, n$. The next step is to understand what is the manifold in which $(x, u, u_{[1]} \ldots, u_{[k]})$ is a local chart. The answer is as follows.

Let M be a manifold, $z \in M$ and $\mu_z = \{f \in C^\infty(M) \mid f(z) = 0\}$ the ideal of the point z. Elements of the quotient algebra $J_z(M) = C^\infty(M)/\mu_z^{k+1}$ (μ_z^s stands for the s-th power of the ideal μ_z) are called *k-th order jets* of functions at the point $z \in M$. The *k-th jet of f at z*, denoted by $[f]_z^k$, is the image of f under the factorization homomorphism $C^\infty(M) \to J_z(M)$. This definition also holds for $k = \infty$ if we put $\mu_z^\infty = \bigcap_{k \in \mathbb{N}} \mu_z^k$. It is easy to see that $[f]_z^k$=$[g]_z^k$ if and only if in a local chart all the derivatives of the functions f and g of order $\leq k$ at the point z are equal.

Two n-dimensional submanifolds $L_1, L_2 \subset E$ are called *tangent with the order k at a common point z* if for any $f \in C^\infty(M)$, $[f|_{L_1}]_z^k = 0$ implies $[f|_{L_2}]_z^k = 0$ and vice versa. Obviously, k-th order tangency is an equivalence relation.

Definition 3.1. The equivalence class of n-dimensional submanifolds of E, which are k-th order tangent to L at $z \in L$, is called the *k-th order jet of L at z* and is denoted by $[L]_z^k$.

The set of all k-jets of n-dimensional submanifolds L of E is naturally supplied with the structure of a smooth manifold, which will be denoted by $J^k(E, n)$. Namely, associate with an n-dimensional submanifold L of E the map

$$j_k(L) : L \to J^k(E, n),\ L \ni z \mapsto [L]_z^k$$

and call a function ϕ on $J^k(E, n)$ *smooth* if $j_k(L)^*(\phi) \in C^\infty(L)$ for all n-dimensional $L \subset E$. The so-defined smooth function algebra will be denoted by $\mathcal{F}_k(E, n)$, i.e., $C^\infty(J^k(E, n)) = \mathcal{F}_k(E, n)$.

Remark 3.1. If $k < \infty$ the above definition of the smooth structure on $J^k(E, n)$ is equivalent to the standard one, which uses charts and atlases (see below). But it becomes essential for $k = \infty$, since the standard "cartographical" approach in this case creates some boring inconveniences.

If L is given by equations $u^i = f^i(x)$, $i = 1, \ldots, n$, in a divided chart and $(x_1^0, \ldots, x_n^0, u_0^1 \ldots, u_0^m)$ are coordinates of z in this chart, then, as it is easy to see, $[L]_z^k$ is uniquely defined by the derivatives

$$u_{\sigma,0}^i = \frac{\partial^{|\sigma|} f^i}{\partial x_\sigma}(x_1^0, \ldots, x_n^0), \quad 1 \leq i \leq m, \quad |\sigma| \leq k, \tag{6.7}$$

and vice versa. So, the numbers x_j^0 together with the numbers $u^i_{\sigma,0}$ may be taken for local coordinates of the point $\theta = [L]^k_z \in J^k(E,n)$. By putting $u^i_\sigma(\theta) = u_{\sigma,0}$, we have

$$j_k(L)^*(u^i_\sigma) = \frac{\partial^{|\sigma|} f^j}{\partial x_\sigma}$$

and we see that the functions u^i_σ, $1 \le i \le m$, $|\sigma| \le k$, together with the functions x_j form a smooth local chart on $J^k(E,n)$.

Thus we see that $(x, u, u_{[1]}, \dots, u_{[k]})$ is a local chart on $J^k(E,n)$ and hence (6.1) is the equation of a submanifold in $J^k(E,n)$. This allows us to interpret the standard definition of PDEs in an invariant coordinate-free manner.

Definition 3.2. A system of PDEs of order k imposed on n-dimensional submanifolds of a manifold E is a submanifold $\mathcal{E}$ of $J^k(E,n)$.

Remark 3.2. $\mathcal{E}$ as a submanifold of $J^k(E,n)$ may have singularities.

3.2 Jet tower Note that E is naturally identified with $J^0(E,n) : z \leftrightarrow [L]^0_z$, and natural projections

$$\pi_{k,l} : J^k(E,n) \to J^l(E,n),\ [L]^k_z \mapsto [L]^l_z,\ l \le k,$$

relate jet spaces of various orders in a unique structure

$$E = J^0(E,n) \overset{\pi_{1,0}}{\longleftarrow} J^1(E,n) \overset{\pi_{2,1}}{\longleftarrow} \cdots \overset{\pi_{k,k-1}}{\longleftarrow} J^k(E,n) \overset{\pi_{k+1,k}}{\longleftarrow} \dots J^\infty(E,n). \quad (6.8)$$

It is easy to see that $J^\infty(E,n)$ is the inverse limit of the system of maps $\{\pi_{k,l}\}$. Also note that $\pi_{k,l} : J^k(E,n) \to J^l(E,n)$ is a fiber bundle. Moreover, $\pi_{k,k-1} : J^k(E,n) \to J^{k-1}(E,n)$ is an *affine bundle* if $k \ge 2$ and $m > 1$ or if $k \ge 3$ and $m = 1$ (see [30, 44]).

Dually to (6.8), smooth function algebras on jet spaces form a telescopic system of inclusions

$$C^\infty(E) = \mathcal{F} \overset{\pi^*_{1,0}}{\rightarrow} \mathcal{F}_1 \overset{\pi^*_{2,1}}{\longrightarrow} \cdots \overset{\pi^*_{k,k-1}}{\longrightarrow} \mathcal{F}_k \overset{\pi^*_{k+1,k}}{\longrightarrow} \dots \mathcal{F}_\infty. \quad (6.9)$$

So, $\mathcal{F}_\infty$ may be viewed as the direct limit of (6.9). By identifying $\mathcal{F}_k$ with $\pi^*_{\infty,k}(\mathcal{F}_k)$ we get the filtered algebra $\mathcal{F}_0 \subset \mathcal{F}_1 \subset \cdots \subset \mathcal{F}_k \subset \dots,\ \mathcal{F}_\infty = \bigcup_{k=0}^\infty \mathcal{F}_k$.

Since two submanifolds of the same dimension are first order tangent at a point z if and only if they have the common tangent space at z, $[L]^1_z$ is naturally identified with T_zL. In this way $J^1(E,n)$ is identified with the Grassmann bundle $\mathrm{Gr}_n(E)$ of n-dimensional subspaces tangent to E , and the canonical projection $\mathrm{Gr}_n(E) \to E$ is identified with $\pi_{1,0}$. In particular, the standard fiber of $\pi_{1,0}$ is the Grassmann manifold $Gr_{n+m,n}$ so that $\pi_{1,0}$ is not an affine bundle. If $m = 1$, then the fiber of $\pi_{2,1}$ is the Lagrangian Grassmannian.

Example 3.1. The equation $\mathcal{E}$ of minimal surfaces in the 3-dimensional Euclidean space E^3 is a hypersurface in $J^2(E^3, 2)$. The projection $\pi_{2,1} : \mathcal{E} \to J^1(E^3, 2)$ is a nontrivial bundle whose fiber is the 2-dimensional torus. So, rigorously speaking, (6.4) is not the equation of minimal surfaces but a local piece of it.

This and many other similar examples show that, generally, (6.1) is just a local coordinate-wise description of a PDE.

3.3 Classical symmetries of PDEs The language of jet spaces not only gives a due conceptual rigor to the traditional theory of PDEs but it also simplifies many technical aspects of it and makes transparent and better workable various basic constructions. This will be shown in the course of the subsequent exposition. But now we shall illustrate this point by explaining how "point transformations" acts on PDEs.

First, we observe that (6.3) is just a local coordinate-wise description of a diffeomorphism $F : E \to E$. Now the question we are interested in is: how does F act on jets? The answer is obvious: F induces the diffeomorphism

$$F_{(k)} : J^k(E,n) \to J^k(E,n),\ [L]_z^k \mapsto [F(L)]_{F(z)}^k, \tag{6.10}$$

called the k-lift of F. This immediately leads to formulate the definition of a "classical" (= "point") symmetry of a PDE.

Definition 3.3. A classical/point symmetry of a PDE $\mathcal{E} \subset J^k(E,n)$ is a diffeomorphism $F : E \to E$ such that $F_{(k)}(\mathcal{E}) = \mathcal{E}$.

Similarly one can define lifts of "infinitesimal point transformations", i.e., vector fields on E. Recall that if X is a vector field on E and $F_t : E \to E$ is the flow it generates, then

$$X = \left.\frac{d(F_t^*)}{dt}\right|_{t=0} \quad \text{with } F_t^* : C^\infty(E) \to C^\infty(E).$$

Then the lift $X_{(k)}$ of X to $J^l(E,n)$ is defined as

$$X_{(k)} = \left.\frac{d((F_t)^*_{(k)})}{dt}\right|_{t=0}.$$

Definition 3.4. A vector field X on E is an *infinitesimal classical/point symmetry of a PDE* $\mathcal{E} \in J^k(E,n)$ if $X_{(k)}$ is tangent to $\mathcal{E}$.

Nonlinear partial differential operators are also easily defined in terms of jets.

Definition 3.5. A *(nonlinear) partial differential operator* $\square$ of order k sending n-dimensional submanifolds of E to E' is defined as the composition $\Phi \circ j_k$ with $\Phi : J^k(E,n) \to E'$ being a (smooth) map, i.e., $\square(L) = \Phi(j_k(L))$.

For instance, functions on $J^l(E,n)$ are naturally interpreted as (nonlinear) differential operators of order l.

The above definitions and constructions are easily specified to fiber bundles. Namely, if $\pi : E \to M$, dim $M = n$, is a fiber bundle, then the k-th order jet $[s]^k_x$ of a local section of it $s : U \to E$ (U is an open in M) at a point $x \in M$ is defined as $[s]^k_x = [s(U)]^k_{s(x)}$. These specific jets form an everywhere dense open subset in $J^k(E,n)$ denoted by $J^k(\pi)$ and called the *k-order jet bundle of π*. The substitute of maps $j_k(L)$ in this context are maps $j_k(s) \overset{\text{def}}{=} j_k(s(U))$. Additionally, we have natural projections $\pi_k = \pi \circ \pi_{k,0} : J^k(\pi) \to M$. If π is a vector bundle, then π_k is a vector bundle too, and the equation $\mathcal{E} \subset J^k(\pi)$ is *linear* if $\mathcal{E}$ is a linear sub-bundle of π_k, etc. For further details concerning the "fibered" case, see [29, 70, 44].

4 Higher order contact structures and generalized solutions of NPDEs

4.1 Higher order contact structures Now we are going to reformulate the standard definition of a solution of a PDE in a coordinate-free manner. Put $L_{(k)} = \operatorname{Im} j_k(L)$ for an n-dimensional submanifold of E. Obviously, $L_{(k)}$ is an n-dimensional submanifold L of $J^k(E,n)$, which is projected diffeomorphically onto L via $\pi_{k,0}$.

Definition 4.1. L is a *solution in the standard sense of a PDE* $\mathcal{E} \subset J^k(E,n)$ if $L_{(k)} \subset \mathcal{E}$.

If $u^i = f^i(x), i = 1,\dots,n$ are local equations of L, then

$$u^i_\sigma = \frac{\partial^{|\sigma|} f^i}{\partial x^\sigma}(x), \quad i = 1,\dots,n, \quad |\sigma| \le n,$$

are local equations of $L_{(k)}$ in $J^k(E,n)$. This shows that the coordinate-free definition 4.1 coincides with the standard one. Also, we see that the $L_{(k)}$s form a very special class of n-dimensional submanifolds in $J^k(E,n)$. This class is not intrinsically defined, and hence Definition 4.1 is not *intrinsic*. For this reason it is necessary to supply $J^k(E,n)$ with an additional structure, which allows to distinguish the submanifolds $L_{(k)}$ from others. Such a structure is a distribution on $J^k(E,n)$ defined as follows.

Definition 4.2. The minimal distribution $\mathcal{C}^k : J^k(E,n) \ni \theta \mapsto \mathcal{C}^k_\theta \subset T_\theta(J^k(E,n))$ on $J^k(E,n)$ such that all the $L_{(k)}$ are integral submanifolds of it, i.e., $T_\theta(L_{(k)}) \subset \mathcal{C}^k_\theta$, $\forall \theta \in L_{(k)}$, is called the *$k$-th order contact structure* or the *Cartan distribution* on $J^k(E,n)$.

It directly follows from the definition that

$$\mathcal{C}^k_\theta = \operatorname{span}\{T_\theta(L_{(k)}) \text{ for all } L \text{ such that } L_{(k)} \ni \theta\}. \tag{6.11}$$

Due to the importance of the subspaces $T_\theta(L_{(k)}) \subset T_\theta(J^k(E,n))$ we shall call them *R-planes* (at θ). By construction any R-plane at θ belongs to $\mathcal{C}^k_\theta$. The following simple fact is very important and will be used in various constructions further on.

Lemma 4.1. Let $\theta = [L]^k_z = [N]^k_z$ and $\theta' = \pi_{k,k-1}(\theta)$. Then $T_{\theta'}(L_{(k-1)}) = T_{\theta'}(N_{(k-1)})$ and hence the R-plane $R_\theta = T_{\theta'}(L_{(k-1)})$ is uniquely defined by θ. Moreover, the correspondence $\theta \mapsto R_\theta$ between points of $J^k(E,n)$ and R-planes at points of $J^{k-1}(E,n)$ is biunique.

This lemma allows to identify the fiber $\pi^{-1}_{k,k-1}(\theta')$ with the variety of all R-planes at θ and hence $J^k(E,n)$ with the variety of R-planes at points of $J^{k-1}(E,n)$.

Below we list some basic facts concerning the Cartan distribution and R-planes (see [60, 30, 70]).

Proposition 4.1.

1. $\mathcal{C}^k_\theta = (d_\theta\pi_{k,k-1})^{-1}(R_\theta)$ with $d_\theta\pi_{k,k-1} : T_\theta(J^k(E,n)) \to T_{\theta'}(J^{k-1}(E,n))$ being the differential of $\pi_{k,k-1}$ at θ. In particular, $d_\theta\pi_{k,k-1}(\mathcal{C}^k_\theta) \subset \mathcal{C}^{k-1}_{\theta'}$.
2. In local coordinates the Cartan distribution is given by the equations

$$\omega^i_\sigma \overset{\text{def}}{=} du^i_\sigma - \sum_j u^i_{\sigma+1_j} dx_j = 0, \quad |\sigma| < k, \text{ where } (\sigma + 1_j)_i = \sigma_i + \delta_{ij},$$

or, dually, is generated by the vector fields

$$D^k_i = \frac{\partial}{\partial x_i} + \sum_{j,|\sigma|<k} u^j_{\sigma+1_i} \frac{\partial}{\partial u^j_\sigma}, \quad i = 1,\dots,n, \text{ and } \frac{\partial}{\partial u^j_\sigma}, \ |\sigma| = k.$$

3. $$\dim \mathcal{C}^k_\theta = m \binom{n+k-1}{k} + n, \text{ if } \ 0 \le k < \infty; \quad \dim \mathcal{C}^\infty_\theta = n.$$

4. Tautologically, a point $\theta = [L]^\infty_z \in J^\infty(E,n)$ is the inverse limit of $\theta_k = \pi_{\infty,k}(\theta) = [L]^k_z$, $k \to \infty$. Then $\mathcal{C}^\infty_\theta$ is the inverse limit of the chain

$$\cdots \xleftarrow{d_{\theta_k}\pi_{k,k-1}} \mathcal{C}^k_{\theta_k} \xleftarrow{d_{\theta_{k+1}}\pi_{k+1,k}} \mathcal{C}^{k+1}_{\theta_{k+1}} \xleftarrow{d_{\theta_{k+2}}\pi_{k+2,k+1}} \cdots$$

different nature, which have the same dimension. One of the simplest examples of this kind is the differential system $(J^k(E,1), \mathcal{C}^k)$ with $\dim E = 2$, $k > 1$, for which locally maximal integral submanifolds of types 0 (fibers or the projection $\pi_{k,k-1}$) and 1 (R-manifolds) are all 1-dimensional. Moreover, integral submanifolds of type 0 are irrelevant/"parasitic" in the context of the theory of differential equations.

Secondly, an equation $\mathcal{E} \subset J^k(E,n)$ is called *rigid* if the k-th order contact manifold $(J^k(E,n), \mathcal{C}^k)$ can be recovered if $(\mathcal{E}, \mathcal{C}^k_{\mathcal{E}})$ as an abstract differential system is only known. For instance, if the codimension of $\mathcal{E}$ in $J^k(E,n)$ is less than the difference of dimensions of locally maximal integral submanifolds of types 0 and 1, then $\mathcal{E}$ is, as a rule, rigid. Indeed, in this case integral submanifolds of $(\mathcal{E}, \mathcal{C}^k_{\mathcal{E}})$ of absolutely maximal dimension are intersections of fibers of $\pi_{k,k-1}$ with $\mathcal{E}$. In other words, these are fibers of the projection $\pi_{k,k-1}|_{\mathcal{E}} : \mathcal{E} \to J^{k-1}(E,n)$. If, additionally, this projection is surjective, then $J^{k-1}(E,n)$ is recovered as the variety of integral submanifolds of $(\mathcal{E}, \mathcal{C}^k_{\mathcal{E}})$ of maximal dimension. Next, under some weak condition projections of spaces $\mathcal{C}^k_{\mathcal{E},\theta}$, $\theta \in \mathcal{E}$, on the so-interpreted jet space $J^{k-1}(E,n)$ span the distribution $\mathcal{C}^{k-1}$. In this way $(J^{k-1}(E,n), \mathcal{C}^{k-1})$ is recovered from $(\mathcal{E}, \mathcal{C}^k_{\mathcal{E}})$ and, finally, $(J^k(E,n), \mathcal{C}^k)$ is recovered from $(J^{k-1}(E,n), \mathcal{C}^{k-1})$ as the variety of R-planes on $J^{k-1}(E,n)$ according to Proposition 4.1, (1). Thus converting rigid equations into differential systems is counterproductive, since this procedure creates non-necessary additional problems. In this connection it is worth mentioning that the most important PDEs in geometry, mechanics and physics we deal with are determined or slightly overdetermined systems of PDEs, like Maxwell or Einstein equations, and hence are rigid.

Even more important arguments, which do not speak in favor of differential systems, come from the fact that the calculus of differential forms is a small part of a much richer structure formed by *natural functors of differential calculus* and *objects representing them*. For instance, indispensable for formal integrability theory diff- and jet-Spencer complexes are examples of this kind (see [50, 54, 59, 44, 30, 70]). Finally, our distrust of differential systems is supported by the fact that practical computations of symmetries, conservation laws and other quantities characterizing PDEs become much more complicated in terms of differential systems.

4.5 Singularities of generalized solutions The concept of generalized solutions for NPDEs, which is important in itself, naturally leads to an important part of a general theory of PDEs, namely, the theory of singularities of generalized solutions. Below we shall outline some key points of this theory.

Let $N \subset J^k(E,n)$ be an R-manifold. A point $\theta \in N$ is called *singular of type* s if the kernel of the differential $d_\theta \pi_{k,k-1}$ restricted to $T_\theta N$ is of dimension $s > 0$. Otherwise, θ is called *regular*. It should be stressed here that "singular" refers to singularities of the map $\pi_{k,k-1}|_N$, while, by definition, N is a smooth submanifold. According to Proposition 4.1, (6), N is of the form $L_{(k)}$ in a neighborhood of any regular point.

Put $F_\theta = (\pi_{k,k-1})^{-1}(\pi_{k,k-1}(\theta))$ (the fiber of $\pi_{k,k-1}$ passing through θ) and $VC^k_\theta = C^k_\theta \cap T_\theta(F_\theta)$. The *bend* of N at a point $\theta \in N$ is

$$B_\theta N \overset{\text{def}}{=} \ker d_\theta \left(\pi_{k,k-1} \mid_N\right) = T_\theta N \cap T_\theta(F_\theta) \subset VC^k_\theta.$$

Also, we shall call an *s-bend* (at $\theta \in J^k(E,n)$) an s-dimensional subspace of VC^k_θ, which is of the form $B_\theta N$ for some R-manifold N. Bends are very special subspaces in VC^k_θ. A remarkable fact is that s-dimensional bends are classified by s-dimensional Jordan algebras of a certain class over $\mathbb{R}$, which contains all unitary algebras (see [64, 68]).

> *PDEs differ from each other by the types of singularities which their generalized solutions admit.*

For instance, 2-dimensional Jordan algebras associated with 2-dimensional bends are 2-dimensional unitary algebras and hence are isomorphic to one of the following three algebras

$$\mathbb{C}_\epsilon = \{a + b\zeta \mid a, b \in \mathbb{R},\ \zeta^2 = \epsilon 1\} \quad \text{with} \quad \epsilon = \pm 1 \text{ or } 0.$$

Obviously, $\mathbb{C}_- = \mathbb{C}$ and $\mathbb{C}_+ = \mathbb{R} \oplus \mathbb{R}$ (as algebras). An equation in two independent variables is *elliptic* (resp., *parabolic* or *hyperbolic*) if its generalized solutions possess singularities of type $\mathbb{C}_-$ (resp., $\mathbb{C}_0$, $\mathbb{C}_+$) only. Geometrically, singularities corresponding to the algebra $\mathbb{C}$ are Riemann ramifications, while bicharacteristics of hyperbolic equations reflect the fact that $\mathbb{C}_+$ splits into the direct sum $\mathbb{R} \oplus \mathbb{R}$.

Obviously, the simplest singularities correspond to the algebra $\mathbb{R}$. They present a kind of folding and can be *analytically detected* in terms of non-uniqueness of Cauchy data. A similar analytic approach is hardly possible for more complicated algebras. This explains why analogues of the classical subdivision of PDEs in two independent variables into elliptic, parabolic and hyperbolic ones are not yet known. This fact emphasizes once again that only analytical methods for PDEs, even linear ones, are not sufficient and the geometrical approach is indispensable.

The description of singularities that solutions of a given PDE admit is naturally settled as follows. Let Σ be a type of s-bends, which may be identified with the corresponding Jordan algebra. If N is an R-manifold, then

$$N_\Sigma = \{\theta \in N \mid B_\theta N \text{ is of the type } \Sigma\}$$

is the locus of its singular points of type Σ. Generally, $\dim N_\Sigma = n - s$. If N is a solution of a PDE $\mathcal{E}$, then N_Σ must satisfy an auxiliary system of PDEs, which we denote by $\mathcal{E}_\Sigma$. For "good" equations $\mathcal{E}_\Sigma$ is, generally, a nonlinear, undetermined system of PDEs in $n - s$ independent variables.

4.6 The reconstruction problem So, any PDE is not a single but is surrounded by an "aura" of subsidiary equations, which put in evidence the internal structure of its solutions. The importance of these equations becomes especially clear in the light of the *reconstruction problem*:

Whether the behavior of singularities of solutions of a PDE $\mathcal{E}$ uniquely determines the equation itself or, equivalently, whether it is possible to reconstruct $\mathcal{E}$ assuming that the $\mathcal{E}_\Sigma$s are known.

In a physical context this question sounds as

Whether the behavior of singularities of a field (medium, etc.) completely determines the behavior of the field (medium, etc.) itself?

A remarkable example of this kind is the deduction of Maxwell's equations from the elementary laws of electricity and magnetism (Coulomb, ..., Faraday) (see [35]).

The reconstruction problem can be solved positively for hyperbolic NPDEs on the basis of equations $\mathcal{E}_{\mathrm{FOLD}}$ that describe singularities corresponding to the algebra $\mathbb{R}$. The equations describing wave fronts of solutions of a linear hyperbolic PDE $\mathcal{E}$ are part of the system $\mathcal{E}_{\mathrm{FOLD}}$.

Example 4.1. *Fold-type singularities for the equation* $u_{xx} - \frac{1}{c^2}u_{tt} - mu^2 = 0$.
Consider wave fronts of the form $x = \varphi(t)$ and put

$$g = u|_{\text{wave front}}, \quad h = u_x|_{\text{wave front}}.$$

Then we have

$$\begin{cases} \ddot{g} + (cm)^2 g = \pm 2c\dot{h} \\ 1 - \frac{1}{c^2}\dot{\varphi}^2 = 0 \Leftrightarrow \dot{\varphi} = \pm c \end{cases} \Leftarrow \left[\begin{array}{c} \text{Equations describing } \mathit{the} \\ \text{behavior of fold-type} \\ \text{singularities} \end{array} \right.$$

The second of these equations is of eikonal type and describes the space-time shapes of singularities. On the contrary, the first equation describes a "particle" in the "field" h. If this field is constant $\Leftrightarrow$ $\dot{h} = 0$, then the first equation represents a harmonic oscillator of frequency $\nu = mc$.

Example 4.2. *Fold-type singularities for the Klein–Gordon equation*

$$(\partial_t^2 - \vec{\nabla}^2 + m^2)u = 0.$$

Consider wave fronts of the form $t = \varphi(x_1, x_2, x_3)$ and g and h as in Example 4.1

$$\mathcal{E}_{FOLD} = \begin{cases} (\vec{\nabla}\varphi)^2 = 1 \leftarrow \quad \text{eikonal type equation} \\ \nabla^2 h + m^2 h - g - (\nabla^2\varphi)g = 2\vec{\nabla}\varphi \cdot \vec{\nabla} g \leftarrow ??? \end{cases}$$

The physical meaning of the second of these equations is unclear.

Example 4.3. The classical Monge–Ampère equations are defined as equations of the form

$$S(u_{xx}u_{yy} - u_{xy}^2) + Au_{xx} + Bu_{xy} + Cu_{yy} + D = 0$$

with S, A, B, C, D being functions of x, y, u, u_x, u_y (see [33]). As was already observed by S. Lie this class of equations is invariant with respect to contact transformations. This fact forces to think that Monge–Ampère equations are distinguished by some "internal" property. This is the case, and Monge–Ampère equations are completely characterized by the fact that the reconstruction problem for these equations is equivalent to a problem in contact geometry (see [8, 39]).

The reader will find in [36] further details and examples concerning the auxiliary singularities equations. Some exact generalized solutions of Einstein equations (the "square root" of the Schwarzshild solution, etc.) are described in [49].

4.7 Quantization as a reconstruction problem Let $\mathcal{E}$ be a PDE, whose solutions admit fold-type singularities. Then we have the following series of interconnected equations:

$$\mathcal{E} \Longrightarrow \mathcal{E}_{\mathrm{FOLD}} \Longrightarrow \mathcal{E}_{\mathrm{eikonal}} \Longrightarrow \mathcal{E}_{\mathrm{char}}. \tag{6.12}$$

Here $\mathcal{E}_{\mathrm{eikonal}}$ is the equation from the system $\mathcal{E}_{\mathrm{FOLD}}$ that describes space-time shape ("wave front") of fold-type singularities. It is a Hamilton–Jacobi equation (see Examples 4.1 and 4.2). In its turn $\mathcal{E}_{\mathrm{char}}$ is the system of ODEs that describes *characteristics* of $\mathcal{E}_{\mathrm{eikonal}}$. In the context where space-time coordinates are independent variables, $\mathcal{E}_{\mathrm{eikonal}}$ is a Hamiltonian system whose Hamiltonian is the main symbol of $\mathcal{E}$. Now we see that the correspondence

$$\mathrm{CHAR} : \mathcal{E}\ (\mathrm{PDE}) \Longrightarrow \mathcal{E}_{\mathrm{char}}\ (\text{Hamiltonian system of ODEs}) \tag{6.13}$$

is parallel to the correspondence between quantum and classical mechanics

$$\mathrm{BOHR} : (\text{Schrödinger's PDE}) \Longrightarrow (\text{Hamiltonian ODEs}). \tag{6.14}$$

Moreover, the correspondence (6.13) is at the root of the famous "optics-mechanics analogy", which guided E. Schrödinger in his discovery of the "Schrödinger equation" (see Schrödinger's Nobel lecture [47]).

It is remarkable that in "Cauchy data" terms, the correspondence (6.13) was known already to T. Levi-Civita and he tried to put it at the foundations of quantum mechanics (see [34]). From what is known today this attempt was doomed to failure. However, the idea that quantization is something like the reconstruction problem explains well why numerous quantization procedures proposed up to now form a kind of recipe book not based on some universal principles. Indeed, from this point of view the quantization looks like an attempt to restore the whole system $\mathcal{E}_{\mathrm{FOLD}}$ on the basis of knowledge of $\mathcal{E}_{\mathrm{char}}$ only. This is manifestly impossible, since $\mathcal{E}_{\mathrm{char}}$ depends only on the main symbol of $\mathcal{E}$. On the other hand, the above outlined solution singularity theory admits some interesting generalizations and refinements, which not only keep alive the Levi-Civita idea but even make it more attractive.

4.8 Higher order contact transformations and the Erlangen program The above interpretation of PDEs as submanifolds of higher order contact manifolds is the first step toward a "conceptualization" of the standard approach to PDEs. It is time now to test its validity through the philosophy of the Erlangen program. First of all, this means that we have to describe the symmetry group of higher contact geometries, i.e., the group of *higher contact transformations*.

Definition 4.4. A diffeomorphism/transformation $\Phi : J^k(E,n) \to J^k(E,n)$ is called a *k-order contact* if for any $X \in \mathcal{C}^k$, $\Phi(X) \in \mathcal{C}^k$ or, equivalently, $d_\theta \Phi(\mathcal{C}^k_\theta) = \mathcal{C}^k_{\Phi(\theta)}$, $\forall \theta \in \Phi$.

If Φ is a k-th order contact, then, obviously, it preserves the class of locally maximal integral submanifolds of type s. In particular, it preserves the fibers of the projection $\pi_{k,k-1}$ and hence the locally maximal integral submanifolds of type n that are transversal to these fibers. But the latter are locally of the form $L_{(k)}$ (Proposition 4.1, (6)). This proves that the differential of Φ sends R-planes into R-planes. By identifying these R-planes with points of $J^{k+1}(E,n)$ we see that Φ induces a diffeomorphism $\Phi_{(1)}$ of $J^{k+1}(E,n)$. More exactly, if $\theta \in J^{k+1}(E,n)$ and $\theta' = \pi_{k+1,k}(\theta)$, then $(d_{\theta'}\Phi)(R_\theta)$ is an R-plane and hence is of the form R_ϑ for a $\vartheta \in J^{k+1}(E,n)$. Then we put $\Phi(\theta) = \vartheta$. Moreover, it directly follows from Proposition 4.1, (1), that $\Phi_{(1)}$ is a $(k+1)$-order contact and the diagram

$$\begin{array}{ccc} J^{k+1}(E,n) & \stackrel{\Phi_{(1)}}{\longrightarrow} & J^{k+1}(E,n) \\ \downarrow \pi_{k+1,k} & & \downarrow \pi_{k+1,k} \\ J^k(E,n) & \stackrel{\Phi}{\longrightarrow} & J^k(E,n) \end{array}$$

commutes. By continuing this process we, step by step, construct contact transformations

$$\Phi_{(l)} : J^{k+l}(E,n) \stackrel{F_{(1)}}{\longrightarrow} J^{k+l}(E,n), \quad \Phi_{(l)} \stackrel{\text{def}}{=} (\Phi_{(l-1)})_{(1)}.$$

Theorem 4.4. *Let* $\Phi : J^k(E,n) \to J^k(E,n)$, $k > 0$, *be a k-order contact transformation. Then* $\Phi = \Psi_{(l)}$ *(resp.,* $\Phi = \Psi_{(l-1)}$ *) where* Ψ *is a diffeomorphism of E if* $m > 1$ *(resp., a contact transformation of* $J^1(E,n)$ *if* $m = 1$*).*

A proof of this fundamental result for the classical symmetry theory can be easily deduced from the fact explained above that a k-th order contact transformation preserves fibers of $\pi_{k,k-1}$ and hence induces a $(k-1)$-th order contact transformation of $J^{k-l}(E,n)$. For $m = 1$ this was proven by Lie and Bäcklund (see [60, 29]).

If one takes Definition 3.2 for a true definition of PDEs, then the definition of a symmetry of a PDE should be

Definition 4.5. A symmetry of a PDE $\mathcal{E} \subset J^k(E,n)$ is

(1) a k-th order contact transformation $\Phi : J^k(E,n) \to J^k(E,n)$ such that $\Phi(\mathcal{E}) = \mathcal{E}$ (à la S. Lie);

(2) a diffeomorphism $\Psi : \mathcal{E} \to \mathcal{E}$ preserving the distribution $\mathcal{C}^k_\mathcal{E}$ (à la É. Cartan).

The rigidity theory shows that Definitions (1) and (2) are equivalent for rigid PDEs, i.e., for almost all PDEs of practical interest. Moreover, by Theorem 4.4, Definitions 4.5 and 3.3 are equivalent in this case too.

Remark 4.1. There are analogues of Theorem 4.4 and Definition 4.5 for infinitesimal k-order contact transformations and symmetries. They do not add anything new to our discussion, and we shall skip them.

In the light of the "Erlangen philosophy" the result of Theorem 4.4 looks disappointing. Indeed, it tells us that the group of k-order contact transformations coincides with the group of first order transformations. So, higher order contact geometries are governed by the same group as the classical one. This does not meet a natural expectation that transformations of higher order geometries should form some larger groups. Hence, by giving credit to this philosophy, we are forced to conclude that

> *Definition* 3.2 *or what is commonly meant by a differential equation is not a conceptual definition but should be considered just as a description of an object, whose nature must be still discovered.*

So, the question of what object is hidden under this description is to be investigated. One rather evident hint is to examine the remaining case $k = \infty$. This is psychologically difficult, since $J^{\infty}(E, n)$ being an infinite-dimensional manifold of a certain kind does not possess any "good" topology or norm, etc. which seem indispensable for the existence of a "good" differential calculus on it. Another hint comes from the principle "chercher la symétrie". For instance, if $\mathcal{E}$ (resp., $\square$) is a linear equation (resp., a linear differential operator) with constant coefficients, then $\square$ sends solutions of $\mathcal{E}$ to the solutions. For this reason $\square$ may be considered as a symmetry of $\mathcal{E}$, finite or infinitesimal. Symmetries of this kind are not, generally, classical and their analytical description involves partial derivatives of any order. Hence one may expect that something similar takes place for general PDEs, and we are going to show that this is the case.

5 From integrable systems to diffieties and higher symmetries

5.1 New experimental data: integrable systems The discovery in the late 1960s of some remarkable properties of the now famous Korteweg-de Vries equation and later of other *integrable systems* brought to light various new facts, which had no conceptual explanation in terms the classical symmetry theory. In particular, any such equation is included in an infinite series of similar equations, the *hierarchy*, which are interpreted as commuting Hamiltonian flows with respect to an, in a sense, infinite-dimensional Poisson structure. For this reason equations of this hierarchy may be considered as infinitesimal symmetries of each other. Moreover, they involves derivatives of any order and hence are outside the classical theory (see [71]). Therefore, attempts to include these non-classical symmetries in common with classical symmetries frames directly leads to infinite jets.

5.2 Infinite jets and infinite order contact transformations Recall that the Cartan distribution $\mathcal{C}^\infty$ on $J^\infty(E,n)$ is (paradoxically!) n-dimensional and completely integrable (Proposition 4.1, (5)). A consequence of this fact is that locally maximal integral submanifolds of $\mathcal{C}^\infty$ are of the same type in sharp contrast with finite-order contact geometries (Proposition 4.2). This is a weighty argument in favor of infinite jets. After that we have to respond to the question of whether the group of infinite-order contact transformations is broader than the group of classical ones. More exactly, we ask whether there are infinite-order contact transformations that are not of the form $\Phi_{(\infty)}$ where Φ is a finite-order contact transformation (see Theorem 4.4). Here $\Phi_{(\infty)}$ stands for the direct limit of the $\Phi_{(l)}$s. The answer is positive: this (local) group consists of all *invertible differential operators* (in the generalized sense outlined above) acting on n-dimensional submanifolds of E. These operators involve partial derivatives of arbitrary orders and in this sense they justify the credit given to infinite jets. We shall skip the details (see [61]), since the same question about infinite order infinitesimal symmetries is much more interesting from the practical point of view and at the same time it reveals some unexpected a priori details, which become essential for further discussion.

Recall that an infinitesimal symmetry of a distribution $\mathcal{C}$ on a manifold M is a vector field $X \in D(M)$ such that $[X,Y] \in \mathcal{C}$ if $Y \in \mathcal{C}$ (symbolically, $[X,\mathcal{C}] \subset \mathcal{C}$). Infinitesimal symmetries form a subalgebra in $D(M)$ denoted $D_{\mathcal{C}}(M)$. The flow generated by a field $X \in D_{\mathcal{C}}(M)$ moves, if it is globally defined, (maximal) integral submanifolds of $\mathcal{C}$ into themselves. If it is not globally defined this flow moves only sufficiently small pieces of integral submanifolds. In this sense we can speak of a local flow in the "space of (maximal) integral submanifolds of $\mathcal{C}$".

If the distribution $\mathcal{C}$ is integrable/Frobenius, then it may be interpreted as a foliation whose leaves are its locally maximal integral submanifold. In this case $\mathcal{C}$ is an ideal in $D_{\mathcal{C}}(M)$. If $N \subset M$ is a leaf of $\mathcal{C}$, then any $Y \in \mathcal{C}$ is tangent to N and, therefore, the flow generated by Y leaves N invariant, i.e., any leaf of $\mathcal{C}$ slides along itself under the action of this flow. We may interpret this fact by saying that the local flow generated by Y on the "space of all leaves of $\mathcal{C}$" is trivial. This is, obviously, no longer so if $Y \in D_{\mathcal{C}}(M) \setminus \mathcal{C}$. Hence the flow generated by Y in the "space of all leaves of $\mathcal{C}$" is uniquely defined by the coset $[Y \mod \mathcal{C}]$, and the quotient Lie algebra

$$\operatorname{Sym} \mathcal{C} \stackrel{\text{def}}{=} \frac{D_{\mathcal{C}}(M)}{\mathcal{C}}, \tag{6.15}$$

called the *symmetry algebra of* $\mathcal{C}$, is naturally interpreted as the algebra of *vector fields on the "space of leaves" of* $\mathcal{C}$. It should be stressed that it would be rather counterproductive to try to give a rigorous meaning to the "space of leaves". On the contrary, the above interpretation of the quotient algebra (6.15) is very productive and may be interpreted as the smile of the Cheshire Cat.

Now we shall apply the above construction to the distribution $\mathcal{C}^\infty$ and introduce for this special case the following notation:

$$\mathcal{C}D(J^\infty(E,n)) = \mathcal{C}^\infty, \quad D_{\mathcal{C}}(J^\infty(E,n)) = D_{\mathcal{C}^\infty}(J^\infty(E,n)), \quad \varkappa = \operatorname{Sym}\mathcal{C}^\infty. \tag{6.16}$$

The Lie algebra $\varkappa$ will play a prominent role in our subsequent investigation. At the moment we know that it is the "algebra of vector fields on the space of all locally maximal integral submanifolds of $J^\infty(E,n)$". As a first step we have to describe $\varkappa$ in coordinates. However, in order to do that with due rigor we have to clarify before what is differential calculus on infinite-dimensional manifolds of this kind. It is rather obvious that the usual approaches based on "limits", "norms", etc. cannot be applied to this situation. Therefore, we need the following digression.

5.3 On differential calculus over commutative algebras Let A be a unitary, i.e., commutative and with unit, algebra over a field $\boldsymbol{k}$ and P and Q be some A-modules.

Definition 5.1. $\Delta : P \longrightarrow Q$ is a *linear differential operator (DO)* of order $\leq m$ if Δ is $\boldsymbol{k}$-linear and $[a_0, [a_1, \ldots, [a_m, \Delta] \ldots]] = 0$, $\forall a_0, a_1, \ldots, a_m \in A$.

Elements $a_i \in A$ figuring in the above multiple commutator are understood as the multiplication by a_i operators.

If $A = C^\infty(M)$, $P = \Gamma(\pi)$, $Q = \Gamma(\eta)$ with π, η being some vector bundles, then Definition 5.1 is equivalent to the standard one. The "logic" of differential calculus is formed by *functors of differential calculus* together with their natural transformations and representing them objects in a *differentially closed* category of A-modules [54, 60, 59]. In particular, this allows one to construct analogues of all known structures in differential geometry, say, tensors, connections, de Rham and Spencer cohomology, an so on, over an arbitrary unitary algebra. The reader will find in [42] an elementary introduction to this subject based on a *physical motivation*.

By applying this approach to the filtered algebra $\mathcal{F}_\infty = \{\mathcal{F}_i\}$ (see (6.9)) we shall get all necessary instruments to develop differential calculus on spaces $J^\infty(E.n)$ and, more generally, on *diffieties* (see below). The informal interpretation of the filtered algebra $\mathcal{F}_\infty$ as the smooth function algebra on the "cofiltered manifold" $J^\infty(E.n)$ helps to keep the analogy with the calculus on smooth manifolds under due control. In coordinates an element of $\mathcal{F}_\infty$ looks as a function of a finite number of variables x_i and u^j_σ. This reflects the fact that any "smooth function" on $J^\infty(E.n)$ is, by definition, a smooth function on a certain $J^k(E.n)$, $k < \infty$, pulled back onto $J^\infty(E.n)$ via $\pi_{\infty,k}$. Therefore, the filtered structure of $\mathcal{F}_\infty$ is essential and differential operators $\Delta : \mathcal{F}_\infty \longrightarrow \mathcal{F}_\infty$ (in the sense of definition 5.1) must respect it. This means that $\Delta(\mathcal{F}_k) \subset \mathcal{F}_{k+s}$ for some s. In particular, a vector field on $J^\infty(E.n)$ is defined as a derivation of $\mathcal{F}_\infty$, which respects, in this sense, the filtration. In coordinates such a vector field looks as an infinite series

$$X = \sum_i \alpha_i \frac{\partial}{\partial x_i} + \sum_{j,\sigma} \beta_\sigma^j \frac{\partial}{\partial u_\sigma^j}, \quad \phi_i, \psi_\sigma^j \in \mathcal{F}_\infty. \tag{6.17}$$

The $\mathcal{F}_\infty$-module of vector fields on $J^\infty(E.n)$ will be denoted by $D(J^\infty(E.n))$.

5.4 The algebra $\varkappa$ in coordinates Since $\mathcal{C}^\infty$ is an $\mathcal{F}_\infty$-module generated by the vector fields D_i (Proposition 4.1, (5)), it is convenient to represent a vector field $X \in D(J^\infty(E.n))$ in the form

$$X = \sum_i \psi_i D_i + \sum_{j,\sigma} \varphi_\sigma^j \frac{\partial}{\partial u_\sigma^j}, \quad \psi_i, \varphi_\sigma^j \in \mathcal{F}_\infty \tag{6.18}$$

where the first summation, which belongs to $\mathcal{C}^\infty$, is the *horizontal* part of X, while the second one is its *vertical* part. This splitting of a vector field into horizontal and vertical parts is unique but depends on the choice of coordinates. Obviously, the coset $[X \mod \mathcal{C}^\infty]$ is uniquely characterized by the vertical part of X.

Below we use the notation $D_\sigma \stackrel{\text{def}}{=} D_1^{\sigma_1} \cdot \dots \cdot D_n^{\sigma_n}$ for a multiindex $\sigma = (\sigma_1, \dots, \sigma_n)$.

Proposition 5.1.

1. $\varkappa$ is an $\mathcal{F}_\infty$-module and $\partial/\partial u^1, \dots, \partial/\partial u^m$ is its local basis in the chart U with coordinates $(\dots, x_i, \dots, u_\sigma^j, \dots)$;
2. the correspondence
$$(\mathcal{F}_\infty^m)_U \ni \varphi = (\varphi^1, \dots, \varphi_m) \Leftrightarrow \partial_\varphi = \sum D_\sigma(\varphi_i) \frac{\partial}{\partial u_\sigma^i} \in \varkappa_U$$
is an isomorphism of $\mathcal{F}_\infty$-modules localized to the chart U;
3. the Lie algebra structure $\{\cdot,\cdot\}$ in $\varkappa_U$ is given by the formula
$$\{\varphi, \psi\} = \partial_\varphi(\psi) - \partial_\psi(\varphi), \quad [\partial_\varphi, \partial_\psi] = \partial_{\{\varphi,\psi\}};$$
4. $(f, \partial_\varphi) \mapsto \partial_{f\varphi}$ is the $(\mathcal{F}_\infty)_U$-module product in $\varkappa_U$.

The vector fields ∂_φs locally representing elements of the module $\varkappa$ are called *evolutionary derivations*, and φ is called the *generating function* of ∂_φ. The bracket $\{\cdot,\cdot\}$ introduced for the first time in [58] (see also [61, 29]) is a generalization of both the Poisson and the contact brackets. Indeed, these are particular cases where $m = 1$ and the generating functions depend only on the x_is and on the first derivatives and in the contact case also of u. If Y is a vector field on E $(m > 1)$ or a contact vector field on $J^1(E,n)$ $(m = 1)$ and $Y_{(\infty)}$ is its lift to $J^\infty(E,n)$, then $Y_{(\infty)} \in D_\mathcal{C}(J^\infty(E,n))$

and the composition $Y \mapsto Y_{(\infty)} \mapsto [Y_{(\infty)} \mod \mathcal{C}D(J^\infty(E,n))] \in \varkappa$ is injective. Therefore, infinitesimal point and contact transformations are naturally included in $\varkappa$. Their generating functions depends only on x, u and first derivatives, and we see that the Lie algebra $\varkappa$ is much larger than the algebras of infinitesimal point and contact transformations. Hence the passage to infinite jets is in fairly good accordance with the "Erlangen philosophy". But in order to benefit from this richness of infinite order contact transformations we must bring PDEs in the context of infinite order contact geometry. But in that case we cannot mimic Definition 3.2, since, in sharp contrast with finite order jet spaces, an arbitrary submanifold of $S \subset J^\infty(E,n)$ cannot be interpreted as a PDE. Indeed, the restriction of $\mathcal{C}^\infty$ to S is, generally, not n-dimensional, while we need n-dimensional integral submanifolds to define the solutions. So, we must concentrate on those submanifolds S to which $\mathcal{C}^\infty$ is tangent, i.e., such that $\mathcal{C}^\infty_\theta \subset T_\theta S$, $\forall \theta \in S$. These are obtained by means of the *prolongation procedure*.

5.5 Prolongations of PDEs and diffieties Let $\mathcal{E} \subset J^k(E,n)$ be a PDE in the sense of Definition 3.2 and $N \subset \mathcal{E}$ be its solution (Definition 4.3). Then, obviously, $T_\theta N \subset \mathcal{E}_\theta$, $\forall \theta \in N$. Therefore, if $\mathcal{E}$ admits a solution passing through a point $\theta \in \mathcal{E}$, then there is at least one R-plane at θ, which is tangent to $\mathcal{E}$. Since any R-plane is of the form R_ϑ, $\pi_{k+1,k}(\vartheta) = \theta$, the variety of all R-planes tangent to $\mathcal{E}$ is identified with the submanifold (probably, with singularities)

$$\mathcal{E}_{(1)} \stackrel{\text{def}}{=} \{\vartheta \in J^{k+1}(E,n) \mid R_\vartheta \text{ is tangent to } \mathcal{E}\} \subset J^{k+1}(E,n).$$

Therefore, tautologically, a solution of $\mathcal{E}$ passes only through points of $\pi_{k+1,k}(\mathcal{E}_{(1)}) \subset \mathcal{E}$. In other words, a solution of $\mathcal{E}$ is automatically a solution of $\pi_{k+1,k}(\mathcal{E}_{(1)})$. Hence by substituting $\pi_{k+1,k}(\mathcal{E}_{(1)})$ for $\mathcal{E}$ we eliminate "parasitic" points. Moreover, by construction, if $L_{(k)} \subset \mathcal{E}$, then $L_{(k+1)} \subset \mathcal{E}_{(1)}$ and vice versa. Hence $\mathcal{E}$ and $\mathcal{E}_{(1)}$ have common "usual" solutions but $\mathcal{E}_{(1)}$ is without "parasitic" points of $\mathcal{E}$. By continuing this process of elimination of "parasitic" points we inductively construct successive *prolongations* $\mathcal{E}_{(r)} \stackrel{\text{def}}{=} \left(\mathcal{E}_{(r-1)}\right)_{(1)}$ of $\mathcal{E}$. In this way we get an infinite series of equations, which have common "usual" solutions:

$$\mathcal{E} = \mathcal{E}_{(0)} \stackrel{\pi_{k+1,k}}{\longleftarrow} \mathcal{E}_{(1)} \stackrel{\pi_{k+2,k+1}}{\longleftarrow} \mathcal{E}_{(2)} \stackrel{\pi_{k+3,k+2}}{\longleftarrow} \dots, \quad \text{with} \quad \mathcal{E}_{(r)} \subset J^{k+r}. \tag{6.19}$$

The inverse limit $\mathcal{E}_\infty$ of the sequence (6.19) called the *infinite prolongation* of $\mathcal{E}$ is a submanifold of $J^\infty(E,n)$ (in the same sense as the latter) and one of the results of the *formal theory* of PDEs tells:

Proposition 5.2. (see [23, 50, 44, 30]) If the distribution $\mathcal{C}^\infty$ is tangent to a submanifold $S \subset J^\infty(E,n)$, then $S = \mathcal{E}_\infty$ for a PDE $\mathcal{E}$.

In coordinates, prolongations of $\mathcal{E}$ are described as follows

$$\mathcal{E}_{(2)} = \left\{ \mathcal{E}_{(1)} = \left\{ \begin{array}{c} \mathcal{E} = \{F_s(x,u,\dots,u^j_\sigma,\dots) = 0,\ s = 1,\dots,l\} \\ D_i F_s = 0 \end{array} \right\} \atop D_i D_j F_s = 0 \right\} \tag{6.20}$$

$$\dots$$

$$\Downarrow$$

$$\mathcal{E}_\infty = \{D_\sigma F_s = 0, \forall s, \sigma\}$$

Remark 5.1. $\mathcal{E}_\infty$ may be empty.

The algebra $\mathcal{F}_\infty(\mathcal{E}) \stackrel{\text{def}}{=} \mathcal{F}|_{\mathcal{E}_\infty}$ plays the role of the smooth function algebra on $\mathcal{E}_\infty$. It is a filtered algebra

$$\mathcal{F}_0(\mathcal{E}) \subset \dots \subset \mathcal{F}_s(\mathcal{E}) \subset \dots \mathcal{F}_\infty(\mathcal{E}) \text{ with } \mathcal{F}_s(\mathcal{E}) = \mathrm{Im}(C^\infty(\mathcal{E}_{(s)}) \stackrel{\pi^*_{\infty,k+s}}{\longrightarrow} \mathcal{F}_\infty(\mathcal{E})). \tag{6.21}$$

As in the case of infinite jets differential calculus on $\mathcal{E}_\infty$ is understood as differential calculus over the filtered algebra $\mathcal{F}_\infty(\mathcal{E})$.

Thus we have constructed the central object of the general theory of PDEs.

Definition 5.2. The pair $(\mathcal{E}_\infty, \mathcal{C}^\infty_\mathcal{E})$ with $\mathcal{C}^\infty_\mathcal{E} \stackrel{\text{def}}{=} \mathcal{C}^\infty\,|_{\mathcal{E}_\infty}$ is called the *diffiety* associated with $\mathcal{E}$.

The distribution $\mathcal{C}^\infty_\mathcal{E}$ is n-dimensional, since $\mathcal{C}^\infty$ is tangent to $\mathcal{E}_\infty$. The projection $\pi_{\infty,k}$ establishes a one-to-one correspondence between integral submanifolds of $\mathcal{C}^\infty_\mathcal{E}$ and those of $\mathcal{C}^k\,|_\mathcal{E}$, which are transversal to fibers of $\pi_{k,k-1}$. Thus, n-dimensional integral submanifolds of $\mathcal{C}^\infty_\mathcal{E}$ are identified with non-singular solutions of $\mathcal{E}$.

The following interpretation, even though absolutely informal, is a very good guide in the task of deciphering the native language that NPDEs speak and, therefore, in terms of which they can be only understood adequately:

> *The diffiety associated with a PDE $\mathcal{E}$ (in the standard sense of this term) is the* space of all solutions of $\mathcal{E}$.

Remark 5.2. The reader may have already observed that nontrivial generalized solutions of $\mathcal{E}$ cannot be interpreted as integral submanifolds of $\mathcal{C}^\infty_\mathcal{E}$ and hence the diffiety $(\mathcal{E}, \mathcal{C}^\infty_\mathcal{E})$ is not the "space of all solutions of $\mathcal{E}$". However, this is not a conceptual defect, since this diffiety can be suitably completed.

As a rule, diffieties are infinite-dimensional. Diffieties of finite dimension are foliations, probably, with singularities. Diffieties associated with determined and overdetermined systems of ordinary differential equations (ODEs) are 1-dimensional foliations on finite-dimensional manifolds. On the contrary, diffieties associated with underdetermined systems of ODEs are infinite-dimensional. A good part of control theory is naturally interpreted as the structural theory of this kind of diffieties (see [13]).

5.6 Higher infinitesimal symmetries of PDEs Now having in hands the concept of diffiety we can extend the classical symmetry theory described above by including in it the new already mentioned "experimental data" that come from the theory of integrable systems. To this end it is sufficient to apply the same approach we have used to understand what are infinite-order infinitesimal contact transformations.

As before, by abusing language, we shall denote the $\mathcal{F}_\infty(\mathcal{E})$-module of vector fields on $\mathcal{E}_\infty$ belonging to $\mathcal{C}_{\mathcal{E}}^\infty$ by the same symbol $\mathcal{C}_{\mathcal{E}}^\infty$. Since the distribution $\mathcal{C}^\infty$ is tangent to $\mathcal{E}_\infty$, vector fields D_is are also tangent to $\mathcal{E}_\infty$. For this reason restrictions of the D_i to $\mathcal{E}_\infty$ are well-defined, Denote them by $\bar{D}_i$. The Lie algebra of infinitesimal transformations preserving the distribution $\mathcal{C}_{\mathcal{E}}^\infty$ is

$$D_{\mathcal{C}}(\mathcal{E}_\infty) \stackrel{\text{def}}{=} \{X \in D(\mathcal{E}_\infty) \mid [X, Y] \in \mathcal{C}_{\mathcal{E}}^\infty, \ \forall Y \in \mathcal{C}_{\mathcal{E}}^\infty\}. \tag{6.22}$$

Now the Lie algebra of *infinitesimal higher symmetries* of a PDE $\mathcal{E}$ is defined as

$$\boxed{\operatorname{Sym}\mathcal{E} = \frac{D_{\mathcal{C}}(\mathcal{E}_\infty)}{\mathcal{C}_{\mathcal{E}}^\infty}} \tag{6.23}$$

This definition merits some comments. First, we use the adjective "higher" to stress the fact that generating functions of elements of the algebra $\operatorname{Sym}\mathcal{E}$ may depend, contrary to the classical symmetries, on arbitrary order derivatives. Next, in conformity with the above interpretation of the diffiety $(\mathcal{E}, \mathcal{C}_{\mathcal{E}}^\infty)$, the informal interpretation of Definition (6.23) is :

> *Elements of the Lie algebra* $\operatorname{Sym}\mathcal{E}$ *are* vector fields on the "space of all solutions of $\mathcal{E}$".

The importance of this interpretation is that it forces the question:

> *What are tensors, differential operators, PDEs, etc. on the "space of all solutions of* $\mathcal{E}$*".*

Later we shall give some examples and indications on how to define and use this kind of objects. These objects form the thesaurus of *secondary calculus*, which is a natural language of the general theory of PDEs (see [66, 27, 29]).

Finally, note that higher symmetries are not genuine vector fields as in the classical theory but just some cosets of them modulo $\mathcal{C}_{\mathcal{E}}^\infty$. For this reason their action on functions on $\mathcal{E}_\infty$ is not even defined. This at first glance discouraging fact leads to the bifurcation point: either to give up or to understand what are *functions on the* *"space of solutions of $\mathcal{E}$"*. Since, as we shall see, Definition (6.23), works well, the first alternative should be discarded, while the second one will lead us to discover *differential forms on the "space of solutions of $\mathcal{E}$"*.

5.7 Computation of higher symmetries Though elements of $\varkappa$ are cosets of vector fields modulo $\mathcal{E}_\infty$ we can say that $\chi = [X] \in \varkappa$ is tangent to $\mathcal{E}_\infty$ if the vector field X is tangent to $\mathcal{E}_\infty$. Since $\mathcal{C}_\infty$ is tangent to $\mathcal{E}_\infty$, this definition is correct. If $\mathcal{E}_\infty$ is locally given by equations (6.19) and χ by the evolutionary derivation $\ni_\varphi$, then χ is tangent to $\mathcal{E}_\infty$ if and only if $\ni_\varphi(D_\sigma(F_s))\,|_{\mathcal{E}_\infty} = 0$, $\forall \sigma, s$. Since $\ni_\varphi$ and the D_i commute these conditions are equivalent to $\ni_\varphi(F_s)\,|_{\mathcal{E}_\infty} = 0$, $\forall s$, or, in short, to $\ni_\varphi(F)\,|_{\mathcal{E}_\infty} = 0$ with $F = (F_1, \dots, F_\ell)$. The bidifferential operator $(\varphi, F) \mapsto \ni_\varphi(F)$ may be rewritten in the form $\ni_\varphi(F) = \ell_F(\varphi)$ with

$$\ell_F = \begin{pmatrix} \sum_\sigma \frac{\partial F_1}{\partial u^1_\sigma} D_\sigma & \dots & \sum_\sigma \frac{\partial F_1}{\partial u^m_\sigma} D_\sigma \\ \vdots & & \vdots \\ \sum_\sigma \frac{\partial F_l}{\partial u^1_\sigma} D_\sigma & \dots & \sum_\sigma \frac{\partial F_l}{\partial u^m_\sigma} D_\sigma \end{pmatrix} \tag{6.24}$$

and ℓ_F is called the *universal linearization operator*. Being tangent to $\mathcal{E}_\infty$ the fields D_i can be restricted to $\mathcal{E}_\infty$. It follows from (6.24) that ℓ_F can also be restricted to $\mathcal{E}_\infty$. This restriction will be denoted by $\overline{\ell_F}$. Thus, by definition, $\overline{\ell_F}(G\,|_{\mathcal{E}_\infty}) = \ell_F(G)\,|_{\mathcal{E}_\infty}$, $\forall G$. In these terms the condition of tangency of χ to $\mathcal{E}_\infty$ reads

$$\boxed{\overline{\ell_F}(\bar\varphi) = 0, \quad \bar\varphi = \varphi\,|_{\mathcal{E}_\infty}} \quad \Longrightarrow \quad \boxed{\operatorname{Sym}\mathcal{E} = \ker \overline{\ell_F}} \tag{6.25}$$

Hence the problem of the computation of the infinitesimal symmetries of a PDE $\mathcal{E}$ is reduced to the resolution of Equation (6.25). This equation is not a usual PDE, since it is imposed on functions depending on an unlimited number of variables. Nevertheless, it is not infrequent that it can be exactly solved. For instance, this method allows not only to easily rediscover "classical" hierarchies associated with well known integrable systems but also to find various new ones (see [29, 25]).

The interpretation of higher symmetries as vector fields on the "space of solutions of $\mathcal{E}$" leads to the question : What are the trajectories of this field? The equation of trajectories of χ is very natural:

$$u_t = \varphi(x, u, \dots, u^i_\sigma, \dots) \quad \text{with} \quad \varphi = (\varphi_1, \dots, \varphi_m). \tag{6.26}$$

Equation (6.26) is the exact analogue of the classical equations

$$x_i = a_i(x), \quad i = 1, \dots, m, \quad x = (x_1, \dots, x_m), \tag{6.27}$$

which describe trajectories of the vector field $X = \sum_{i=1}^m a_i \partial/\partial x_i$. An essential difference between Equations (6.26) and (6.27) is that the initial data uniquely determine solutions of (6.27), while it is not longer so for (6.26). Indeed, the uniqueness for the partial evolution equation is guaranteed by some addition to the initial conditions, for instance, the boundary ones. For this reason a "vector field" $\chi \in \varkappa$ does not generate a flow on the "space of solutions of $\mathcal{E}$."

A very important consequence of this fact is that in this new context the classical relation between Lie algebras and Lie groups breaks down. Consequently, absolute priority should be given to infinitesimal symmetries, not to the finite ones.

One of the most popular applications of symmetry theory takes an especially simple form if expressed in terms of generating functions. Namely, imagine for a while that the flow generated by $\chi \in \varkappa$ exists. Then, according to (6.26), "stable points" of this flow are solutions of the equation $\varphi = 0$. In other words, these "stable points" are solutions of the last equation. If $\varphi^1, \dots, \varphi^l$ are generating functions of some symmetries of $\mathcal{E}$, then the solutions of the system

$$\begin{cases} F = 0 \\ \varphi^1 = 0 \\ \quad \vdots \\ \varphi^l = 0 \end{cases} \tag{6.28}$$

represent those solutions of $\mathcal{E}$ that are stable in the above sense with respect to the "flows" generated by $\varphi^1, \dots, \varphi^l$. The system (6.28) is well overdetermined and for this reason can be exactly solved in many cases. For instance, famous multi-soliton solutions of the KdV equation are solutions of this kind.

5.8 What are partial differential equations? The fact that we have built a self-consistent and well working theory of symmetries for PDEs based on diffieties gives a considerable reason to recognize diffieties as objects of the category of PDEs. Another argument supporting this idea is as follows.

Take any PDE, say,

$$u_{xx}u_{tt}^2 + u_{tx}^2 + (u_x^2 - u_t)u = 0. \tag{6.29}$$

This is a hypersurface $\mathcal{E} \subset J^2(E,2)$, $\dim E = 3$. The equivalent system of first order PDEs is

$$\begin{cases} u_x = v \\ u_t = w \\ v_x w_t^2 + v_t w_x + (v^2 - w)u = 0. \end{cases} \tag{6.30}$$

This is a submanifold $\mathcal{E}' \subset J^1(E',2)$, $\dim E' = 5$, of codimension 3. $\mathcal{E}$ and $\mathcal{E}'$ live in different jet spaces and have different dimensions. For this reason their classical symmetries cannot even be compared. On the other hand, associated with $\mathcal{E}$ and $\mathcal{E}'$ diffieties are naturally identified and hence have the same (higher) symmetries. Therefore, this fact may be interpreted by saying that (6.29) and (6.30) are different descriptions of the same object, namely, of the associated diffiety.

Another example illustrating priority of diffieties is the factorization problem. Namely, if G is a Lie algebra of classical symmetries of an equation $\mathcal{E}$, then the question is: Can $\mathcal{E}$ be factorized by the action of G and what is the resulting "quotient equation"? In terms of diffieties the answer is almost obvious : this is the equation $\mathcal{E}'$ such that $\mathcal{E}_\infty \backslash G = \mathcal{E}'_\infty$. On the contrary, it is not very clear how to answer this question in terms of the usual approach.

Example 5.1. Let G be the group of translations of the Euclidean plane. Obviously, these translations are symmetries of the Laplace equation $u_{xx} + u_{yy} = 0$. Then the corresponding quotient equation is again the Laplace equation.

There are many other examples manifesting that

> *A PDE as a mathematical object is a* diffiety, *while what is usually called a PDE is just one of many possible "identity cards" of it.*

It should be stressed that the diffiety associated with a system of PDEs (in the usual sense of this word) is the exact analogue of the algebraic variety associated to a system of algebraic equations. Indeed, if a system of algebraic equations is $f_1 = 0, \ldots, f_r = 0$, then the ideal defining the corresponding variety is algebraically generated by the polynomials f_i. In the case of a PDE $\mathcal{E} = \{F_i = 0\}$ the ideal defining $\mathcal{E}_\infty$ is algebraically generated not only by the functions F_i but also by all their differential consequences $D_\sigma(F_i)$ (see (6.20)). Viewed from this side algebraic geometry is seen as the zero-dimensional case of the general theory of PDEs.

6 On the internal structure of diffieties

On the surface, a diffiety $\mathcal{O} = (\mathcal{E}, \mathcal{C}^\infty_{\mathcal{E}})$ looks as a simple enough object like a foliation. All foliations of a given finite dimension and codimension are locally equivalent. On the contrary, the situation drastically changes when the codimension becomes infinite. Therefore, the problem of how to extract all the information on the equation $\mathcal{E}$, which is encoded in the "poor" Frobenius distribution $\mathcal{C}^\infty_{\mathcal{E}}$, naturally arises and becomes central. To gain a first insight into the problem we consider as a simple model a Frobenius distribution $\mathcal{D}$, or, equivalently, a foliation, on a finite-dimensional manifold M.

6.1 The normal complex of a Frobenius distribution Let $\mathcal{D}$ be an r-dimensional Frobenius distribution on a manifold M. The quotient $C^\infty(M)$-module $\mathcal{N} = D(M)/\mathcal{D}$ is canonically isomorphic to $\Gamma(\nu)$ where ν is the normal bundle to $\mathcal{D}$, i.e., the bundle whose fiber over $x \in M$ is $T_xM/\mathcal{D}_x$. Put $\hat{Y} = [Y \mod \mathcal{D}] \in \mathcal{N}$ for $Y \in D(M)$ and

$$\nabla_X(\hat{Y}) = \widehat{[X, Y]} \quad \text{for} \quad X \in \mathcal{D}.$$

It is easy to see that $\nabla_{fX} = f\nabla_X$, $\nabla_X(f\hat{Y}) = X(f)\hat{Y} + f\nabla_X(\hat{Y})$ if $f \in C^\infty(M)$ and $[\nabla_X, \nabla_{X'}] = \nabla_{[X,X']}$. These formulas tell that the correspondence $\nabla : X \mapsto \nabla_X$ is a flat $\mathcal{D}$-*connection*. This means that this construction can be restricted to a leaf $\mathcal{L}$

of the foliation associated with $\mathcal{D}$ and this restriction is a flat connection $\nabla^{\mathcal{L}}$ in the bundle ν normal to $\mathcal{D}$ restricted to $\mathcal{L}$. Recall that with a flat connection is associated a de Rham-like complex i (see [11]), which for $\nabla^{\mathcal{L}}$ is

$$0 \longrightarrow \mathcal{N}_{\mathcal{L}} \stackrel{\nabla^{\mathcal{L}}}{\longrightarrow} \Lambda^1(\mathcal{L}) \otimes_{C^\infty(\mathcal{L})} \mathcal{N}_{\mathcal{L}} \stackrel{\nabla^{\mathcal{L}}}{\longrightarrow} \cdots \stackrel{\nabla^{\mathcal{L}}}{\longrightarrow} \Lambda^r(\mathcal{L}) \otimes_{C^\infty(\mathcal{L})} \mathcal{N}_{\mathcal{L}} \longrightarrow 0 \quad (6.31)$$

where the covariant differential is abusively denoted also by $\nabla^{\mathcal{L}}$ and $\mathcal{N}_{\mathcal{L}} = \Gamma(\nu|_{\mathcal{L}})$. This complex is, in fact, the restriction to $\mathcal{L}$ of the complex.

$$0 \longrightarrow \mathcal{N} \stackrel{\nabla}{\longrightarrow} \Lambda^1_{\mathcal{D}} \otimes_{C^\infty(M)} \mathcal{N} \stackrel{\nabla}{\longrightarrow} \cdots \stackrel{\nabla}{\longrightarrow} \Lambda^r_{\mathcal{D}} \otimes_{C^\infty(M)} \mathcal{N} \longrightarrow 0 \qquad (6.32)$$

where $\Lambda^i_{\mathcal{D}} = \Lambda^i(M)/\mathcal{D}\Lambda^i(M)$ with

$$\mathcal{D}\Lambda^i(M) = \{\omega \in \Lambda^i(M) \mid \omega(X_1, \ldots, X_i) = 0,\ \forall X_1, \ldots, X_i \in \mathcal{D}\}.$$

The terms of the complex (6.32) are $\mathcal{N}$-valued differential forms on $\mathcal{D}$, i.e., $\rho(X_1, \ldots, X_s) \in \mathcal{N}$ if $X_1, \ldots, X_s \in \mathcal{D}$. The *covariant differential* ∇ is defined as

$$\nabla(\rho)(X_1, \ldots, X_{s+1}) = \textstyle\sum_{i=1}^{s+1}(-1)^{i-1}\nabla_{X_i}(\rho(X_1, \ldots, \widehat{X_i}, \ldots, X_{s+1})) +$$

$$\textstyle\sum_{i<j}(-1)^{i+j}\rho([X_i, X_j], X_1 \ldots, X_i, \ldots, X_j, \ldots, X_{s+1}).$$

Pictorially, this situation may be seen as a "foliation" of the complex (6.32) by complexes (6.31). The i-th cohomology of complexes (6.31) and (6.32) will be denoted by $H^i(\nabla^{\mathcal{L}})$ and $H^i(\nabla)$, respectively. We also have a natural restriction map $H^i(\nabla) \to H^i(\nabla^{\mathcal{L}})$ in cohomology.

Formally, the above construction remains valid for any Frobenius distribution and hence can be applied to diffieties. In order to duly specify the complex (6.32) to this particular case we need a new construction from differential calculus over commutative algebras.

6.2 Modules of jets Let A be an unitary algebra and let P, Q be A-modules. Denote by $\mathrm{Diff}_k(P, Q)$ the totality of DOs of order $\leq k$ considered as a left A-module, i.e., $(a, \square) \mapsto a\square$, $a \in A$, $\square \in \mathrm{Diff}_k(P, Q)$. Consider a subcategory $\mathcal{K}$ of the category of A-modules such that $\mathrm{Diff}_k(P, Q) \in \mathrm{Ob}\,\mathcal{K}$ if $P, Q \in \mathrm{Ob}\,\mathcal{K}$. For a fixed P we have the functor $Q \mapsto \mathrm{Diff}_k(P, Q)$. We say that a pair composed of an A-module $\mathcal{J}^k_{\mathcal{K}}(P)$ and a k-th order DO $j_k = j_k^{P,\mathcal{K}} : P \to \mathcal{J}^k_{\mathcal{K}}(P)$ represents this functor in the category $\mathcal{K}$ if the map $\mathrm{Hom}_A(\mathcal{J}^k_{\mathcal{K}}(P), Q) \ni h \mapsto h \circ j_k \in \mathrm{Diff}_k(P, Q)$ is an isomorphism of A-modules. Under some weak condition on $\mathcal{K}$, which we skip, the *representing object* $(\mathcal{J}^k_{\mathcal{K}}(P), j_k)$ exists and is unique up to isomorphism. $\mathcal{J}^k_{\mathcal{K}}(P)$ is called the module of *k-th order jets of* P (in $\mathcal{K}$). Thus for a DO $\square \in \mathrm{Diff}_k(P, Q)$ there is a unique A-module homomorphism $h_\square : \mathcal{J}^k_{\mathcal{K}}(P) \to Q$ such that $\square = h_\square \circ j_k$.

As an A-module, $\mathcal{J}^k_{\mathcal{K}}(P)$ is generated by elements $j_k(p)$, $p \in P$. A natural transformation of functors $\mathrm{Diff}_l(P, \cdot) \mapsto \mathrm{Diff}_k(P, \cdot)$, $l \leq k$, induces a homomorphism $\pi_{k,l} = \pi^P_{k,l} : \mathcal{J}^k_{\mathcal{K}}(P) \to \mathcal{J}^l_{\mathcal{K}}(P)$ of A-modules such that $j_l = \pi_{k,l} \circ j_k$. This allows to define the inverse limit of pairs $(\mathcal{J}^k_{\mathcal{K}}(P), j_k)$ called the module of *infinite jets* of P and denoted by $(\mathcal{J}^\infty_{\mathcal{K}}(P), j_\infty = j^{P,\mathcal{K}}_\infty)$. Natural projections $\pi_{\infty,k} : \mathcal{J}^\infty_{\mathcal{K}}(P) \to \mathcal{J}^k_{\mathcal{K}}(P)$ come from the definition. These maps supply $\mathcal{J}^\infty_{\mathcal{K}}(P)$ with a decreasing filtration

$$\mathcal{J}^\infty_{\mathcal{K}}(P) \supset \ker(\pi_{\infty,0}) \supset \ker(\pi_{\infty,1}) \supset \cdots \supset \ker(\pi_{\infty,k}) \supset \cdots \tag{6.33}$$

Finally, we stress that $\mathcal{J}^k_{\mathcal{K}}(P)$ and all related constructions essentially depend on $\mathcal{K}$.

Any operator $\square \in \mathrm{Diff}_r(P, Q)$ induces a homomorphism

$$h^r_\square : \mathcal{J}^{k+r}_{\mathcal{K}}(P) \to \mathcal{J}^r_{\mathcal{K}}(Q), \quad r \geq 0.$$

Namely, the composition $P \xrightarrow{\square} Q \xrightarrow{j_r} \mathcal{J}^r_{\mathcal{K}}(Q))$ is a DO of order $\leq k + r$. Thus, it can be presented in the form $h_{j_r \circ \square} \circ j_{k+r}$, and we put

$$h^r_\square \overset{\text{def}}{=} h_{j_r \circ \square} : \mathcal{J}^{k+r}_{\mathcal{K}}(P) \to \mathcal{J}^r_{\mathcal{K}}(Q). \tag{6.34}$$

The inverse limit of the homomorphisms $h^r_\square$ defines a homomorphism of filtered modules

$$h^\infty_\square : \mathcal{J}^\infty_{\mathcal{K}}(P) \to \mathcal{J}^\infty_{\mathcal{K}}(Q)$$

which shifts filtration (6.33) by $-k$.

If $Q = \mathcal{J}^k_{\mathcal{K}}(P)$ and $\square = j_k$, then the above construction gives natural inclusions

$$\iota_{k,r} \overset{\text{def}}{=} h^r_{j_r \circ j_k} : \mathcal{J}^{k+r}_{\mathcal{K}}(P) \hookrightarrow \mathcal{J}^r_{\mathcal{K}}(\mathcal{J}^k_{\mathcal{K}}(P)).$$

The inverse limit of these inclusions is

$$\iota_\infty : \mathcal{J}^\infty_{\mathcal{K}}(P) \hookrightarrow \mathcal{J}^\infty_{\mathcal{K}}(\mathcal{J}^\infty_{\mathcal{K}}(P)).$$

Now we shall describe constructively the above *conceptually* defined modules of jets for *geometrical modules* over the algebra $A = C^\infty(M)$. Recall that an A-module P is geometrical if all its elements p such that $p \in \mu_z \cdot P$, $\forall z \in M$, are equal to zero (see [42]). Here $\mu_z = \{f \in C^\infty(M) \mid f(z) = 0\}$. The category of geometrical A-modules will be denoted by $\mathcal{G}$ and we shall write simply $\mathcal{J}^k(P)$ for $\mathcal{J}^k_{\mathcal{G}}(P)$.

Put $\mathcal{J}^k = \mathcal{J}^k(A)$ and note that $\mathcal{J}^k$ is a unitary algebra with the product $(f_1 j_k(g_1)) \cdot (f_2 j_k(g_2)) = f_1 f_2 j_k(g_1 g_2)$, $f_i, g_i \in A$. In particular, $\mathcal{J}^k$ is a bimodule. Namely, left (standard) and right multiplications by $f \in A$ are defined as $(f, \theta) \mapsto f\theta$ and $(f, \theta) \mapsto \theta j_k(f)$, respectively. $\mathcal{J}^k$ supplied with the right A-module structure will be denoted by $\mathcal{J}^k_>$.

We have (see [42]).

Proposition 6.1.

1. Let α_k be the vector bundle whose fiber over $z \in M$ is $J_z(M)$ (see Section 3.1). Then $\mathcal{J}^k = \Gamma(\alpha_k)$.
2. $\mathcal{J}^k(P) = \mathcal{J}^k_> \otimes_A P$.

6.3 Jet-Spencer complexes The k-th jet-Spencer complex of P denoted by $\mathcal{S}_k(P)$ is defined as

$$0 \to \mathcal{J}^k(P) \xrightarrow{S_k} \mathcal{J}^{k-1}(P) \otimes_A \Lambda^1(M) \xrightarrow{S_k} \cdots \xrightarrow{S_k} \mathcal{J}^{k-n}(P) \otimes_A \Lambda^n(M) \to 0 \tag{6.35}$$

with $S_k(j_{k-s}(p) \otimes \omega) = j_{k-s-1}(p) \otimes d\omega$, $\omega \in \Lambda^s(M)$. Here $n = \dim M$ and we assume that $\mathcal{J}^s(P) = 0$ if $s < 0$.

Differentials of Spencer complexes are 1-st order DOs. For $k \geq l$, the homomorphisms

$$\pi_{k-s,l-s} \otimes \mathrm{id}\,|_{\Lambda^s(M)} : \mathcal{J}^{k-s}(P) \otimes_A \Lambda^s(M) \longrightarrow \mathcal{J}^{l-s}(P) \otimes_A \Lambda^s(M), \quad s = 0, 1, \ldots, n,$$

define a cochain map $\sigma_{k,l} : \mathcal{S}_k(P) \longrightarrow \mathcal{S}_l(P)$. In particular, we have the following sequence of Spencer complexes

$$0 \leftarrow \mathcal{S}_0(P) \xleftarrow{\sigma_{1,0}} \mathcal{S}_1(P) \xleftarrow{\sigma_{2,1}} \cdots \xleftarrow{\sigma_{k,k-1}} \mathcal{S}_k(P) \xleftarrow{\sigma_{k+1,k}} \cdots \tag{6.36}$$

The infinite jet-Spencer complex $\mathcal{S}_\infty(P)$ is defined as the inverse limit of (6.36) together with natural cochain maps $\sigma_{\infty,k}$. As in the case of jets the complex $\mathcal{S}_\infty(P)$ is filtered by subcomplexes $\ker(\sigma_{\infty,k})$.

An operator $\Box \in \mathrm{Diff}_r(P, Q)$ induces a cochain map of Spencer complexes

$$\sigma_\Box^k : \mathcal{S}_k(P) \longrightarrow \mathcal{S}_{k-r}(Q), \tag{6.37}$$

which acts on the s-th term of $\mathcal{S}_k(P)$ as

$$h_\Box^{k-s} \otimes \mathrm{id}\,|_{\Lambda^s(M)} : \mathcal{J}^{k-s}(P) \otimes_A \Lambda^s(M) \longrightarrow \mathcal{J}^{k-s-r}(Q) \otimes_A \Lambda^s(M) \tag{6.38}$$

(see (6.34)).

Conceptually, the k-th jet-Spencer complex is a *acyclic resolvent* for the *universal* k-order differential operator j_k. Namely, we have

Proposition 6.2. If $P = \Gamma(\xi)$ with ξ being a vector bundle over M, then

1. $\mathcal{S}_k(P)$ is acyclic in positive dimensions $\Leftrightarrow$ $H^i(\mathcal{S}_k) = 0$ if $i > 0$.
2. $H^0(\mathcal{S}_k) = P$ and 0-cocycles are $j_k(p) \in \mathcal{J}^k(P)$, $p \in P$.

Jet-Spencer complexes are *natural*, since they can be defined over arbitrary unitary (graded) algebras (see [59]). In other words, they are *compatible* with homomorphisms of these algebras. In particular, they restrict to submanifolds. For our purposes, we need to describe this procedure.

Let $N \subset M$ be a submanifold. In the notation of Proposition 6.2 we put $P_N = \Gamma(\xi|_N)$ and $j_k^N : P_N \to \mathcal{J}^k(P_N)$ to distinguish this jet-operator on N from j_k :

$P \to \mathcal{J}^k(P)$. Since $C^\infty(N)$-modules can also be considered as $C^\infty(M)$-modules, the composition $\square$

$$P \overset{restriction}{\longrightarrow} P_N \overset{j_k^N}{\longrightarrow} \mathcal{J}^k(P_N)$$

is a k-th order DO over $C^\infty(M)$. The homomorphism of $C^\infty(M)$-modules $h^\square$: $\mathcal{J}^k(P) \to \mathcal{J}^k(P_N)$ associated with $\square$ is, by definition, the *restriction operator*. Now, by tensoring this restriction operator with the well-known restriction operator for differential forms we get the restriction operator for terms of $\mathcal{S}_k(P)$. Finally, by passing to the inverse limit we get the restriction operator for $\mathcal{S}_\infty(P)$.

6.4 Foliation of Spencer complexes by a Frobenius distribution Take the notation of Subsection 6.1 and denote by $\mathcal{D}\Lambda^i(M)$ (resp., $\mathcal{D}\mathcal{J}^k(P)$) the totality of all differential forms (resp., jets) whose restrictions to all leaves of $\mathcal{D}$ are trivial. Similarly, $\mathcal{D}\mathcal{S}_k(P)$ stands for the maximal subcomplex of $\mathcal{S}_k(P)$ such that its restrictions to leaves of $\mathcal{D}$ are trivial. *Horizontal* (with respect to $\mathcal{D}$) differential forms and jets are elements of the quotient modules

$$\bar{\Lambda}^i_{\mathcal{D}}(M) \overset{\text{def}}{=} \Lambda^i(M)/\mathcal{D}\Lambda^i(M), \quad \bar{\mathcal{J}}^k(P) \overset{\text{def}}{=} \mathcal{J}^k(P)/\mathcal{D}\mathcal{J}^k(P).$$

Similarly, the *horizontal jet-Spencer complex* is

$$\bar{\mathcal{S}}_k(P) \overset{\text{def}}{=} \mathcal{S}_k(P)/\mathcal{D}\mathcal{S}_k(P).$$

Restrictions of all the above horizontal objects to leaves of $\mathcal{D}$ are naturally defined. Conversely, a horizontal differential form (resp., jet-Spencer complex) may be viewed as a family of differential forms (resp., jet-Spencer complex) defined on single leaves. In other words, if $\mathcal{L}$ runs through all leaves of $\mathcal{D}$, then the $\Lambda^i(\mathcal{L})$ foliate $\bar{\Lambda}^i_{\mathcal{D}}(M)$, and similarly for jets and Spencer complexes.

Accordingly, the exterior differential d as well as the Spencer differential S_k factorize to $\bar{\Lambda}^*(M)$ and $\bar{\mathcal{S}}_k(P)$, since $\mathcal{D}\Lambda^i(M)$ and $\mathcal{D}\mathcal{S}_k(P)$ are stable with respect to d and S_k, respectively. These quotient differentials will be denoted by $\bar{d}$ and $\bar{S}_k$, respectively. In this way we get the *horizontal de Rham* and *Spencer complexes* and hence *horizontal de Rham* and *Spencer cohomology*.

Remark 6.1. For simplicity, in the above definition of horizontal objects we have used leaves of $\mathcal{D}$. It fact, with a longer formal procedure this can be done explicitly in terms of the distribution $\mathcal{D}$.

6.5 The normal complex of a diffiety and more symmetries First, we shall describe the normal complex for $J^\infty(E,n)$. Denote the normal bundle to $\mathcal{C}^k$ by ν_k, $1 \le k \le \infty$ and put $\varkappa_k \overset{\text{def}}{=} \Gamma(\nu_k) = D(J^k(E,n))/\mathcal{C}^k$, $\varkappa_{l,k} \overset{\text{def}}{=} \Gamma(\pi^*_{k,l}(\nu_l)) = \pi^*_{k,l}(\varkappa_l)$, $l \le k$.

Proposition 6.3.

1. $\varkappa = \varkappa_{1,\infty}$, $\varkappa_{k,\infty} = \bar{\mathcal{J}}^{k-1}(\varkappa)$ and $\varkappa_\infty = \bar{\mathcal{J}}^\infty(\varkappa)$;
2. the normal to $\mathcal{C}^\infty$ complex is isomorphic to $\bar{\mathcal{S}}_\infty(\varkappa)$;
3. $H^0(\bar{\mathcal{S}}_\infty) = \varkappa$, $H^i(\bar{\mathcal{S}}_\infty) = 0$, $i \neq 0$.

Assertion (3) in this proposition is a pro-finite consequence of Proposition 6.2. The following important interpretation is a consequence of this and the second assertions of Proposition 6.3:

> *$\varkappa$ is the zero-th cohomology space of the complex normal to the infinite contact structure $\mathcal{C}^\infty$.*

Now we can describe the normal complex to $\mathcal{C}^\infty_\mathcal{E}$ by restricting, in a sense, Proposition 6.3 to $\mathcal{E}_\infty$. First, to this end, we need a conceptually satisfactory definition of the universal linearization operator (6.24), which was defined coordinate-wisely in Subsection 5.7.

The equation $\mathcal{E} \subset J^k(E,n)$ may be presented in a coordinate-free form as $\Phi = 0$, $\Phi \in \Gamma(\xi)$ with ξ being a suitable vector bundle over $J^k(E,n)$. If $P = \Gamma(\pi^*_{\infty,k}(\xi))$ and $\Phi_\infty = \pi^*_{\infty,k}(\Phi)$, then $\mathcal{E}_\infty = \{\bar{j}_\infty(\Phi_\infty) = 0\}$. Additionally, assume that P is supplied with a connection ∇. If $Y \in \mathcal{C}^\infty$, then, as it is easy to see, $\nabla_Y(\Phi)|\mathcal{E}_\infty = 0$. For this reason the following definition is reasonable.

$$\ell_\mathcal{E}(\chi) \overset{\text{def}}{=} \nabla_X(\Phi)|\mathcal{E}_\infty \quad \text{with} \quad \chi = [X \mod \mathcal{C}^\infty_\mathcal{E}]. \tag{6.39}$$

The operator $\ell_\mathcal{E} : \varkappa \to P$ does not depend on the choices of Φ and ∇. It defines the cochain map of jet-Spencer complexes

$$\sigma^\infty_{\ell_\mathcal{E}} : \mathcal{S}_\infty(\varkappa) \longrightarrow \mathcal{S}_\infty(P) \tag{6.40}$$

(see (6.37)).

Proposition 6.4. Let $\mathcal{N}_\mathcal{E}$ be the normal to the distribution $\mathcal{C}^\infty_\mathcal{E}$ complex on $\mathcal{E}_\infty$. Then

1. $\mathcal{N}_\mathcal{E}$ is isomorphic to the complex $\ker \sigma^\infty_{\ell_\mathcal{E}}$.
2. The cohomology $H^i(\mathcal{N}_\mathcal{E})$ of the complex $\mathcal{N}_\mathcal{E}$ is trivial if $i > n$.
3. $\operatorname{Sym} \mathcal{E} = H^0(\mathcal{N}_\mathcal{E}) = \ker \ell_\mathcal{E}$.

Assertions (2) and (3) of this proposition are consequences of the first one, which allows to compute the cohomology of $\mathcal{N}_\mathcal{E}$. Moreover, assertion (3) is one of many other arguments that motivate the following definition.

Definition 6.1. The *Lie algebra of (higher) infinitesimal symmetries of a PDE $\mathcal{E}$* is the cohomology of the normal complex $\mathcal{N}_\mathcal{E}$.

Accordingly, denote by $\mathrm{Sym}_i\,\mathcal{E}$ the i-th cohomology space of $\mathcal{N}_{\mathcal{E}}$. Thus, the whole Lie algebra of infinitesimal symmetries of $\mathcal{E}$ is *graded*:

$$\boxed{\mathrm{Sym}_*\,\mathcal{E} = \sum_{i=0}^{n} \mathrm{Sym}_i\,\mathcal{E}, \quad \mathrm{Sym}_i\,\mathcal{E} = H^i(\mathcal{N}_{\mathcal{E}})}$$

In particular, $\mathrm{Sym}\,\mathcal{E} = \mathrm{Sym}_0\,\mathcal{E}$ (see Subsection 5.6).

Remark 6.2. The description of the Lie product in $\mathrm{Sym}_*\,\mathcal{E}$ is not immediate and requires some new instruments of differential calculus over commutative algebras. For this reason we shall skip it.

The following proposition illustrates what the algebra $\mathrm{Sym}_*\,\mathcal{E}$ looks like.

Proposition 6.5. If $\mathcal{E}$ is not an overdetermined system of PDEs, then

$$\mathrm{Sym}_0\,\mathcal{E} = \ker \ell_{\mathcal{E}}, \quad \mathrm{Sym}_1\,\mathcal{E} = \mathrm{coker}\ell_{\mathcal{E}} \quad \text{and} \quad \mathrm{Sym}_i\,\mathcal{E} = 0 \quad \text{if} \quad i \neq 0, 1.$$

In this connection we note that a great majority of the PDEs of current interest in geometry, physics and mechanics are not overdetermined. As an exception we mention the system of Yang–Mills equations, which is sightly overdetermined, and for these equations $\mathrm{Sym}_2\,\mathcal{E} \neq 0$.

To conclude this section we would like to emphasize the role of the structure of differential calculus over commutative algebras in the above discussion. While we have used the "experimental data" coming from the theory of integrable systems to discover "by hands" the conceptually simplest part of infinitesimal symmetries of PDEs, i.e., the Lie algebra $\mathrm{Sym}\,\mathcal{E}$, a familiarity with the structures of differential calculus over commutative algebras is indispensable to discover that it is just the zeroth component of the full symmetry algebra $\mathrm{Sym}_*\,\mathcal{E}$, which in its turn is the cohomology of a certain complex.

7 Nonlocal symmetries and once again: what are PDEs?

In the previous section we have constructed a self-consistent symmetry theory, which, from one side, resolves shortcomings of the classical theory discussed in Sections 2–5 and, from another side, incorporates "experimental data" that emerged in the theory of integrable systems. However, one important element of this theory was not taken into account. Namely, we have in mind *nonlocal symmetries*. Roughly speaking, these are symmetries whose generating function depends on variables of the form $D_i^{-1}(u)$. Fortunately, these unusual symmetries can be tamed by introducing only one new notion we are going to describe.

7.1 Coverings of a diffiety Schematically, a diffiety $\mathcal{O}$ is a *pro-finite manifold* $\mathfrak{M}$ supplied with a finite-dimensional pro-finite Frobenius distribution $\mathcal{D} = \mathcal{D}_{\mathcal{O}} : \mathcal{O} = (\mathfrak{M}, \mathcal{D})$. We omit technical details that these data must satisfy.

Recall that a pro-finite manifold is the inverse limit of a sequence of smooth maps

$$M_0 \xleftarrow{\mu_1} M_1 \xleftarrow{\mu_2} \cdots \xleftarrow{\mu_k} M_k \xleftarrow{\mu_{k+1}} \cdots \quad \Leftarrow \mathfrak{M} \tag{6.41}$$

A pro-finite distribution on $\mathfrak{M}$ is the inverse limit via the μ_i of the distributions $\mathcal{D}_i$ on M_i. The associated sequence of homomorphisms of smooth function algebras

$$C^\infty(M_0) \xrightarrow{\mu_1^*} C^\infty(M_1) \xrightarrow{\mu_2^*} \cdots \xrightarrow{\mu_k^*} C^\infty(M_k) \xrightarrow{\mu_{k+1}^*} \cdots \quad \Rightarrow \mathcal{F}_{\mathfrak{M}} \tag{6.42}$$

with $\mathcal{F}_{\mathfrak{M}}$ being the direct limit of homomorphisms μ_k^* is filtered by subalgebras $\mathcal{F}_{\mathfrak{M}}^k \stackrel{\text{def}}{=} \mu_{\infty,k}^* C^\infty(M_k)$ where $\mu_{\infty,k} : \mathfrak{M} \to M_k$ is a natural projection. Differential calculus on $\mathfrak{M}$ is interpreted as the calculus over the filtered algebra $\mathcal{F}_{\mathfrak{M}}$ (see Subsection 5.3). The dimension of $\mathcal{D}$ is interpreted as the "number of independent variables."

A morphism $F : \mathcal{O} \to \mathcal{O}'$ between a diffiety $\mathcal{O} = (\mathfrak{M}, \mathcal{D})$ and a diffiety $\mathcal{O}' = (\mathfrak{M}', \mathcal{D}')$ is, abusing the notation, a map $F : \mathfrak{M} \to \mathfrak{M}'$ such that $F^*(\mathcal{F}_{\mathfrak{M}'}) \subset \mathcal{F}_{\mathfrak{M}}$, F^* is compatible with filtrations and $d_\theta F(\mathcal{D}_\theta) \subset \mathcal{D}'_{F(\theta)}$, $\forall \theta \in \mathfrak{M}$.

Definition 7.1. A surjective morphism $F : \mathcal{O} \to \mathcal{O}'$ of diffieties is called a *covering* if $\dim \mathcal{D} = \dim \mathcal{D}'$ and $d_\theta F$ isomorphically sends $\mathcal{D}_\theta$ to $\mathcal{D}'_{F(\theta)}$, $\forall \theta \in \mathfrak{M}$.

This terminology emphasizes the analogy with the standard notion of a covering in the category of manifolds. Namely, fibers of a covering are zero-dimensional diffieties in the sense that the their structure distributions $\mathcal{D}$ are zero-dimensional. If these fibers are finite-dimensional in the usual sense, then the covering is called *finite-dimensional*.

A covering $F : \mathcal{E}_\infty \to \mathcal{E}'_\infty$ may be interpreted as a (nonlinear) DO, which sends solutions of $\mathcal{E}$ to solutions of $\mathcal{E}'$. More exactly, it associates with a solution of $\mathcal{E}'$ a families of solutions of $\mathcal{E}$. For instance, the famous Cole–Hopf substitution $v = 2u_x/u$ that sends solutions of the heat equation $\mathcal{E} = \{u_t = u_{xx}\}$ to solutions of the Burgers equation $\mathcal{E}' = \{v_t = v_{xx} + vv_x\}$ comes from a 1-dimensional covering of $\mathcal{E}'$. Equivalently, the passage from a PDE to a covering equation is the *inversion of a (nonlinear) DO* on solutions of this PDE. For instance, by inverting the operator $v \mapsto 2v_x/v$ on solutions of the Burgers equation one gets the heat equation.

7.2 Where coverings appear The notion of a covering of a diffiety was introduced by the author (see [62]) as a common basis for various constructions that appeared in PDEs. Below we list and briefly discuss some of them.

1) In the language of diffieties the passage from Lagrange's description of a continuum media to that of Euler is interpreted as a covering. This interpretation allows to apply instruments of secondary calculus to this situation and, as a result, to derive from this fact some important consequences for mechanics of continua.

2) ***Factorization of PDEs.*** If G is a symmetry group of a diffiety $\mathcal{O}$, then under some natural conditions the quotient diffiety $\mathcal{O}\backslash G$ is well-defined and $\mathcal{O} \to \mathcal{O}\backslash G$ is a covering. In particular, if $\mathcal{O} = \mathcal{E}_\infty$, then $\mathcal{O}\backslash G = \mathcal{E}'_\infty$. In such a case $\mathcal{E}'$ is the *quotient equation* of $\mathcal{E}$ by G. A remarkable fact is that the group G in this construction may be an "infinite-dimensional" Lie group like the group $\mathrm{Diffeo}(M)$ of diffeomorphisms of a manifold M, or the group of contact transformations, etc.

3) ***Differential invariants and characteristic classes.*** Let $\pi : E \to M$ be a fiber bundle of geometrical structures of a type $\mathfrak{S}$ on M (see [1]). Then $\mathrm{Char}\,\mathfrak{S} = J^\infty(\pi)\backslash\mathrm{Diffeo}(M)$ is the *characteristic diffiety* for $\mathfrak{S}$-structures. This diffiety is with singularities, which are in turn diffieties with a smaller numbers of independent variables. Functions on $\mathrm{Char}\,\mathfrak{S}$ are scalar differential invariants of $\mathfrak{S}$- structures, their horizontal de Rham cohomology is composed of their characteristic classes, etc.

   Similarly one can define differential invariants and characteristic classes for solutions of *natural* PDEs, i.e., those that are invariant with respect to the group $\mathrm{Diffeo}(M)$ or some more specific subgroups of this group. For instance, Einstein equations and many other equations of mathematical physics are natural. Gel'fand–Fucks characteristic classes are quantities of this kind. The reader will find more details and examples in [65, 39, 53].

4) ***Bäcklund transformations.*** The notion of covering allows to rigorously define Bäcklund transformations. Namely, the diagram

$$\begin{array}{ccc} & \mathcal{E}_\infty & \\ {}^{F'}\swarrow & & \searrow^{F''} \\ \mathcal{E}'_\infty & & \mathcal{E}''_\infty \end{array}$$

   where F' and F'' are coverings presents the Bäcklund transformation $F'' \circ (F')^{-1}$ from $\mathcal{E}'$ to $\mathcal{E}''$ and its inverse $F' \circ (F'')^{-1}$. The importance of this definition lies in the fact that it suggests an efficient and regular method for finding Bäcklund transformations for a given PDE (see [28, 29, 22]). Previously this was a kind of handcraft art. Moreover, it turned out to be possible to prove for the first time nonexistence of Bäcklund transformations connecting two given equations (see [21]). This seems to be an impossible task by using only the standard techniques of the theory of integrable systems.

5) ***Poisson structures and the Darboux lemma in field theory***. The efficiency and elegance of the Hamiltonian approach to the mechanics of systems with a finite number of degrees of freedom motivates the search for its extension to the mechanics of continua and field theory. Obviously, this presupposes a due formalization of the idea of a Poisson structure in the corresponding infinite-dimensional context. Over the past 70–80 years various concrete constructions of the Poisson bracket in field theory were proposed, mainly, by physicists. But the first attempts to build a systematic general theory can be traced back only to the late 1970s. Here we mention B. A. Kuperschmidt's paper [31] where he constructs an analogue of the Poisson structure on the cotangent bundle on infinite jets, and the paper by I. M. Gel'fand and I. Dorfman [16] in the context of "formal differential geometry". A general definition of a Poisson structure on infinite jets was proposed by the author in [57] but its extension to general diffieties appeared to be a not very trivial task.

More precisely, while the necessary definition of multivectors in secondary calculus, sometimes also called variational multivectors, is a natural generalization of Definition 6.1, some technical aspects of the related Schouten bracket mechanism are to be still elaborated. See [66, 26, 18] for further results.

On the other hand, in the context of integrable systems numerous concrete Poisson structures were revealed. Among them the bi-hamiltonian ones deserve a special mention (see [38]). Therefore, the question of their classification arises. In the finite-dimensional case the famous Darboux lemma tells that symplectic manifolds or, equivalently, nondegenerate Poisson structures of the same dimension are locally equivalent. "What is its analogue in field theory?" is a good question, which, at first glance, seems to be out of place as many known examples show. Nevertheless, by substituting "coverings" for "diffeomorphisms" in the formulation of this lemma and observing that these two notions are locally identical for finite-dimensional manifolds we get some satisfactory results. Namely, all Poisson structures explicitly described up to now on infinite jets are obtained from a few models by passing to suitable coverings. See [3] for more details.

7.3 Nonlocal symmetries The first idea about nonlocal symmetries takes its origin at a seemingly technical fact. It was observed that the PDEs forming the KdV hierarchy are obtained from the original KdV equation $\mathcal{E} = \{u_t = uu_x + u + u_{xxx}\}$ by applying the so-called *recursion operator*. This operator

$$\mathcal{R} = D_x^2 + \frac{2}{3}u + \frac{1}{3}u_x D_x^{-1}$$

is not defined rigorously. By applying it to generating functions of symmetries of $\mathcal{E}$ one gets new ones that may depend on $D_x^{-1}u = \int u dx$. A due rigor to this formal trick can be given by passing to a 1-dimensional covering $\mathcal{E}'_\infty \to \mathcal{E}_\infty$ by adding to the standard coordinates on $\mathcal{E}_\infty$ a new one w such that $D_x w = u$ and $D_t w = u_{xx} + \frac{1}{2}u^2$. In this setting the above symmetries of $\mathcal{E}$ depending on $\int dx$, i.e., *nonlocal* ones, become symmetries of $\mathcal{E}'_\infty$ in the sense of Definition 6.1, i.e., *local* ones. This and other similar arguments motivate the following definition.

Definition 7.2. A *nonlocal symmetry* (finite or infinitesimal) of an equation $\mathcal{E}$ is a local symmetry of a diffiety $\mathcal{O}$, which covers $\mathcal{E}_\infty$. If $\tau : \mathcal{O} \to \mathcal{E}_\infty$ is a covering, then symmetries of $\mathcal{O}$ are called *τ-symmetries* of $\mathcal{E}$.

Similarly are defined nonlocal quantities of any kind. For instance, Poisson structures in field theory discussed in Subsection 7.2 are nonlocal with respect to the original PDE/diffiety.

Let $\tau_i : \mathcal{O}_i \to \mathcal{E}_\infty$, $i = 1, 2$, be two coverings of $\mathcal{E}_\infty$. A remarkable fact, which is due to I. S. Krasil'shchik, is that the Lie bracket of a τ_1-symmetry and a τ_2-symmetry can be defined as a τ-symmetry for a suitable covering $\tau : \mathcal{O} \to \mathcal{E}_\infty$ together with coverings $\tau'_i : \mathcal{O} \to \mathcal{O}_i$, $i = 1, 2$, such that $\tau = \tau_i \circ \tau'_i$. The covering τ is not defined uniquely. Nevertheless, this non-uniqueness can be resolved by passing to a common covering for "all parties in question." The Jacobi identity as well as other ingredients of Lie algebra theory can be settled in a similar manner (see [28, 29]). Thus, nonlocal symmetries of a PDE $\mathcal{E}$ form this *strange* Lie algebra, and this fact in turn confirms the validity of Definition 7.2.

Thus this definition incorporates all theoretically or experimentally known candidates for symmetries of a PDE. Moreover, it brings us to a new challenging question:

> *Symmetries of which object are the elements of the above "strange" Lie algebra?*

Indeed, this algebra can be considered not only as the algebra of nonlocal symmetries of the equation $\mathcal{E}$ but also as the symmetry algebra of any equation that covers $\mathcal{E}$. In other words, the question: What are partial differential equations? arises again in this new context. But before we shall take a necessary look at the related problem of construction of coverings.

7.4 Finding of coverings The problem of how to find coverings of a given equation is key from both practical and theoretical points of view. At present we are rather far from its complete solutions. Therefore, below we shall illustrate the situation by sketching a direct method, which works well for PDEs in two independent variables and also supplies us with an interesting experimental material.

Let $\mathcal{E} \subset J^k(E, n)$, $\mathcal{O} = (\mathfrak{M}, \mathcal{D})$ and $\tau : \mathcal{O} \to \mathcal{E}_\infty$ be a covering. A τ-projectable vector field $X \in \mathcal{D}$ is of the form $X = \bar{X} + V$ where $\bar{X} \in \mathcal{C}_\mathcal{E}$ and V is τ-vertical, i.e., tangent to the fibers of τ. Locally τ can be represented as the projection $U \times W \to \mathcal{E}_\infty$ with W being a pro-finite manifold and U a domain in $\mathcal{E}_\infty$. If U is sufficiently small, then the restrictions $\bar{D}_i, i = 1, \ldots, n$, of the total derivatives D_is to U span the distribution $\mathcal{C}_\mathcal{E}|_U$. The vector fields $\widehat{D}_i \in \mathcal{D}$ that project onto the $\bar{D}_i$ span $\mathcal{D}|_{\tau^{-1}(U)}$ and $\widehat{D}_i = \bar{D}_i + V_i$ where V_i is τ-vertical. The Frobenius property of $\mathcal{D}$ is equivalent to

$$0 = [\widehat{D}_i, \widehat{D}_j] \Leftrightarrow [\bar{D}_i, V_j] - [\bar{D}_j, V_i] + [V_i, V_j] = 0, \quad 1 \le i < j \le n. \qquad (6.43)$$

By inverting this procedure we get a method to search for coverings of $\mathcal{E}_\infty$. Namely, take a pro-finite manifold W with coordinates $w_1, w_2, \dots$ and vector fields $V_i = \sum_r a_r \partial/\partial w_s$ on $U \times W$ with indeterminate coefficients $a_r \in C^\infty(U \times W)$. Any choice of these coefficients satisfying relations (6.43) defines a Frobenius distribution $\mathrm{span}\{\widehat{D}_1, \dots, \widehat{D}_n\}$, which covers $\mathcal{E}_\infty$. Therefore, by resolving Equations (6.43) with respect to the a_r we get local coverings of $\mathcal{C}_\mathcal{E}$. Many exact solutions of these equations for concrete PDEs of interest can be found for $n = 2$ and they reveal a very interesting structure, which we illustrate with the following example.

Example 7.1. For the KdV equation $\mathcal{E} = \{u_t = uu_x + u_{xxx}\}$ we may take $t, x, u, u_x, u_{xx}, \dots$ for coordinates on $\mathcal{E}_\infty$. Then the vector fields

$$D_x \overset{\text{def}}{=} \bar{D}_1 = \frac{\partial}{\partial x} + \sum_{s=0}^{\infty} u_{s+1} \frac{\partial}{\partial u_s}, \quad D_t \overset{\text{def}}{=} \bar{D}_2 = \frac{\partial}{\partial t} + \sum_{s=0}^{\infty} D_x^s(u_3 + uu_1)\frac{\partial}{\partial u_s},$$

with $u_s = u_{x\dots x}$ (s-times) span $\mathcal{C}_\mathcal{E}$. Put $V_x = V_1$, $V_t = V_2$ and look for solutions of *(6.43)* assuming that $a_r = a_r(u, u_1, u_2, w_1, w_2, \dots)$ for simplicity. The result is worth to be reported on in details. We have

$$\begin{aligned} V_x &= u^2 A + uB + C, \\ V_t &= 2uu_2 A + u_2 B - u_1^2 A + u_1[B, C] + \tfrac{2}{3}u^3 A + \\ &\quad + \tfrac{1}{2}(B + [B, [C, B]]) + u[C, [C, B]] + D \end{aligned} \tag{6.44}$$

with A, B, C, D being some fields on W such that

$$\begin{aligned} &[A, B] = [A, C] = [C, D] = 0, \quad [B, D] + [C, [C, [C, B]]] = 0, \\ &[B, [B, [B, C]]] = 0, \quad [A, D] + \tfrac{3}{2}[B, [C, [C, B]]] = 0. \end{aligned} \tag{6.45}$$

This result tells that if we consider the Lie algebra generated by four elements A, B, C, D, which are subject to the relations *(6.45)*, then any representation of this algebra by vector fields on a manifold W gives a covering of $\mathcal{E}_\infty$ associated with the vector fields *(6.44)*.

Remark 7.1. The Lie algebra defined by the relations (6.45) "mystically" appeared for the first time in the paper by H. D. Wahlquist and F. B. Estabrook [72], in which they introduced the so-called prolongation structures. The fact that it is, as explained above, a necessary ingredient in the construction of coverings is due to the author.

The reader will find many other examples of this kind together with related non-local symmetries, conservation laws, recursion operators, Bäcklund transformations, etc. in [28, 29, 25].

7.5 But what really are PDEs? Now we can turn back to the question posed at the end of Subsection 7.3. Recall that the possibility to commute nonlocal symmetries of a PDE $\mathcal{E}$ living in different coverings of $\tau_i : \mathcal{O}_i \to \mathcal{E}_\infty$, $i = 1, \dots, m$, is ensured by the existence of a common covering diffiety $\mathcal{O}$, i.e., a system of coverings $\tau_i' : \mathcal{O} \to \mathcal{O}_i$ such that $\tau = \tau_i \circ \tau_i'$. Thus, in order to include into consideration all nonlocal symmetries we must consider "all" coverings $\tau_\alpha : \mathcal{O}_\alpha \to \mathcal{E}_\infty$ of $\mathcal{E}_\infty$ as well as coverings $\mathcal{O}_\alpha \to \mathcal{O}_\beta$. In this way we come to the category **Cobweb** $\mathcal{E}$ of coverings of $\mathcal{E}_\infty$. Then it is natural to call the *universal covering* of $\mathcal{E}$ the *terminal object* of **Cobweb** $\mathcal{E}$. Denote this hypothetical universal covering by $\tau_\mathcal{E} : \mathcal{O}_\mathcal{E} \to \mathcal{E}_\infty$, $\mathcal{O}_\mathcal{E} = (\mathfrak{M}_\mathcal{E}, \mathcal{D}_\mathcal{E})$. Now it is easy to see that **Cobweb** $\mathcal{E} =$ **Cobweb** $\mathcal{E}'$ if and only if there is a common covering diffiety $\mathcal{O}$, $\mathcal{E}_\infty \leftarrow \mathcal{O} \to \mathcal{E}'_\infty$. In other words, $\mathcal{E}$ and $\mathcal{E}'$ are related by a Bäcklund transformation (see Subsection 7.2). Recalling that coverings present inversions of differential operators we can trace the following analogy with algebraic geometry:

- *Affine algebraic variety associated with an algebraic equation* $\Rightarrow$ $\mathcal{E}_\infty$.
- *Birational transformations connecting two affine varieties* $\Rightarrow$ Bäcklund transformations.
- *The field of rational functions on an affine variety* $\Rightarrow$ $\mathcal{O}_\mathcal{E}$.

This analogy becomes a tautology if one considers algebraic varieties as PDEs in *zero independent variables*. Indeed, any DO in this case is of zero order, i.e., multiplications by a function, and hence the inversion of such a DO is the division by this function.

Unfortunately, the universal covering understood as a terminal object of a category is not sufficiently constructive to work with. However, we have some indication of how to proceed. From the theoretical side, the indication is to look for an analogue of the fundamental group in the category of diffieties in order to construct the universal covering. By taking into account that we deal with infinitesimal symmetries it would be more adequate to look for the *infinitesimal fundamental group*, i.e., for the *fundamental Lie algebra* of the diffiety $\mathcal{E}_\infty$. On the other hand this idea is on an "experimental" ground. Namely, the Lie algebra associated with a Wahlquist–Estabrook prolongation structure (see Example 7.1) is naturally interpreted as the universal algebra for a special class of coverings.

In this connection a very interesting result by S. Igonin should be mentioned. In [21] he constructed an object which possesses basic properties of the fundamental algebra for a class of PDEs in two independent variables. Moreover, on this basis he succeeded to prove the non-existence of Bäcklund transformations connecting some integrable PDEs, for instance, the KdV equation and the Krichever–Novikiov equation.

Thus the question: What are PDEs? continues to resist well, and the reader may see that this is a highly nontrivial conceptual problem. Yet though universal coverings of diffieties (if they exist!) point at a plausible answer, a good bulk of work should be done in order to put these ideas on a firm ground.

8 A couple of words about secondary calculus

In these pages we, first, tried to attract attention to two intimately related questions: "what are symmetries of an object?" and "what is the object itself?". They form something like an electro-magnetic wave when one of them induces the other and vice versa. Probably, this dynamical form is the most adequate adaptation of the background ideas of the Erlangen program to realities of present-day mathematics. The launch of such a wave in the area of nonlinear partial differential equations was the inestimable contribution of S. Lie to modern mathematics as it is now clearly seen in the hundred-years retrospective.

In the above picture of the post-Lie phase of propagation of this wave we did not touch such fundamental questions as what are general tensor fields, connections, differential operators, etc. on the "space of all solutions" of a given PDE, i.e., on the corresponding diffiety. They all together form what we call *secondary calculus*. It turns out that any natural notion or construction of the standard "differential mathematics" has an analogue in secondary calculus, which is referred to by adding the adjective "secondary". In these terms (higher) symmetries of a PDE $\mathcal{E}$ are nothing but *secondary vector fields* on $\mathcal{E}_\infty$. Surprisingly, all secondary notions are cohomology classes of suitable natural complexes of differential operators, one of which, the jet-Spencer complex, was discussed in Section 6. For the whole picture see [66].

To illustrate this point we shall give some details on secondary differential forms. They constitute the first term of the $\mathcal{C}$*-spectral sequence*, which is defined as follows. Let $\mathcal{O} = (\mathfrak{M}, \mathcal{D})$ be a diffiety and $\mathcal{D}\Lambda(\mathcal{O}) = \oplus_{i\geq 0}\mathcal{D}\Lambda^i(\mathcal{O})$ the ideal of differential forms on $\mathfrak{M}$ vanishing on the distribution $\mathcal{D}$. This ideal is differentially closed and its powers $\mathcal{D}^k\Lambda(\mathcal{O})$ form a decreasing filtration of $\Lambda(\mathcal{O})$. The $\mathcal{C}$-spectral sequence $\{E_r^{p,q}(\mathcal{O}), d^{p,q}\mathcal{O}\}$ is the spectral sequence associated with this filtration. By definition, the space of secondary differential forms of degree p is the graded object $\oplus_{q=0}^n E_1^{p,q}(\mathcal{O})$ and d_1 is the secondary exterior differential. Note that a smooth fiber bundle may be naturally viewed as a diffiety and the corresponding $\mathcal{C}$-spectral sequence is identical to the Leray–Serre spectral sequence of this bundle.

Nontrivial terms of the $\mathcal{C}$-spectral sequence are all in the strip $0 \leq q \leq n,\ p \geq 0$ with $n = \dim \mathcal{D}$, and $E_1^{0,q}(\mathcal{O}) = \bar{H}^q(\mathcal{O})$ (horizontal de Rham cohomology of $\mathcal{O}$, see Subsection 6.4). Below we write simply $E_r^{p,q}$ for $E_r^{p,q}(\mathcal{O})$ if the context does not allow a confusion. Also recall that $\mathcal{C}$-differential DOs are those that admit restrictions to integral submanifolds of $\mathcal{D}$.

The following proposition illustrates the fact that the calculus of variations is just an element of the calculus of secondary differential forms.

Proposition 8.1. Let $\mathcal{O} = J^\infty(E, n)$. Then

1. If $E_1^{p,q}$ is nontrivial, then either $p = 0$ or $q = n$ (*"one line theorem"*).
2. $E_1^{0,q} = \bar{H}^q(J^1(E,n))$, if $q < n$, and $E_1^{0,n}$ is composed of variational functionals $\int \omega \bar{d}x_1 \wedge \cdots \wedge \bar{d}x_n$.

3. $d_1^{0,n}$ is the *Euler operator* of the calculus of variations:

$$E_1^{0,n} = \bar{H}^n(J^\infty(E,n)) \ni \int \omega \bar{d}x_1 \wedge \cdots \wedge \bar{d}x_n \overset{d_1^{0,n}}{\longmapsto} \ell_\omega^*(1) \in \widehat{\varkappa}$$

where $\widehat{\varkappa} \overset{\text{def}}{=} \mathrm{Hom}_{\mathcal{F}}(\varkappa, \bar{\Lambda}^n(J^\infty(E,n)))$ and ℓ_ω^* stands for the adjoint to ℓ_ω $\mathcal{C}$-differential operator.

4. $E_1^{2,n} = \mathcal{C}\,\mathrm{Diff}^{\mathrm{alt}}(\varkappa, \widehat{\varkappa}) = \{$skew-self-adjoint $\mathcal{C}$-differential operators from $\varkappa$ to $\widehat{\varkappa}\}$, and

$$d_1^{1,n} : \widehat{\varkappa} \ni \Psi \longmapsto \ell_\Psi^* - \ell_\Psi^* \in \mathcal{C}\,\mathrm{Diff}^{\mathrm{alt}}(\varkappa, \widehat{\varkappa}).$$

5. $E_2^{p,n} = H^{p+n}(J^1(E,n))$ and, in particular, the complex $\{E_1^{p,n}, d_1^{p,n}\}_{p\geq 0}$ is locally acyclic.

The reader will find a similar description of the terms $E_1^{p,n}$ and the differentials $d_1^{p,n}$ for $p > 2$ in [63, 66].

If $\mathcal{O} = \mathcal{E}_\infty$, then the terms $E_1^{0,q}(\mathcal{O})$ present various conserved quantities of the equation $\mathcal{E}$. For instance, the Gauss electricity conservation law is an element of $E_1^{0,2}(\mathcal{E}_\infty)$ for the system of Maxwell equations $\mathcal{E}$. The term $E_1^{0,n-1}(\mathcal{E}_\infty)$ is composed of standard *conservation laws* of a PDE $\mathcal{E}$, which are associated with *conserved densities*. In this connection we have

Proposition 8.2. Let $\mathcal{E}$ be a determined system of PDEs and $\mathcal{CL}(\mathcal{E}) \overset{\text{def}}{=} E_1^{0,n-1}(\mathcal{E}_\infty)$ the vector space of conservation laws for $\mathcal{E}$. Then

1. If $E_1^{p,q}$ is nontrivial, then either $p = 0$ or $q = n-1, n$ *("two lines theorem")*.
2. $\ker d_1^{0,n-1} = H^{n-1}(\mathcal{E})$ *(trivial conservation laws)*.
3. $E_1^{1,n-1} = \ker \ell_{\mathcal{E}}^*$ and $E_1^{1,n} = \operatorname{coker} \ell_{\mathcal{E}}^*$.

$\Upsilon = d_1^{0,n-1}(\Omega)$ is called the *generating function* of a conservation law $\Omega \in \mathcal{CL}(\mathcal{E})$. Assertion (2) of Proposition 8.2 tells that a conservation law is uniquely defined by its generating function up to a trivial one. Moreover, by assertion (3) of this proposition, generating functions are solutions of the equation $\ell_{\mathcal{E}}^*\Upsilon = 0$, and this is the most efficient known method for finding conservation laws (see [67, 29, 27]).

Propositions 8.1 and 8.2 unveil the nature of the classical Noether theorem. Namely, by assertions (3) and (4) of Proposition 8.1, the Euler–Lagrange equation $\mathcal{E}$ corresponding to the Lagrangian $\int \omega \bar{d}x_1 \wedge \cdots \wedge \bar{d}x_n$ is $\Psi = 0$ with $\Psi = \ell_\omega^*(1)$ and $\ell_{\mathcal{E}} = \ell_{\mathcal{E}}^*$. In other words, Euler–Lagrange equations are *self-adjoint*. Thus, in this case the equation $\ell_{\mathcal{E}}^* = 0$ whose solutions are generating functions of conservation laws of $\mathcal{E}$ (assertion (3) of Proposition 8.2) coincides with the equations $\ell_{\mathcal{E}} = 0$ whose solutions are generating functions of symmetries of $\mathcal{E}$ (formula (6.25)). Moreover, we see that this relation between symmetries and conservation laws takes place for a much larger class than the Euler–Lagrange class of PDEs, namely, the class of *conformally self-adjoint* equations: $\ell_{\mathcal{E}}^* = \lambda \ell_{\mathcal{E}}$, $\lambda \in \mathcal{F}_{\mathcal{E}}$.

All natural relations between vector fields and differential forms such as Lie derivatives, insertion operators, etc. survive at the level of secondary calculus in the form of some relations between the horizontal jet-Spencer cohomology and the first term of the $\mathcal{C}$-spectral sequence. Also, a morphism of diffieties induces a pull-back homomorphism of $\mathcal{C}$-spectral sequences. In particular, this allows to define *nonlocal conservation laws* of a PDE $\mathcal{E}$ as conservation laws of diffieties that cover $\mathcal{E}_\infty$. These are just a few of numerous facts that show high self-consistence of secondary calculus and its adequacy for needs of physics and mechanics.

Remark 8.1. The $\mathcal{C}$-spectral sequence was introduced by the author in [56]. It was preceded by some works by various authors on the inverse problem of calculus of variations and the resolvent of the Euler operator (or the Lagrange complex). These works may now be seen as results about the $\mathcal{C}$-spectral sequence for $\mathcal{O} = J^\infty(\pi)$ (see, for instance, [31, 52]). If $\mathcal{E} \subset J^k(\pi)$, then the first term of the $\mathcal{C}$-spectral sequence for $\mathcal{E}_\infty \subset J^\infty(\pi)$ acquires the second differential coming from the spectral sequence of the fiber bundle $\pi_\infty : J^\infty(\pi) \to M$ and it becomes the *variational bicomplex* associated with $\mathcal{E}$. This local interpretation of the $\mathcal{C}$-spectral sequence is due to T. Tsujishita [53], who described these two differentials in a semi-coordinate manner.

9 New language and new barriers

In the preceding pages we were trying to show that a pithy general theory of PDEs exists and to give an idea about the new mathematics that comes into light when developing this theory in a systematic way. Even now this young theory provides many new instruments allowing the discovery of new features and facts about well-known and for long time studied PDEs in geometry, mechanics and mathematical physics. The theory of singularities of solutions of PDEs sketched in Section 4 is an example of this to say nothing about symmetries, conservation laws, hamiltonian structures and other more traditional aspects. Moreover, numerous possibilities, which are within one's arm reach, are still waiting to be duly elaborated simply because of a lack of workmen in this new area. This situation is to a great extent due to a *language barrier*, since

> *the specificity of the general theory of PDEs is that it cannot be systematically developed in all its aspects on the basis of the traditionally understood differential calculus.*

Indeed, one very soon loses the way by performing exclusively direct manipulations with coordinate-wise descriptive definitions of objects of differential calculus, especially if working on such infinite-dimensional objects as diffieties. By their nature, these descriptive definitions cannot be applied to various situations when some kind of singularities or other nonstandard situations occur naturally. Not less important is that descriptive definitions give no idea about natural relations between objects of differential calculus. Typical questions that can in no way be neglected when dealing with the foundations of the theory of PDEs are: "What are tensor fields on manifolds with singularities, or on pro-finite manifolds, or what are tensor fields respecting a specific structure on a smooth manifold", etc. This kind of questions becomes much more delicate when working with diffieties.

All these questions can be answered by analyzing why and how the traditional differential calculus of Newton and Leibniz became a natural language of classical physics (including geometry and mechanics). Since the fundamental paradigm of classical physics states that existence means observability and vice versa, the first step in this analysis must be a due mathematical formalization of the observability mechanism in classical physics.

We do that by assuming that from a mathematical point of view a classical physical laboratory is the *unitary algebra* A over $\mathbb{R}$ *generated by measurement instruments* installed in this laboratory and called the *algebra of observables*. A state of an observed object is interpreted as a homomorphism $h : A \to \mathbb{R}$ of $\mathbb{R}$-algebras ($\equiv$ "readings of all instruments"). Hence the variety of all states of the system is identified with the real spectrum $\mathrm{Spec}_{\mathbb{R}} A$ of A. The validity of this formalization of the classical observation mechanism is confirmed by the fact that all aspects of classical physics are naturally and, even more, elegantly expressed in terms of this language. Say, one of the simplest necessary concepts, namely, that of velocity of an object at a state $h \in \mathrm{Spec}_{\mathbb{R}} A$ is defined as a tangent vector to $\mathrm{Spec}_{\mathbb{R}} A$ at the "point" h, i.e., as an $\mathbb{R}$-linear map $\xi : A \to \mathbb{R}$ such that $\xi(ab) = h(a)\xi(b) + h(b)\xi(a)$. Thus, velocity is a particular first order DO over the algebra A of observables in the sense of Definition 5.1. In this case $P = A$ and $Q = \mathbb{R}$ as an $\mathbb{R}$-vector space with the $\mathbb{R}$-module product $a \star r \overset{\text{def}}{=} h(a)r$, $a \in A$, $r \in \mathbb{R}$. The reader will find other simple examples of this kind in an elementary introduction to the subject [42].

Thus, by formalizing the concept of a classical physical laboratory as a commutative algebra, we rediscover differential calculus in a new and much more general form. The next question is: "What is the structure of this new language and what are its informative capacities?" In the standard approach the zoo of various structures and constructions in modern differential and algebraic geometry, mechanics, field theory, etc. that are based on differential calculus seems not to manifest any regularity. Moreover, numerous questions like "why do skew symmetric covariant tensors, i.e., differential forms, possess a natural differential d, while the symmetric ones do not" cannot be answered within this approach. On the contrary, in the framework of differential calculus over commutative algebras all these "experimental materials" are nicely organized within a scheme composed of *functors of differential calculus connected by natural transformations* and *the objects that represent them* in various categories of modules over the ground algebra.

The reader will find in a series of notes [69] various examples illustrating what one can discover by analyzing the question "what is the conceptual definition of covariant tensors". From the last three notes of this series he can also get an idea on the complexity of the theory of iterated differential forms and, in particular, tensors, in secondary calculus.

It should be especially mentioned that new views, instruments and facts coming from the general theory of PDEs and related mathematics offer not only new perspectives for many branches of contemporary mathematics and physics but at the same time put in question some popular current approaches and expectations ranging from algebraic geometry to QFT. Unfortunately, there is too much to say in order to present the necessary reasons in a satisfactory manner.

We conclude by stressing that

> *The complexity and the dimension of problems in the general theory of PDEs are so high that a new organization of mathematical research similar to that in experimental physics is absolutely indispensable.*

Unfortunately, the dominating mentality and the "social organization" of the modern mathematical community seems not to be sufficiently adequate to face this challenge.

Acknowledgements. The author is very grateful to A. B. Sossinsky who moved him to write a popularization paper dedicated to the modern general theory of PDEs and addressed to a general audience, and to the Editors for their kind invitation to include such a paper in the present volume. Additional thanks to Katya Vinogradova for correcting the author's "continental English" and to Athanase Papadopoulos for useful suggestions which improved the exposition.

Bibliography

[1] D. V. Alekseevski, V. V. Lychagin and A. M. Vinogradov, *Basic ideas and concepts of differential geometry*. Encyclopedia of Math. Sciences, 28 (1991), Springer-Verlag, Berlin.

[2] V. I. Arnold, *Mathematical Methods of Classical Mechanics*. Springer-Verlag, 1989.

[3] A. M. Astashov and A. M. Vinogradov, On the structure of Hamiltonian operator in field theory, *J. Geom. Phys.* 3 (1986), no. 2, 263–287.

[4] G. I. Barenblatt, *Similarity, Self-Similarity and Intermediate Asymptotics*. Consultants Bureau, New York, 1979

[5] G. W. Bluman and J. D. Cole, *Similarities Methods for Differential Equations*. Appl. Math, Sci., No. 13, Springer-Verlag, New York, 1974.

[6] Bryant, Chern, Gardner, Goldschmidt and Griffiths, *Exterior Differential Systems*. Mathematical Sciences Research Institute Publications, Springer-Verlag.

[7] É. Cartan, *Les systèmes différentiels extérieurs et leurs applications géométriques*. Hermann, Paris, 1945.

[8] D. Catalano-Ferraioli and A. M. Vinogradov, *Differential Invariants of Generic Parabolic Monge–Ampère equations*, J. Phys. A: Math. Theor., vol. 45, no. 26, 2012.

[9] R. Courant and D. Hilbert, *Metoden der Mathematischen Physik*. Springer, Heidelberg, 1924.

[10] G. Darboux, *Leçons sur la théorie générale des surfaces et les applications géométriques du calcul infinitésimal*. Vol. 1–4, Gauthier-Villars, Paris, 1887–1896.

[11] A. De Paris and A. M. Vinogradov, *Fat Manifolds and Linear Connections*. World Sci., 2009.

[12] Ch. Ehresmann, *Introduction à la théorie des structures infinitésimales et des pseudo-groupes de Lie*. Geometrie Differentielle, Colloq. Inter. du Centre Nat. de la Recherche Scientifique, Strasbourg, 1953, 97–127.

[13] M. Fliess, J. Levin, Ph. Martin and P. Rouchon, *Nonlinear control and diffieties with an application to physics*. Contemporary Mathematics, vol. 219, 81–92, 1998.

[14] A. R. Forsyth, *Theory of differential equations*. Vol. 5,6, Cambridge University Press, 1900.

[15] C. F. Gauss, *Disquisitiones generales circa superficies curvas*, Commentationes Societatis Regiae Scientiarum Gottingesis Recentiores, vol. VI, 1827 (Engl. transl, "General Investigations of Curved Surfaces", Raven Press, New York, 1965.

[16] I. M. Gel'fand and I. Dorfman, Hamiltonian operators and algebraic structures related to them. *Funct. Anal. Appl.* 13 (1979), 13–30.

[17] I. M. Gel'fand and G. E. Shilov, "Generalized functions", vol. 1–5, Nauka, Moscow, 1959 (Russian); English Transl., Acad. Press, 1966–1968.

[18] V. A. Golovko, I. S. Krasil'shchik and A. M. Verbovetsky, Variational Poisson–Nijenhuis structures on partial differential equations. *Theor. Math. Phys.* 154, no. 2, 268–282, 2008 (in Russian).

[19] Ed. M. Henneaux, I. S. Krasil'shchik and A. M. Vinogradov, *Secondary Calculus and Cohomological Physics*. Proc. Conf. "Secondary Calculus and Cohomological Physics", August 24–31, 1997, Moscow; Contemporary Mathematics, 219 (1998).

[20] N. H. Ibragimov, *Transformation groups applied to mathematical physics*. Riedel, Dordrecht, 1985.

[21] S. Igonin, Coverings and fundamental algebras for partial differential equations. *J. Geom. Phys.*, vol. 56, no. 6, 2006.

[22] S. Igonin, Higher jet prolongation Lie algebras and Backlund transformations for (1+1)-dimensional PDEs, arXiv:1212.2199 [nlin.SI] 10 Dec 2012.

[23] M. Janet, *Les systèmes d'équations aux derivée partielles*. Mém. Sci. Math., Gauthier-Villars, Paris, 1927.

[24] E. Kähler, *"Introduction to the theory of systems of differential equations,"* B. G. Teubner, Leipzig–Berlin, 1934.

[25] I. S. Krasil'shchik and P. Kersten, *Symmetries and Recursion Operators for Classical and Supersymmetric Differential Equations*. Kluwer Acad. Publ., 2000.

[26] I. S. Krasil'shchik, P. Kersten and A. M. Verbovetsky, Hamiltonian operators and ℓ^*-coverings, *J. Geom. Phys.* 50, no. 4, 273–302, 2004.

[27] I. S. Krasil'shchik and A. M. Verbovetsky, *Homological methods in equations of mathematical physics*. Open Education, Opava, 1998, 150 pp.

[28] I. S. Krasil'shchik and A. M. Vinogradov, Nonlocal trends in the geometry of differential equations: symmetries, conservation laws, and Bäcklund transformations, *Acta Appl. Math.*, 1989, vol. 15, no. 1–2, 161–209. Also in: *Symmetries of Partial Differential Equations*. Ed. by Vinogradov A. M., Kluwer Acad. Publ., Dordrecht, Boston, London, 1989, 161–209.

[29] I. S. Krasil'shchik and A. M. Vinogradov, eds., *Symmetries and Conservation Laws for Differential Equations of Mathematical Physics*. AMS, Translations of Mathematical Monograph series, 182 (1999).

[30] I. S. Krasil'shchik, V. V. Lychagin and A. M. Vinogradov, *Geometry of Jet Spaces and Nonlinear Differential Equations*. Advanced Studies in Contemporary Mathematics, 1 (1986), Gordon and Breach, New York, London. xx+441 pp.

[31] B. A. Kuperschmidt, *Geometry of jet bundles and the structure of Lagrangian and Hamiltonian formalism, in Geometric Methods in Mathematical Physics*. G. Kaiser and J. E. Marsden, eds., Lect. Notes in Math., Springer-Verlag, 1980, 162–218.

[32] M. Kuranishi, *Lectures on exterior differential systems*. Tata Inst. Fund. Res., Bombay, 1962.

[33] A. Kushner, V. Lychagin and V. Rubtsov, *Contact Geometry and Nonlinear Differential Equations*. Cambridge Univ., Press, Cambridge, 2007.

[34] T. Levy-Civita, *Caratteristiche dei Sistemi Differenziali e Propagazione Ondosa*. Zanichelli, Bologna, 1931.

[35] R. K. Luneburg, *Mathematical Theory of Optics*. Univ. of California Press, Berkeley – Los Angeles, 1964.

[36] F. Lizzi, G. Marmo, G. Sparano and A. M. Vinogradov, Eikonal type equations for geometrical singularities of solutions in field theory, *J. Geom. Phys.* 14, 211–235, 1994.

[37] V, V, Lychagin, Singularities of multivalued solutions of nonlinear differential equations and nonlinear phenomena. *Acta Appl. Math.* vol. 3, 1985, 135–173.

[38] F. Magri, A simple model of the integrable Hamiltonian Equation. *J. Math. Phys.* 19 (1978), 1156–1162.

[39] M. Marvan, A. M. Vinogradov and V. A. Yumaguzhin, *Differential Invariants of Generic Hyperbolic Monge–Ampère Equations*. Cent. Eur. J. Math., vol. 5, no. 1, 2007.

[40] V. P. Maslov, Operational methods, Nauka, Moscow, 1973 (Russian); (English translation) MIR, Moscow, 1976.

[41] G. Monge, *Application de l'analyse à la géométrie*. Bachelier, Paris, 1850.

[42] J. Nestruev, *Smooth manifolds and observables*. Graduate Texts in Math., vol. 220, Springer, 2003.

[43] L. V. Ovsiannikov, Gruppovoy analiz differenzial'nyh uravneniy, Nauka, Moskva, 1978 (Russian); English transl., *Group analysis of differential equations*. Academic Press, New York, 1982.

[44] J.-F. Pommaret, *Systems of partial differential equations and Lie pseudogroups*, Gordon and Breach, New York, 1978.

[45] P. K. Rashevsky, *Geometrical Theory of Partial Differential Equations* (in Russian). OGIZ, 1947.

[46] Ch. Riquier, *Les Systemes D'équations aux Derivees Partielles*, Gauthier-Villars, Paris, 2011.

[47] E. Schrödinger, *The fundamental idea of wave mechanics, Nobel Lectures – Physics 1922–1941*. Elsevier Publishing Company, Amsterdam, 1965.

[48] L. Schwartz, *Théorie des distributions*. Hermann, 2 vols., 1966.

[49] G. Sparano, G. Vilasi and A. M. Vinogradov, Vacuum Einstein Metrics with Bidimensional Killing Leaves. II. Global Aspects, *Diff. Geom. Appl.*, vol. 17, 15–35, 2002.

[50] D. C. Spencer, Overdetermined systems of partial differential equations, *Bull. Amer. Math. Soc.* 75 (1969), 179–239.

[51] S. L. Sobolev, *Applications of Functional Analysis in Mathematical Physics*. Leningrad: Leningrad University Press, 1950 (Russian); (English transl.), Some Applications of Functional Analysis in Mathematical Physics, AMS Transl. of Math. Monographs ser., 2008.

[52] W. M. Tulczyjew, The Lagrange complex, *Bull. Soc. Math. France*, 105 (1977), 419–431.

[53] T. Tsujishita, On variation bicomplexes associated to differential equations. *Osaka J. Math.* 19 (1982), 311–363.

[54] A. M. Vinogradov, The logic of the theory of linear differential operators (Russian). *Dokl. Akad. Nauk*, 1972, vol. 205, no. 5, 1025–1028; English transl. in Soviet Math. Dokl., 1972, vol. 13, no. 4, 1058–1062.

[55] A. M. Vinogradov, Multivalued solutions and a principle of classification of non-linear differential equation (Russian). *Dokl. Akad. Nauk*, 1973, vol. 210, no. 1, 11–14; English transl. in Soviet Math. Dokl., 1973, 14, no. 3, 661–665.

[56] A. M. Vinogradov, A spectral sequence associated with a non-linear differential equation, and the algebro-geometric foundations of Lagrangian field theory with constraints (Russian). *Dokl. Akad. Nauk SSSR*, 1978, vol. 238, no. 5, 1028–1031; Engl. transl. in Soviet Math. Dokl., 1978, vol. 19, 144–148.

[57] A. M. Vinogradov, Hamiltonian structures in field theory (Russian). *Dokl. Akad. Nauk SSSR*, 1978, vol. 241, no. 1, 18–21; English transl. in Soviet Math. Dokl., 1978, vol. 19, no. 4, 790–794.

[58] A. M. Vinogradov, The theory of higher infinitesimal symmetries of nonlinear partial differential equations. *Dokl. Acad. Nauk SSSR*, vol. 248, 274–278, 1979; English transl. in Soviet Math. Dokl., vol. 20, 985–990, 1979.

[59] A. M. Vinogradov, Some new homological systems associated with differential calculus over commutative algebras (Russian). *Uspechi Mat. Nauk*, 1979, vol. 34, no. 6, 145–150; English transl. in Russian Math. Surveys, 1979, vol. 34, no. 6, 250–255.

[60] A. M. Vinogradov, Geometry of nonlinear differential equations, *J. Sov. Math.* 17(1981), 1624–1649.

[61] A. M. Vinogradov, Local symmetries and conservation laws, *Acta Appl. Math.* vol. 2, 1984.

[62] A. M. Vinogradov, Category of nonlinear differential equations, *Springer Lect. Notes in Math.*, vol. 1108 (1984), 77–102.

[63] A. M. Vinogradov, The $\mathcal{C}$-spectral sequence, Lagrangian formalism and conservation laws. Part I. *J. Math. Anal. Appl.*, 100 (1984), 1–40. Part II, *Ibid.*, 41–129.

[64] A. M. Vinogradov, Geometric singularities of solutions of nonlinear partial differential equations. *Proc. Conf. Differential geometry and its applications*, Brno, 1986, Math. Appl. (East European Ser.), , Reidel, Dordrecht-Boston, MA, 1987, vol. 27, 359–379.

[65] A. M. Vinogradov, Scalar differential invariants, diffieties, and characteristic classes, In *Francaviglia M.* (Ed.), Mechanics, Analysis and Geometry: 200 Years after Lagrange, Elsvier, Amsterdam, 1991, 379–416.

[66] A. M. Vinogradov, *Cohomological analysis of partial differential equations and secondary calculus*. AMS, series: Transl. Math. Monographs, 204 (2001).

[67] A. M. Vinogradov, ed., *Symmetries of Partial Differential Equations: Conservation Laws-Applications-Algorithms*. Kluwer Acad. Publ., Dordrecht, 1989.

[68] A. M. Vinogradov, Singularities of solutions of partial differential equations. in preparation.

[69] A. M. Vinogradov and L. Vitagliano, *Iterated differential forms*. I-VI, Doklady Mathematics, I (Tensors), 2006, vol. 73, no. 2, 169–171; II (Riemannian Geometry Revisited), 2006, vol. 73, no. 2, 182–184; III (Integral Calculus), 2007, vol. 75, no. 2, 177–180; IV (The $\mathcal{C}$-Spectral Sequence), 2007, vol. 75, no. 3, 403–406; V (The $\Lambda_{k-1}\mathcal{C}$-Spectral Sequence on Infinite Jets), 2007, vol. 76, no. 2, 673–677; VI (The $\Lambda_{k-1}\mathcal{C}$-Spectral Sequence on Infinitely Prolonged Equations), 2007, vol. 76, no. 2, 692–695.

[70] A. M. Vinogradov, I. S. Krasilshchik and V. V. Lychagin, *Introduction to geometry of nonlinear differential equations*, (in Russian). Nauka, Moscow, 1986, 336 pp.

[71] V. E. Zakharov and L. D. Faddeev, *Korteweg–de-Vries equation: A completely integrable Hamiltonian system*. Funkts. Anal. Prilozh., vol. 5, no. 4, 1971.

[72] H. D. Wahlquist and F. B. Estabrook, Prolongation structures of nonlinear evolution equations. *J. Math. Phys.* 16 (1975), 1–7.

Chapter 7

Transformation groups in non-Riemannian geometry

Charles Frances

Contents

1 Introduction

The aim of this chapter is to survey some aspects of the links unifying a geometric structure $\mathcal{S}$ on a smooth manifold M, and its group of automorphisms $\mathrm{Aut}(M, \mathcal{S})$, namely, the group of diffeomorphisms of M preserving $\mathcal{S}$. This domain of research, which could be called theory of geometric transformation groups, takes its roots in the pioneering ideas of S. Lie, and developed all along the past century, to blossom into a vast area where a wide range of mathematics meet: Lie theory of course, but also differential geometry, dynamical systems, geometry of foliations, ergodic theory, algebraic actions, etc.

One of the main motivations to study Lie group actions preserving a geometric structure comes from the fact that for a lot of interesting structures, the automorphism group itself is always a Lie transformation group.[1] What we call "fact" here is actually the result of a series of theorems. One of the earliest is due to S. Myers and N. Steenrod, who proved that the isometry group of a Riemannian manifold is always a Lie group.

Theorem 1.1 ([32]). *Let (M, g) be an n-dimensional Riemannian manifold. Then the isometry group* $\mathrm{Iso}(M, g)$*, endowed with the compact open topology, is a Lie transformation group of dimension at most $\frac{n(n+1)}{2}$. Moreover this group is compact as soon as M is compact.*

From the proof of the theorem, one gets also a local result, which is that on each open subset $U \subset M$, the Lie algebra of Killing fields, namely, the vector fields defined on U and generating local flows of isometries, is finite dimensional. The dimension is again at most $\frac{n(n+1)}{2}$.

The geometric structures for which the two following facts always hold:

1. the automorphism group is a Lie transformation group;
2. the dimension of the Lie algebra of local Killing fields is finite

will be called *rigid* in this paper. Actually, there is a more precise and very general definition of rigid geometric structures introduced by Gromov in [25], which turns out to imply points (1) and (2) above. To avoid too much technicalities, we will keep our "rough" definition here.

The Myers–Steenrod theorem was followed by many other works which increased the list of geometric structures which are known to be rigid. Thus, pseudo-Riemannian metrics, affine connections, projective structures, conformal structures in dimension ≥ 3 turned out to be rigid.

On the other hand, there are structures, like for instance symplectic structures, which are not rigid. Any Hamiltonian flow on a symplectic manifold (M, ω) acts by symplectomorphisms, so that there are far too much symplectic automorphisms for Conditions (1) and (2) to be satisfied.

Observe also that some structures like complex ones display an intermediate behavior. By a theorem of Bochner and Montgomery, the automorphism group of a compact complex manifold is a Lie transformation group, so that for *compact* structures, Condition (1) is satisfied. Nevertheless, Condition (2) clearly always fails, and we won't retain complex structures as being rigid.

Assume now that we focus on a certain class of geometric structures, which are known to be rigid. For instance, we study Riemannian metrics, or Lorentzian ones, or affine connections. Given a manifold M endowed with a structure $\mathcal{S}$ belonging to our given class, it is natural to ask what kind of Lie group $\mathrm{Aut}(M, \mathcal{S})$ can be. More precisely,

[1]The precise definition of a Lie transformation group will be given at the beginning of Section 2.

Question 1.2. *What are the possible Lie groups that can be the automorphism group* $\mathrm{Aut}(M, \mathcal{S})$ *of a structure* $(M, \mathcal{S})$ *of the class we are considering, when* M *is a compact manifold?*

For instance, Theorem 1.1 says that the isometry group of a compact Riemannian manifold must be a compact Lie group. Conversely, it is shown in [38] that any compact Lie group may be realized as the full isometry group of a compact manifold. Hence, Question 1.2 is completely settled for Riemannian structures on compact manifolds.

Observe that Question 1.2 is meaningless if we don't put any global assumption on M, like compactness. To see this, let us consider any Lie group H, and put on its Lie algebra a Euclidean scalar product. Pushing this scalar product by left translations yields a left-invariant Riemannian metric on H. Thus, without any compactness assumption, we see that any Lie group may appear as a subgroup of isometries of a Riemannian manifold.

Answering Question 1.2 for other structures than Riemannian metrics is generally a hard problem, which is solved completely in very few cases. The subject got a renewed impulse thanks to the very influential works of Gromov and Zimmer in the eighties (see in particular the monumental [25], as well as [52]). We will present in Sections 3 and 4 definitive results on this problem, and partial ones in Section 5.

One of the difficulties to tackle Question 1.2 is that for most rigid geometric structures, there is no analogue of the Myers–Steenrod theorem. One can indeed exhibit instances of compact structures with a noncompact automorphism group. On the other hand, it is expected that such occurrences are rather unusual. A good illustration is the following result, proved by J. Ferrand, and independently (in a weaker form) by M. Obata in the early 1970s.

Theorem 1.3 ([33][18]). *A compact connected Riemannian manifold* (M, g) *of dimension* $n \geq 2$ *having a noncompact group of conformal diffeomorphisms must be conformally diffeomorphic to the standard sphere* $\mathbf{S}^n$.

This result is extremely strong, since the mere assumption of noncompactness of the automorphism group is enough to single out only one space: the model space of compact conformal Riemannian structures.

In the survey [15], it is vaguely conjectured that compact rigid geometric structures with a large automorphism group should be peculiar enough to be classified. Rather than a conjecture, we should speak of a principle, which is indeed illustrated by a lot of beautiful results, as Theorem 1.3. The heuristics is as follows. Rigid geometric structures, *generically*, do not admit any symmetry at all (local or global). The reader could take the example of a "generic" Riemannian metric as an illustration. The presence of *a lot* of symmetries is thus extremely unlikely. At the end of the spectrum, one has the highly non-generic case of structures which are homogeneous, or locally homogeneous, if we just consider local symmetries. These structures are few, but generally beautiful and play a prominent role in the theory. We will be especially interested in results showing that rather mild assumptions on the automorphism group force local homogeneity.

Beside noncompactness, other notions of "largeness" for the automorphism group are of interest. For instance, the condition can be put on the action of the group itself. Noncompactness amounts to a nonproperness assumption, but stronger dynamical conditions can be relevant: ergodicity, existence of a dense orbit etc. In this regard, a very striking result was obtained by Gromov in [25]. It is often called *open dense orbit theorem.*

Theorem 1.4 ([25]). *Let* $(M, \mathcal{S})$ *be a rigid geometric structure of algebraic type. Assume that the group* $\mathrm{Aut}(M, \mathcal{S})$ *acts on* M *with a dense orbit. Then there exists a dense open subset* $U \subset M$ *which is locally homogeneous.*

We won't define here the expression "algebraic type" in the above statement. The theorem applies to all the examples of rigid structures we have been considering so far: pseudo-Riemannian metrics, affine connections, etc. This is again a nice illustration of the principles presented above, even though the conclusion is slightly weaker than the one expected: the local homogeneity of the structure holds only on a dense open set. Passing from local homogeneity on this dense open set to the whole manifold is in general extremely difficult (actually it might be false in full generality), but is possible in certain cases [16], [5].

The organization of the chapter is as follows. We will first discuss several notions of geometric structures in Section 2, focusing on Cartan geometries and $G-$structures of finite type, for which the automorphism group is always a Lie transformation group. In Sections 3 and 4, we will present classification theorems for the automorphism group of conformal Riemannian structures, and Lorentzian metrics. Section 5 will be devoted to other geometric structures, for which the picture is less precise. We will emphasize results of Zimmer yielding valuable information about the structure of the Lie algebra of automorphisms of $G-$structures, as well as some developments in the realm of Cartan geometries, focusing on pseudo-Riemannian conformal structures.

2 Rigid geometric structures

Until now, our use of expressions such as "geometric structure" or "rigid structure" was rather informal. In any case, it is always rather arbitrary to adopt a definition of what a geometric structure is. Nevertheless, it is nowadays broadly accepted that a decisive step toward the modern way of viewing geometry was achieved by Klein, who was the first to consider a geometric structure as a manifold M acted upon (transitively) by a group G, and the study of all properties which are invariant under the group action. The modern formulation is that of a *Klein geometry*, as a homogeneous space G/P, where G is a Lie group and P a closed Lie subgroup of G. Whereas this point of view was extremely successful to unify the "classical" geometries such as Euclidean, hyperbolic, spherical, projective geometries, etc. it was later on considered as too restrictive for a general definition of geometric structure. As the mere example of Riemannian geometry shows, homogeneous structures are clearly the exception, not the rule. Thus, other attemps were made during the twentieth century to find broader definitions of geometric structures. We present briefly two of them below, which will be of interest for our purpose. Especially, those structures will have the remarkable property that *their automorphism group is always a Lie transformation group.*

Recall at this point that if M is a smooth manifold, a subgroup $H \subset \mathrm{Diff}^{\infty}(M)$ is a Lie transformation group if it can be endowed with a Lie group structure for which the action $H \times M \to M$ is smooth. Moreover, one requires that any flow of diffeomorphisms included in H is a 1-parameter group for the Lie group structure of H.

2.1 Cartan geometries A first broad generalization of Klein's definition of geometry was introduced by É. Cartan under the name "*espaces généralisés*". It is actually the most natural extension of Klein's point of view, since it gives a precise meaning of what a "curved analogue" of a Klein geometry is. To understand the notion of a Cartan geometry, we start with a homogeneous space $\mathbf{X} = G/P$, and we reverse slightly Klein's point of view, looking for geometric data on $\mathbf{X}$ whose automorphism group is exactly G. By geometric data, we mean here something like a tensor, or a connection, on the space $\mathbf{X}$, or on some space naturally built from $\mathbf{X}$.

The nice thing is that there is a general answer to this question, whatever the homogeneous space $\mathbf{X}$ is. Let us indeed consider the group G itself as a P-principal fiber bundle over G/P, and put on G the so called Maurer–Cartan form ω^{MC}. This is the 1-form on G, with values in the Lie algebra $\mathfrak{g}$, such that for every left-invariant vector field X, $\omega^{MC}(X) = X(e)$. It is not very hard to check that the automorphisms of the P-bundle $G \to G/P$, which moreover preserve ω^{MC}, are exactly the left translations on the group G. So, on our homogeneous space $\mathbf{X}$, we get a nice differential geometric structure, namely, the fiber bundle $G \to \mathbf{X}$, and the 1-form ω^{MC}, which is natural in the sense that its automorphism group is exactly G.

The next step is to generalize this picture to an arbitrary manifold M of the same dimension as $\mathbf{X}$. This is easily done by considering the following data:

- A P-principal fiber bundle $\hat{M} \to M$.
- A 1-form ω on $\hat{M}$ with values in the Lie algebra $\mathfrak{g}$, which mimics the Maurer–Cartan form. In particular, one requires that at every point $\hat{x}$ of $\hat{M}$, the map $\omega : T_{\hat{x}}\hat{M} \to \mathfrak{g}$ is a linear isomorphism. One also requires equivariance properties of ω with respect to the actions of P on $\hat{M}$ and $\mathfrak{g}$, but we won't mention them here. The reader who wants to know more about Cartan geometries is referred to the very comprehensive [37] and [9].

The triple $(\hat{M}, M, \omega)$ is called *a Cartan geometry modelled on* $\mathbf{X}$, and the form ω a *Cartan connection.* A Cartan geometry $(\hat{M}, M, \omega)$ modelled on $\mathbf{X} = G/P$ is often referred to as a *curved analogue* of the Klein geometry $\mathbf{X}$. The precise meaning of this sentence is made clear by the following remark. The Maurer–Cartan form on the

Lie group G satisfies a property, known as the Maurer–Cartan equation. If X and Y are two vector fields on G, then the equation reads:

$$d\omega^{MC}(X,Y) + [\omega^{MC}(X), \omega^{MC}(Y)] = 0. \tag{7.1}$$

Now, for a general Cartan geometry $(\hat{M}, M, \omega)$ modelled on $\mathbf{X} = G/P$, one can introduce the *curvature form* K, which is defined, for any pair (X, Y) of vector fields on $\hat{M}$ as

$$K(X,Y) = d\omega(X,Y) + [\omega(X), \omega(Y)].$$

In general, the curvature form is not zero, and actually, one shows that the curvature form vanishes identically if and only if the Cartan geometry $(\hat{M}, M, \omega)$ is *locally* equivalent to the model $(G, \mathbf{X}, \omega^{MC})$. Hence, spaces locally modelled on the homogeneous space $\mathbf{X}$ (which are commonly called $(G, \mathbf{X})$−structures) are just *flat* Cartan geometries modelled on $\mathbf{X}$.

The notion of a Cartan geometry is an elegant way of formalizing what is a curved Klein geometry. But one drawback of the definition is that it involves an abstract fiber bundle $\hat{M}$ over our manifold M, whereas one would rather like to work with geometric data directly available on M. Thus, if one fixes once and for all the model homogeneous space $\mathbf{X} = G/P$, two natural questions arise:

1. Given a Cartan geometry, modelled on $\mathbf{X}$, on a manifold M, can we interpret the data $(M, \hat{M}, \omega)$ on $\mathbf{X}$ in terms of geometric data $\mathcal{S}$ on M (such as tensors, connections, etc.).
2. Conversely, if such a set of geometric data $\mathcal{S}$ is given on M, can we build a P-principal fiber bundle $\pi : \hat{M} \to M$, which is natural with respect to $\mathcal{S}$, as well as a Cartan connection $\omega : T\hat{M} \to \mathfrak{g}$, so that the procedure described in point (1), when applied to the triple $(M, \hat{M}, \omega)$, yields back $\mathcal{S}$. One would like moreover to find suitable normalization conditions which make the connection ω unique.

A model homogeneous space $\mathbf{X} = G/P$ being given, we say that *the equivalence problem* is solved for Cartan geometries modelled on $\mathbf{X}$ if we can give a positive answer Problems (1) and (2). This is for instance the case for Cartan geometries modelled on the Euclidean space $\mathbf{E}^n = \mathrm{O}(n) \ltimes \mathbb{R}^n/\mathrm{O}(n)$. Such a geometry on a manifold yields a Riemannian metric g on M. Conversely, the existence of the Levi-Civita connection allows to build a Cartan connection ω on $\hat{M}$, the $\mathrm{O}(n)$−bundle of orthonormal frames associated to g. The fact that the Levi-Civita connection is the only torsion-free connection compatible with the metric ensures the uniqueness of this "normal" Cartan connection ω.

Classical examples of Cartan geometries The most interesting Cartan geometries to consider are of course those for which the equivalence problem is solved. We give examples below, but the reader should keep in mind that except for very few cases, solving the equivalence problem and proving the existence of a normal Cartan connection is not an easy matter at all. Since the pioneering works of É. Cartan ([10]), some deep advances on this problem were done in [13], [39], [8] among others.

1. Pseudo-Riemannian structures of type (p,q) are Cartan geometries modelled on type (p,q) Minkowski space $\mathbf{E}^{p,q} = \mathrm{O}(p,q) \ltimes \mathbb{R}^n / \mathrm{O}(p,q)$.
2. In dimensions $p+q \geq 3$, conformal classes of type (p,q) metrics are Cartan geometries (this was proved by É. Cartan himself in the Riemannian case). The model space $\mathbf{X}$ is the pseudo-Riemannian conformal analogue of the round sphere $\mathbf{S}^n$. It is called *Einstein's universe of signature* (p,q), denoted $\mathbf{Ein}^{p,q}$, and is the product $\mathbb{S}^p \times \mathbb{S}^q$ endowed with the conformal class of the product metric $-g_{\mathbb{S}^p} \oplus g_{\mathbb{S}^q}$ (where $g_{\mathbb{S}^m}$ stands for the round Riemannian metric on $\mathbb{S}^m$). As a homogeneous space under its conformal group, it can be written as $\mathbf{Ein}^{p,q} = \mathrm{O}(p+1,q+1)/P$, where P is the stabilizer of a null direction. In Riemannian signature $p=0$, $\mathbf{Ein}^{0,q}$ is just the round sphere.
3. Nondegenerate CR structures (structures appearing on real hypersurfaces with nondegenerate Levi form in complex manifolds) are Cartan geometries modelled on the boundary of the complex hyperbolic space, namely, the homogeneous space $\mathbf{X} = \mathrm{PSU}(p+1,q+1)/P$, where P is the stabilizer of a null direction.
4. Affine connections on an n-dimensional manifold M are Cartan geometries modelled on the affine space $\mathbf{A}^n = \mathrm{GL}(n,\mathbb{R}) \ltimes \mathbb{R}^n / \mathrm{GL}(n,\mathbb{R})$.
5. Projective classes of affine connections in dimension n are Cartan geometries modelled on the n-dimensional projective space $\mathbf{RP}^n = \mathrm{PGL}(n+1,\mathbb{R})/P$.
6. Conformal, CR and projective structures are instances of *parabolic geometries*, which are Cartan geometries modelled on a homogeneous space $\mathbf{X} = G/P$, for which G is a simple Lie group and P a parabolic subgroup. Thanks to the works [39], [8], the equivalence problem is solved for almost all parabolic geometries, meaning that there is a one-to-one correspondence between parabolic geometries with suitable normalizations made on the Cartan connection, and certain geometric data on the manifold.

Rigidity of Cartan geometries Let M be a manifold endowed with a Cartan geometry $\mathcal{S}$ modelled on some homogeneous space $\mathbf{X}$. By $\mathcal{S}$, we mean the triple $(\hat{M}, M, \omega)$. One defines naturally an automorphism of this geometry as a bundle automorphism $\hat{f} : \hat{M} \to \hat{M}$ satisfying $\hat{f}^*\omega = \omega$. When we deal with geometries for which the equivalence principle holds, and if $\mathcal{S}$ stands as well for the geometric data on M equivalent to the triple $(\hat{M}, M, \omega)$, then the automorphisms are exactly the diffeomorphisms $f : M \to M$ satisfying $f^*\mathcal{S} = \mathcal{S}$ (in the sense that such f lift naturally to automorphisms $\hat{f}$ of $\hat{M}$ in the previous sense). There is also a notion of Killing field of $\mathcal{S}$, as a vector field generating local flows of automorphisms.

A very important and nice feature of the automorphism group of a Cartan geometry is the

Theorem 2.1. *Let M be a manifold endowed with a Cartan geometry $\mathcal{S}$. Then the automorphism group* $\mathrm{Aut}(M,\mathcal{S})$ *is a Lie transformation group.*

Moreover, on each open subset of M, the Lie algebra of Killing fields has dimension at most $\dim \mathfrak{g}$.

Let us indicate why this theorem holds. The main point is that the Cartan connection ω defines naturally a *parallelism* (one says also a *framing*) on $\hat{M}$, namely, a family $\hat{X}_1, \dots, \hat{X}_s$ of vector fields of $\hat{M}$, where $s = \dim \hat{M} = \dim \mathfrak{g}$, such that at every $\hat{x} \in \hat{M}$, $(\hat{X}_1(\hat{x}), \dots, \hat{X}_s(\hat{x}))$ is a basis of $T_{\hat{x}}\hat{M}$. To build this parallelism, it is enough to consider a basis $(X_1, \dots, X_s)$ of $\mathfrak{g}$, and to define $\hat{X}_i = \omega^{-1}(X_i)$. Hence $\mathrm{Aut}(M, \mathcal{S})$ is identified with a closed subgroup of the group of automorphisms of our parallelism, the latter being itself a Lie transformation group, when endowed with the compact open topology. To see this, let us first observe that the group of diffeomorphisms preserving a parallelism $\mathcal{P}$ on a manifold must act freely. Indeed, if such an automorphism fixes a point x_0, then it must fix pointwise any curve γ through this point, such that γ' has constant coordinates in the frame field defining the parallelism. But a straigthforward application of the inverse mapping theorem shows that the set of such curves fills in an open neighborhood of x_0. Hence, the set of fixed points of our automorphism is open, and since it is of course closed, the automorphism is trivial. We infer for instance that a Killing field of a parallelism having a zero must be identically zero. This yields the bound on the dimension of the Lie algebra of local Killing fields in Theorem 2.1.

Another consequence of the freeness of the action is that the automorphism group of a parallelism can always be identified with any of its orbits. One has then to show that this identification is a homeomorphism and the orbits are closed smooth manifolds, which is a little bit more involved (details can be found in [26, Theorem 3.2]).

The above argument shows that the topology on $\mathrm{Aut}(M, \mathcal{S})$, making it a Lie transformation group, is the one inherited after identifying $\mathrm{Aut}(M, \mathcal{S})$ with one of its orbits on $\hat{M}$. In general, $\hat{M}$ is a subbundle of the bundle of m-jets of frames over M. Thus a sequence (f_k) converges in $\mathrm{Aut}(M, \mathcal{S})$ when (f_k), together with its m first derivatives, converges uniformly on compact subsets of M. In some cases, extra arguments show that this topology is actually the one induced by the compact-open topology on $\mathrm{Homeo}(M)$.

2.2 G-structures Beside Cartan geometries, there is another family of geometric structures, called G-structures, which was studied a lot by differential geometers.

Let us consider a smooth n-dimensional manifold M, and the bundle $R(M)$ of frames of M (this is a $\mathrm{GL}(n, \mathbb{R})$-principal bundle). Let G be a closed subgroup of the linear group $\mathrm{GL}(n, \mathbb{R})$. One defines *a G-structure on the manifold M* as a G-subbundle $\hat{M}$ of the bundle $R(M)$. This means that at each point of M, we select a subclass of distinguished frames (or, equivalently, we select a distinguished class of charts at each point of M).

There is a natural notion of automorphism for a G-structure: this is a diffeomorphism $f : M \to M$ whose action on $R(M)$ preserves the subbundle $\hat{M}$ defining the G-structure. In the same way, one defines what is an isomorphism between two

G-structures. Unlike for Cartan geometries, there is no direct notion of curvature for a G-structure, but there is still the notion of a *flat* (one says also *integrable*) G-structure, as one which is locally isomorphic to $\mathbb{R}^n \times G \subset \mathbb{R}^n \times \mathrm{GL}(n,\mathbb{R})$ (where the product $\mathbb{R}^n \times \mathrm{GL}(n,\mathbb{R})$ is identified with the frame bundle of $\mathbb{R}^n$).

1. For $G = \mathrm{O}(p,q)$, a G-structure on a manifold M is equivalent to the choice of a pseudo-Riemannian metric g of type (p,q). The subbundle $\hat{M}$ of the frame bundle defining the G-structure is then merely the bundle of orthonormal frames.
2. For $G = \mathbb{R}^* \times \mathrm{O}(p,q)$, where the factor $\mathbb{R}^*$ denotes the homothetic maps in $\mathrm{GL}(n,\mathbb{R})$, a G-structure on M is the same as the choice of a conformal class $[g] = \{e^\sigma g \mid \sigma \in C^\infty(M)\}$ of type (p,q) pseudo-Riemannian metrics.
3. Let $\omega = \Sigma dx_i \wedge dy_i$ be the standard symplectic form on $\mathbb{R}^{2n}$ and $\mathrm{Sp}(n,\mathbb{R})$ the group of linear transformations of $\mathbb{R}^{2n}$ preserving ω. An $\mathrm{Sp}(n,\mathbb{R})$-structure on a manifold M is called an *almost symplectic structure.* It is a genuine symplectic structure (*i.e* $d\omega = 0$) if and only if it is flat (this is Darboux's theorem).

Other examples of interesting G-structures are presented in [26, Chapter I].

Let us now make a trivial remark: a $\mathrm{GL}(n,\mathbb{R})$-structure on some n-dimensional manifold yields nothing more than the differentiable structure. Hence, any diffeomorphism of M is an automorphism of the structure. It is thus clear that we cannot expect all G-structures to be rigid (another example is given by symplectic structures, see (3) above). So, a natural question is: *can we determine, among all G-structures, which ones are rigid?*

There is a reasonable answer to this question, and it depends only on data involving the Lie algebra $\mathfrak{g}$. Let us fix the Lie group $G \subset \mathrm{GL}(n,\mathbb{R})$. Let us identify the bundle of frames of $\mathbb{R}^n$ as the product $\mathbb{R}^n \times \mathrm{GL}(n,\mathbb{R})$, and let us consider the *flat G-structure* on $\mathbb{R}^n$ as the subbundle $\mathbb{R}^n \times G \subset \mathbb{R}^n \times \mathrm{GL}(n,\mathbb{R})$. Let us try to exhibit a lot of Killing fields for this G-structure. Recall that by a Killing field, we mean a vector field X whose local flow acts by G-bundle automorphisms of $\mathbb{R}^n \times G$. Let $k \geq 0$ be an integer, let L_i, $i = 1,\dots,n$, be symmetric $(k+1)$-linear forms on $\mathbb{R}^n$, and let X be the polynomial vector field defined by

$$X(x) = \Sigma_{i=1}^n L_i(x,\dots,x)\frac{\partial}{\partial x_i}.$$

What is the condition for X to be a Killing field? Looking at the action of the local flow of X on the frame bundle, a necessary and sufficient condition is that for every $x \in \mathbb{R}^n$, the endomorphism of $\mathbb{R}^n$ given by

$$u \mapsto \begin{pmatrix} L_1(x,\dots,x,u) \\ \vdots \\ L_n(x,\dots,x,u) \end{pmatrix}$$

is an element of $\mathfrak{g}$. Differentiating k times, we find another necessary (and sufficient) condition which is that for every vectors $v_1, \dots, v_k$, the endomorphism

$$u \mapsto \begin{pmatrix} L_1(v_1, \dots, v_k, u) \\ \vdots \\ L_n(v_1, \dots, v_k, u) \end{pmatrix}$$

is in $\mathfrak{g}$.

This motivates the definition of *the k-th extension $\mathfrak{g}_k$ of $\mathfrak{g}$* as the set of symmetric $(k+1)$–linear maps

$$\varphi : \mathbb{R}^n \times \dots \times \mathbb{R}^n \to \mathbb{R}^n,$$

such that the endomorphism $u \mapsto \varphi(v_1, \dots, v_k, u)$ is in $\mathfrak{g}$ for every choice of the k first entries $v_1, \dots, v_k$. One easily checks that $\mathfrak{g}_0 = \mathfrak{g}$, and whenever $\mathfrak{g}_k = 0$, then $\mathfrak{g}_j = 0$ for every $j > k$.

The alternative is then as follows. Either $\mathfrak{g}_k \neq 0$ for every $k \in \mathbb{N}$, and in this case the flat G-structure admits Killing fields of the form $X = \Sigma_{i=1}^n L_i \frac{\partial}{\partial x_i}$, with L_i a polynomial of arbitrary large degree. The algebra of Killing fields is thus infinite-dimensional for the flat structure, and we won't retain G-structures as rigid structures.

More interestingly, if for some $k \in \mathbb{N}$, $\mathfrak{g}_{k+1} = 0$, one can define successive extensions of the frame bundle $R(M)$ as well, such that the last one yields a P-principal bundle $\hat{M}_k$ over M, *endowed with a natural parallelism* (the reader will find details in [26, Chapter I]). Here, P is a Lie group with Lie algebra $\mathfrak{g} \oplus \mathfrak{g}_1 \oplus \dots \oplus \mathfrak{g}_k$, and "natural" means that every automorphism $f : M \to M$ of the G-structure lifts to a bundle automorphism of $\hat{M}_k$ preserving the parallelism.

As for Cartan geometries, one can conclude:

Theorem 2.2 ([26], Theorem 5.1). *Let M be a manifold endowed with a G-structure of finite type $\mathcal{S}$. Then the automorphism group* $\mathrm{Aut}(M, \mathcal{S})$ *is a Lie transformation group.*

One can also show, as for Cartan geometries, that the dimension of the Lie algebra of local Killing fields is finite. Examples of G-structures of finite type are, for instance:

- Pseudo-Riemannian metrics of type (p, q) ($\mathfrak{g} = \mathfrak{o}(p, q)$ and $\mathfrak{g}_1 = 0$).
- Conformal structures of type (p, q) when $p + q \geq 3$ ($\mathfrak{g} = \mathbb{R} \oplus \mathfrak{o}(p, q)$, and $\mathfrak{g}_2 = 0$).
- One can also interpret the notion of affine connection, or projective class of such connections, but it is then necessary to introduce G-structures of higher order, but we won't do that here.

3 The conformal group of a Riemannian manifold

In the previous section, we exhibited quite a large classe of structures, whose automorphism group is a Lie group. For these structures, we can consider Question 1.2 formulated in the introduction: *a given class of structures being fixed, what are the possible Lie groups* $\mathrm{Aut}(M, \mathcal{S})$, *for* M *compact and* $\mathcal{S}$ *belonging to the class considered.*

As we already mentioned, the first complete general result regarding this question was proved for Riemannian structures by Myers and Steenrod in [32]. Their result describes the isometry group of a compact Riemannian manifold (M, g), namely, the group of smooth diffeomorphisms $\varphi : M \to M$ satisfying $\varphi^* g = g$. For the reader's convenience, we recall the statement:

Theorem 3.1 ([32]). *Let* (M, g) *be an* n*-dimensional Riemannian manifold. Then the isometry group* $\mathrm{Iso}(M, g)$, *endowed with the compact open topology is a Lie transformation group of dimension at most* $\frac{n(n+1)}{2}$. *Moreover this group is compact as soon as* M *is compact.*

This is now a direct consequence of Theorem 2.1 (or equivalently 2.2). Observe that the theorem contains an extra information about the topology making $\mathrm{Iso}(M, g)$ a Lie transformation group. It is induced here by the compact open topology, and we saw in Section 2.1 that this fact is generally not straigthforward. This amounts to showing that $\mathrm{Iso}(M, g)$ is closed in the group of homeomorphisms of M, namely, that a C^0 limit of smooth isometries is still smooth. The reason why this holds is that a homeomorphism preserving the distance defined by g must send parametrized geodesics to parametrized geodesics. The smoothness of such a transformation follows.

3.1 The theorem of Obata and Ferrand Let us now start with a Riemannian manifold (M, g) and let us take as a geometric structure the conformal class $[g] = \{e^{\sigma} g \mid \sigma \in C^{\infty}(M)\}$. In dimension at least three, (pseudo)-Riemannian conformal structures are Cartan geometries (and G-structures of finite type as well), hence by Theorem 2.1, the group of conformal diffeomorphisms $\mathrm{Conf}(M, g)$ (namely, the diffeomorphisms preserving the conformal class) is a Lie group. Observe that a conformal diffeomorphism is a transformation preserving angles between curves. Obviously, the group of conformal diffeomorphisms of M contains the isometry group of any metric in the conformal class $[g]$.

This inclusion can be strict, as shows the example of the round n-sphere $\mathbf{S}^n = (\mathbb{S}^n, g_0)$, where g_0 is "the" metric with constant curvature $+1$ on $\mathbb{S}^n$. The isometry group of $\mathbf{S}^n$ is $\mathrm{O}(n+1)$, whereas the conformal group is the Möbius group $\mathrm{PO}(1, n+1)$. The latter is noncompact, showing that one cannot expect a generalization of Theorem 3.1 for Riemannian conformal structures. At first glance, it is not unreasonable to expect the noncompactness of the conformal group to be a rather general phenomenon. But if the reader tries to determine the conformal group of classical spaces in Riemannian geometry (the real projective space $\mathbf{RP}^n$, flat tori, compact hyperbolic manifolds, etc.), he will always find a compact group. Actually, Lichnerowicz conjectured, in the middle of the sixties, that noncompactness of the conformal group for compact Riemannian manifold only occurs for the standard sphere. His guess became a theorem a few years later, thanks to independent works by Ferrand and Obata.

Theorem 3.2 ([33][18]). *A compact connected Riemannian manifold (M, g) of dimension $n \geq 2$ having a noncompact conformal group must be conformally diffeomorphic to the standard sphere $\mathbf{S}^n$.*

Actually, the result obtained by Obata in [33] is weaker, since he made the stronger assumption that the identity component $\mathrm{Conf}^\circ(M)$ is noncompact. For instance, Obata's result did not cover the possibility (which *a posteriori* never occurs) of an infinite discrete conformal group.

Theorem 3.2 settles Question 1.2 for Riemannian conformal structures. Indeed, it says that for a compact Riemannian manifold (M, g), the conformal group $\mathrm{Conf}(M, g)$ is either a compact Lie group, or the Möbius group $\mathrm{PO}(1, n+1)$. The latter possibility only occurs for the standard sphere $\mathbf{S}^n$.

3.2 The idea of the proof of the Ferrand–Obata theorem

We will work in dimension ≥ 3 (for $n = 2$, Theorem 3.2 is a consequence of the uniformization theorem for Riemann surfaces). The proof is made of two distinct steps. The first one is to use the noncompactness assumption on the conformal group to show that (M, g) is *conformally flat*, namely, every sufficiently small open subset of (M, g) is conformally diffeomorphic to an open subset of Euclidean space $\mathbf{E}^n$. The second step uses tools from the theory of $(G, \mathbf{X})$-structures to show that a compact conformally flat manifold which is not the standard sphere has a compact conformal group.

Let us recall that whereas any Riemannian metric on a surface is conformally flat (this was first shown by Gauss in the analytic case), the situation is completely different in dimension ≥ 3. Then, there exists a tensor W on M, that we will call the *conformal curvature*, which vanishes if and only if (M, g) is conformally flat. This tensor W is the Weyl (resp. Cotton) tensor in dimension ≥ 4 (resp. in dimension 3).

On a Riemannian manifold (M, g), the conformal curvature allows to build a conformally invariant "metric" putting $h_g = ||W||_g^\alpha g$, where $|W||_g$ denotes the norm of the tensor W with respect to g, and $\alpha = 1$ (resp. $\alpha = \frac{3}{2}$) when the dimension is ≥ 4 (resp. is 3). Of course, h_g is not really a Riemannian metric because W may vanish at some points. Nevertheless, if we assume that (M, g) is not conformally flat, h_g is not identically zero. Hence we can define a nontrivial, conformally invariant, singular distance

$$d_h(x, y) = \inf_\gamma \int h_g(\gamma', \gamma'),$$

the infimum being taken over all γ's joining x to y.

Now, still assuming that (M, g) is not conformally flat, let us consider the closed subset K on which the conformal curvature vanishes. This is a proper subset of M, and if $K_\epsilon = \{x \in M,\ d_h(x, K) \geq \epsilon\}$, then (K_ϵ, d_h) is, for ϵ sufficiently small, a genuine (*i.e* nonsingular) metric space. Moreover, K_ϵ is invariant by the conformal group $\mathrm{Conf}(M, g)$, and this group acts isometrically with respect to d_h. Using Ascoli's theorem, and elaborating a little bit, one infers that $\mathrm{Conf}(M, g)$ is compact.

This completes the first step: whenever $\mathrm{Conf}(M, g)$ is noncompact, then (M, g) must have *locally* the same conformal geometry as the standard sphere $\mathbf{S}^n$.

The second part of the proof aims at globalizing this local result. Once we know that (M, g) is conformally flat, then we also know since Kuiper (see [27]), that there is a conformal immersion $\delta : \tilde{M} \to \mathbf{S}^n$ (where $\tilde{M}$ stands for the universal cover of M), called *developing map*, as well as a morphism $\rho : \mathrm{Conf}(\tilde{M}, \tilde{g}) \to \mathrm{PO}(1, n+1)$ satisfying the equivariance relation

$$\delta \circ \gamma = \rho(\gamma) \circ \delta. \tag{7.2}$$

The noncompactness of $\mathrm{Conf}(\tilde{M}, \tilde{g})$ implies that of $\rho(\mathrm{Conf}(\tilde{M}, \tilde{g}))$. Now, divergent sequences (g_k) in $PO(1, n-1)$ have (up to extracting a subsequence) a "north-south" type dynamics, meaning that there is an attracting, and a repelling pole for the sequence (g_k). The equivariance relation (7.2) allows to show that some noncompact sequence (f_k) in $\mathrm{Conf}(\tilde{M}, \tilde{g})$ has a repelling pole on $\tilde{M}$, and that there exists a small open set $U \subset \tilde{M}$ on which δ is one-to-one, such that $f_k(U)$ is an increasing sequence, the union of which is $\tilde{M}$, or $\tilde{M}$ minus a point. We infer that δ is one-to-one on $\tilde{M}$, hence (M, g) is conformally equivalent to a quotient Ω/Γ, where $\Omega \subset \mathbf{S}^n$ is an open subset and $\Gamma \subset \mathrm{PO}(1, n-1)$ is a discrete subgroup acting cocompactly on Ω. The normalizer of Γ in $\mathrm{PO}(1, n+1)$ is then always compact (and so is the conformal group of Ω/Γ), except when $\Gamma = \{\mathrm{id}\}$. This forces $\Omega = \mathbf{S}^n$ by cocompactness and (M, g) is conformally diffeomorphic to the round sphere $\mathbf{S}^n$.

3.3 Generalizations to rank-one parabolic geometries Theorem 3.2 has been generalized to the noncompact case, independently by Ferrand in [20] and Schoen [36].

Theorem 3.3 ([20][36]). *Let (M, g) be a connected Riemannian manifold of dimension $n \geq 2$. If the group* $\mathrm{Conf}(M, g)$ *does not act properly on M, then M is conformally diffeomorphic to the standard sphere $\mathbf{S}^n$, or to Euclidean space $\mathbf{E}^n$.*

The proofs of this theorem are much more involved than for the compact case. The methods of Ferrand led to further generalizations of Theorem 3.3 for the group of K-quasi conformal mappings of a Riemannian manifold (see [21]).

The PDE methods used by Schoen in [36] allowed him to obtain the same kind of statement for strictly pseudoconvex CR structures. This is not that surprising if we take the point of view of Cartan geometries, because the model spaces of conformal Riemannian structures and of stricly pseudoconvex CR structures are the boundary at infinity of real and complex hyperbolic spaces respectively, hence have similar properties (for instance the north-south dynamics for divergent sequences of their automorphism group). These structures fit in the larger class of Cartan geometries for which the model space $\mathbf{X}$ is a quotient G/P, where G is a noncompact simple Lie group of rank one, and P a parabolic subgroup. In other words, $\mathbf{X}$ is the Hadamard boundary of a rank-one symmetric space of noncompact type. The Riemannian conformal case corresponds to $G = \mathrm{PO}(1, n+1)$. There are three other types of geometries involved, respectively for $G = \mathrm{PU}(1, n+1)$ (strictly pseudoconvex CR structures), $G = \mathrm{PSp}(1, n+1)$ and $G = F_4^{-20}$. These Cartan geometries are called *rank one parabolic geometries*. It turns out that the results of Obata–Ferrand–Schoen quoted above generalize to all the geometries of this family. Indeed:

Theorem 3.4 ([24]). *Let $(M, \mathcal{S})$ be a rank-one, regular, Cartan geometry modelled on $\mathbf{X}$. If the automorphism group $\mathrm{Aut}(M, \mathcal{S})$ does not act properly on M, then M is isomorphic, as a Cartan geometry, to the model $\mathbf{X}$, or to $\mathbf{X}$ minus a point.*

Of course, one recovers Theorem 3.3 when the model space $\mathbf{X}$ is the conformal sphere $\mathbf{S}^n$.

4 Lorentzian isometries

Like for conformal structures, there is no analogue of Myers–Steenrod's theorem in Lorentzian geometry. Recall that a Lorentzian metric g on a manifold M is a smooth field of indefinite nondegenerate bilinear forms of signature $(1, n-1)$ (namely, $(-, +, \dots, +)$). Finding compact Lorentz manifolds having a noncompact isometry group is not completely obvious, so that we begin by describing relevant examples.

4.1 Flat tori We begin with the simplest example. Let us consider the matrix

$$A = \begin{pmatrix} 2 & 1 \\ 1 & 1 \end{pmatrix}$$

in $\mathrm{SL}(2, \mathbb{R})$. It has two distinct eigenvalues $\lambda_1 < 1$ and $\lambda_2 > 1$, and we choose two associated eigenvectors u and v. Let g be the translation-invariant Lorentzian metric on $\mathbb{R}^2$ given by $g = du\,dv$. Let $\mathbb{T}^2$ be the quotient of $\mathbb{R}^2$ by the lattice $\mathbb{Z}^2$, equipped with the Lorentzian metric $\overline{g}$ induced by g. Because A normalizes $\mathbb{Z}^2$, it induces a Lorentzian isometry $\overline{A}$ of $(\mathbb{T}^2, \overline{g})$, and the group generated by $\overline{A}$ has noncompact closure in the homeomorphisms of $\mathbb{T}^2$ (actually $\overline{A}$ is an Anosov diffeomorphism).

One can elaborate on this example, by considering a quadratic form g on $\mathbb{R}^n$. We assume moreover that g has Lorentzian signature, and its coefficients are rational. We see g as a flat Lorentz metric on $\mathbb{R}^n$. Again, we consider the torus $\mathbb{T}^n = \mathbb{R}^n/\mathbb{Z}^n$, equipped with the metric $\overline{g}$ induced by g. A theorem of Borel and Harish-Chandra ensures that the group $\mathrm{O}(g, \mathbb{Z}) \subset \mathrm{O}(g)$, comprising all linear transformations of $\mathrm{O}(g)$

with integral entries, is a lattice in $\mathrm{O}(g)$. It is in particular noncompact. The isometry group of $(\mathbb{T}^n, \overline{g})$ coincides, as a Lie group, with the (noncompact) semi-direct product $\mathrm{O}(g, \mathbb{Z}) \ltimes \mathbb{T}^n$.

Observe that in those examples, the identity component of the isometry group is compact (a torus). The noncompactness comes from the discrete part of the group.

4.2 Anti-de sitter 3-manifolds Here is a more general procedure to build compact pseudo-Riemannian manifolds with noncompact isometry group. Start with a Lie group G, such that there is on $\mathfrak{g}$ an $\mathrm{Ad}(G)-$ invariant, pseudo-Riemannian scalar product λ. If one pushes λ by left translations on G, one gets a pseudo-Riemannian metric g on G, which is left-invariant by construction, but also right-invariant because λ is $\mathrm{Ad}(G)-$invariant. Assume that G has a cocompact lattice Γ. Then, on G/Γ, there is an induced pseudo-Riemannian metric $\overline{g}$ which is invariant for the left action of G on G/Γ. We thus get a copy of G inside $\mathrm{Iso}(G/\Gamma, \overline{g})$. If G cannot be embedded in any compact Lie group, this forces $\mathrm{Iso}(G/\Gamma, \overline{g})$ to be noncompact.

A nice example of this construction is obtained by choosing G to be a connected noncompact simple Lie group and λ to be the Killing form on $\mathfrak{g}$. The Killing form is defined by $\lambda(X, Y) = \mathrm{Tr}(\mathrm{ad}(X)\,\mathrm{ad}(Y))$. It is always $\mathrm{Ad}(G)-$invariant, and for simple Lie groups, it is nondegenerate. Moreover, it follows from the works of Borel and Harish-Chandra that noncompact simple Lie groups do admit cocompact lattices, and the above mentioned procedure works. Nevertheless, in general, the pseudo-Riemannian metric $\overline{g}$ on G/Γ is not Lorentzian.

The only case leading to Lorentzian metrics is that of $G = \mathrm{SL}(2, \mathbb{R})$ (or more generally when $\mathfrak{g} = \mathfrak{sl}(2, \mathbb{R})$). The manifold $\mathrm{SL}(2, \mathbb{R})$ endowed with the Killing form is a Lorentzian manifold of constant negative curvature, and is called *anti-de Sitter space* $\mathbf{AdS_3}$. For any cocompact lattice $\Gamma \subset \mathrm{SL}(2, \mathbb{R})$, the quotient $\mathrm{SL}(2, \mathbb{R})/\Gamma$, endowed with the induced Lorentzian metric, is a compact anti-de Sitter manifold, the isometry group of which is noncompact.

Let us mention that when Γ is torsion-free, the manifold $\mathrm{SL}(2, \mathbb{R})/\Gamma$ is naturally identified with (a double cover of) the unit tangent bundle $T^1\Sigma$ of the hyperbolic surface $\Sigma = \mathbb{H}^2/\Gamma$. The flows

$$\left\{ \begin{pmatrix} e^{\frac{t}{2}} & 0 \\ 0 & e^{-\frac{t}{2}} \end{pmatrix} \right\}_{t \in \mathbb{R}}$$

and

$$\left\{ \begin{pmatrix} 1 & t \\ 0 & 1 \end{pmatrix} \right\}_{t \in \mathbb{R}}$$

acting on the left on $\mathrm{SL}(2, \mathbb{R})/\Gamma$ can be interpreted respectively as the geodesic and horocyclic flows on $T^1\Sigma$. Thus, geodesic and horocyclic flows on the unit tangent bundle of hyperbolic surfaces are instances of isometric flows for Lorentzian metrics. This gives an idea of the richness of Lorentzian dynamics.

4.3 Warped Heisenberg groups Abelian and simple Lie groups are not the only instances of Lie groups admitting a bi-invariant pseudo-Riemannian metric. There is actually a classification of the possible Lie algebras for such groups (see [31]).

Here is a nice class of examples which will be important for our purpose. We start with the Heisenberg Lie algebra $\mathfrak{heis}(2d+1)$, generated as a vector space by $2d+1$ elements $X_1, Y_1, X_2, Y_2, \dots, X_d, Y_d$ and Z, with the bracket relations

$$[X_i, Y_i] = Z,\ [X_i, Z] = [Y_i, Z] = 0 \text{ for } i = 1, \dots, d.$$

In the sequel, we will denote by $\mathrm{Heis}(2d+1)$ the connected simply connected Lie group having $\mathfrak{heis}(2d+1)$ for Lie algebra (it can be seen as a subgroup of upper-triangular unipotent matrices in $\mathrm{GL}(d+2, \mathbb{R})$).

Let $\mu = (m_1, \dots, m_d) \in \mathbb{Z}_{\pm}$, where $\mathbb{Z}_{\pm}$ denotes the set of d-uples of integers all having the same sign. We introduce an extra element T satisfying the bracket relations

$$[T, X_i] = m_i Y_i, \text{ for } i = 1, \dots, d,$$
$$[T, Y_i] = -m_i X_i, \text{ for } i = 1, \dots, d,$$

and

$$[T, Z] = 0.$$

The Lie algebra $\mathbb{R} \ltimes_\mu \mathfrak{heis}(2k+1)$ spanned by $T, X_1, \dots, X_d, Y_1, \dots, Y_d, Z$ is called *a warped Heisenberg algebra.*

The derivation $\mathrm{ad}(T)$ can be integrated into an action of $\mathbf{S}^1$ by automorphisms of $\mathfrak{heis}(2d+1)$. Hence there exists a group G_μ, called a *warped Heisenberg group*, isomorphic to a semi-direct product $\mathbf{S}^1 \ltimes \mathrm{Heis}(2d+1)$, having $\mathbb{R} \ltimes_\mu \mathfrak{heis}(2k+1)$ for Lie algebra. Let us put on $\mathbb{R} \ltimes_\mu \mathfrak{heis}(2k+1)$ a Lorentzian product λ defined as follows:

- on $\mathrm{Span}(X_1, \dots, X_d, Y_1, \dots, Y_d)$, λ is a Riemannian scalar product, invariant by the action of $\mathrm{Ad}(\mathbf{S}^1)$.
- the scalar product $\lambda(T, Z)$ equals 1, and $\lambda(T, T) = \lambda(Z, Z) = 0$.
- the space $\mathrm{Span}(T, Z)$ is λ-orthogonal to $\mathrm{Span}(X_1, \dots, X_d, Y_1, \dots, Y_d)$.

Then λ is an $\mathrm{Ad}(G_\mu)$–invariant Lorentzian product on $\mathbb{R} \ltimes_\mu \mathfrak{heis}(2k+1)$, which can be pushed by left translations to get a bi-invariant Lorentz metric on G_μ.

The last remark is that if Γ is a cocompact lattice in $\mathrm{Heis}(2d+1)$, then Γ yields a cocompact lattice in G_μ. It follows that the compact manifold G_μ/Γ is endowed with a Lorentzian metric, for which the isometry group is noncompact (the connected component of this group is actually G_μ).

4.4 The classification of isometry algebras We now present the results classifying all the possible connected components for the isometry group of a compact Lorentz manifold. Actually, those results focus on the possible Lie algebras $\mathfrak{Iso}(M, g)$. They are due, independently and almost simultaneously, to Adams-Stuck and Zeghib, following some pioneering works of Zimmer and Gromov.

We already met examples of compact Lorentzian manifolds (M, g) where the Lie algebra $\mathfrak{Iso}(M, g)$ is $\mathfrak{sl}(2, \mathbb{R})$ or some warped Heisenberg algebra $\mathbb{R} \ltimes_\mu \mathfrak{heis}(2k+1)$. These examples can be enriched by the following warped-product procedure. Assume that (N, h) is a Riemannian manifold, (M, g) a Lorentzian one, and let $\beta : M \to (0, \infty)$ be some smooth function. Then on the product manifold $M \times N$, the warped metric $g' = h \oplus \beta g$ is Lorentzian, and it is clear that $\mathrm{Iso}(M, h) \times \mathrm{Iso}(N, g)$ is included in $\mathrm{Iso}(M \times N, g')$. Here is now the classification result we anounced at the begining of this section.

Theorem 4.1 ([1] [46] [47]). *Let (M, g) be a compact Lorentzian manifold. Then the Lie algebra $\mathfrak{Iso}(M, g)$ is isomorphic to a direct sum $\mathfrak{k} \oplus \mathfrak{a} \oplus \mathfrak{s}$, where $\mathfrak{k}$ is either trivial or the Lie algebra of a compact semisimple group, $\mathfrak{a}$ is abelian, and $\mathfrak{s}$ is either trivial, or of one of the following types:*

1. *The Lie algebra $\mathfrak{sl}(2, \mathbb{R})$.*
2. *The Heisenberg algebra $\mathfrak{heis}(2k + 1)$, for some integer $k \geq 1$.*
3. *A warped Heisenberg algebra $\mathbb{R} \ltimes_\mu \mathfrak{heis}(2k + 1)$, for some integer $k \geq 1$, and $\mu \in \mathbb{Z}_\pm$.*

Conversely, any such algebra is isomorphic to the Lie algebra of the isometry group of some compact Lorentzian manifold.

Beyond this algebraic result, the works [1], [46] give a quite precise picture of the geometry of the manifold (M, g), when the factor $\mathfrak{s}$ is nontrivial. For instance, if the group $\mathrm{SL}(2, \mathbb{R})$ acts faithfully and isometrically on a compact Lorentzian manifold (M, g), then the universal cover $(\tilde{M}, \tilde{g})$ is a warped product of $\widetilde{\mathrm{SL}(2, \mathbb{R})}$ endowed with the Killing form, and some Riemannian manifold (N, h) (this case was actually first proved by Gromov in [25]). The situation for actions of warped Heisenberg groups is also well understood, but less easy to describe.

These results illustrate once again the principle stated in the introduction: a "large" isometry group only occurs for very peculiar geometries.

To conclude this section about Lorentzian isometries, let us quote the following striking result by D'Ambra, the proof of which is a very nice application of Gromov's ideas presented in [25]. It says that some kind of Myers–Steenrod theorem is available in Lorentzian geometry, under some analyticity asumption, and for simply connected manifolds.

Theorem 4.2 ([14]). *Let (M, g) be a compact, analytic, simply connected Lorentzian manifold. Then the group $\mathrm{Iso}(M, g)$ is compact.*

5 Further structure results

In the previous sections, we presented very complete results answering Question 1.2 for peculiar geometric structures. Now, the question is: can we obtain more general statements for entire classes of structures, such as G-structures, or Cartan geometries?

The first significant theorems in this direction were obtained by Zimmer in [52], and also Gromov in [25]. They resulted from a wonderful mix of new ideas, involving ergodic theory and algebraic actions.

5.1 Zimmer's embedding theorem In [52], Zimmer investigated actions of connected, noncompact, simple Lie groups on compact manifolds, preserving a G-structure. The reader who does not know the definition of a simple Lie group should have in mind $\mathrm{SL}(m,\mathbb{R})$, $\mathrm{O}(p,q)$, $\mathrm{SU}(p,q)$, $\mathrm{Sp}(m,\mathbb{R})$ etc. He obtained the following theorem, often called *Zimmer's embedding theorem.*

Theorem 5.1 ([52], Theorem A). *Let H be a connected, noncompact simple Lie group. Assume that H acts faithfully on some n-dimensional manifold M, preserving a G-structure. Assume also that G is an algebraic subgroup of* $\mathrm{SL}'(n,\mathbb{R})$, *the group of linear transformations of $\mathbb{R}^n$ with determinant* ± 1. *Then :*

1. *There is a Lie algebra embedding $\sigma : \mathfrak{h} \to \mathfrak{g}$.*
2. *More precisely, there exists a linear subspace $V \subset \mathbb{R}^n$, a Lie subalgebra $\mathfrak{h}_V \subset \mathfrak{g} \subset \mathfrak{gl}(n,\mathbb{R})$ isomorphic to $\mathfrak{h}$, leaving V invariant, and such that the action of $\mathfrak{h}_V$ on V is conjugate to the linear action of the algebra* $\mathrm{ad}\,\mathfrak{h}$ *on $\mathfrak{h}$.*

Since H is a simple group, the map $\mathrm{ad} : \mathfrak{h} \to \mathrm{End}(\mathfrak{h})$ is one-to-one. It follows that the second point of the theorem implies the first one but, as we will see soon on some examples, it carries more information.

In the above statement, the G-structure we consider is not required to be of finite type. Hence, Zimmer's result applies for instance to symplectic structures which are not rigid in the sense we adopted in this text. Actually, the rigidity comes here from the algebraic assumption (simplicity) on the group H.

When the G-structure is of finite type, the automorphism group is a Lie group (see Section 2.2). The Levi decomposition allows to write the Lie algebra $\mathfrak{aut}(M)$ as a semidirect product $\mathfrak{s} \ltimes \mathfrak{r}$, where $\mathfrak{r}$ is the solvable radical and $\mathfrak{s}$ is a semisimple algebra. Zimmer's theorem puts some restrictions on the semisimple factor $\mathfrak{s}$: Noncompact factors in $\mathfrak{s}$ must embed into $\mathfrak{g}$, hence cannot be "too big", with respect to $\mathfrak{g}$.

5.2 Illustration in the case of isometric actions on Lorentz manifolds Zimmer's Theorem 5.1 predates Theorem 4.1 of almost ten years. It allows to derive quickly results which are now particular cases of Theorem 4.1.

Let us consider a compact manifold M endowed with a Lorentz metric g. Assume that some noncompact, connected, simple Lie group H acts isometrically on (M,g). We already saw that giving a Lorentz metric g on M amounts to giving an $\mathrm{O}(1,n-1)$-structure on M. Because $\mathrm{O}(1,n-1)$ is an algebraic subgroup of $\mathrm{SL}'(n,\mathbb{R})$, Zimmer's embedding theorem applies: there exists a Lie algebra embedding $\sigma : \mathfrak{h} \to \mathfrak{o}(1,n-1)$.

This puts rather strong restrictions on $\mathfrak{h}$. For instance, we infer immediately from the theorem that $\mathrm{SL}(3,\mathbb{R})$ cannot act isometrically on a compact Lorentz manifold. Indeed, such an action would provide a Lie algebra embedding $\sigma : \mathfrak{sl}(3,\mathbb{R}) \to \mathfrak{o}(1,n-1)$. But such an embedding cannot exist, because the real rank of $\mathfrak{sl}(3,\mathbb{R})$ is

2, whereas that of $\mathfrak{o}(1, n-1)$ is 1, and the rank cannot decrease under an embedding of Lie algebras.

Actually, the second point of Theorem 5.1 allows to determine $\mathfrak{h}$ completely. Indeed, it implies that the transformations of $\operatorname{ad}\mathfrak{h}$ are skew-symmetric with respect to some bilinear form B on $\mathfrak{h}$. This form B is just obtained after identifying $\mathfrak{h}$ with the subspace V (given by the statement of Theorem 5.1), and restricting to V a Lorentz scalar product which is $O(1, n-1)$-invariant. In particular, the totally isotropic subspaces of B have dimension at most 1. But on a simple real Lie algebra $\mathfrak{h}$ of non-compact type, one checks easily that any bilinear form B for which the elements of $\operatorname{ad}\mathfrak{h}$ are skew-symmetric must be zero on the root spaces $\mathfrak{h}_\alpha$, and two distinct root spaces $\mathfrak{h}_\alpha, \mathfrak{h}_\beta$ with $\alpha \neq -\beta$ must be B-orthogonal. As a consequence, if the isotropic subspaces of B have dimension at most 1, $\mathfrak{h}$ can have only two roots $\alpha, -\alpha$ with 1-dimensional root spaces. This only happens for the Lie algebra $\mathfrak{sl}(2, \mathbb{R})$. We are thus led to the following

Corollary 5.2. *Let (M, g) be a compact Lorentz manifold and H a connected non-compact simple Lie group acting isometrically on (M, g). Then $\mathfrak{h}$ is isomorphic to $\mathfrak{sl}(2, \mathbb{R})$.*

5.3 The idea of the proof of Zimmer's embedding theorem To illustrate the beautiful methods introduced by Zimmer to prove Theorem 5.1, we give an elementary exposition of the proof in the case of isometric actions on pseudo-Riemannian manifolds.

We are thus considering (M, g) a compact pseudo-Riemannian manifold of type (p, q) (without loss of generality, we will assume $p \leq q$), and H a connected non-compact simple Lie group acting isometrically and faithfully on (M, g).

The first important idea in Zimmer's proof is that an action of a connected Lie group on a manifold, when it preserves a geometric structure, often defines natural equivariant maps to algebraic varieties (called "Gauss maps" in [25]). Let us see this in the case of our isometric action. First, with each element $X \in \mathfrak{h}$, we associate the vector field X^* on M, defined as $X^*(x) = \frac{d}{dt}_{|t=0}(e^{tX}.x)$. We call $\mathcal{S}(\mathfrak{h})$ the space of symmetric bilinear forms on $\mathfrak{h}$, and $\operatorname{Gr}(\mathfrak{h})$ the Grassmannian of subspaces of $\mathfrak{h}$.

A first map we can consider is

$$\alpha : M \to \operatorname{Gr}(\mathfrak{h}),$$

which associates, with each point $x \in M$, the Lie algebra $\mathfrak{h}_x$ of vectors $X \in \mathfrak{h}$ satisfying $X^*(x) = 0$.

A second interesting map $\beta : M \to \mathcal{S}(\mathfrak{h})$ is defined as follows:

$$\beta_x(X, Y) = g_x(X^*(x), Y^*(x)).$$

These maps are natural in the sense that they are H-equivariant, where we make H act on $\mathcal{S}(\mathfrak{h})$ and $\operatorname{Gr}(\mathfrak{h})$ through the representation $\operatorname{Ad} : H \to \operatorname{GL}(\mathfrak{h})$.

Let us now consider the H-invariant open set Ω where the H-orbits have maximal dimension $m_0 \geq 1$, and the map $\alpha \times \beta : \Omega \to \operatorname{Gr}_{n_0}(\mathfrak{h}) \times \mathcal{S}(\mathfrak{h})$, where

$n_0 = \dim H - m_0$. The next key idea in the proof of Zimmer is to notice that, roughly speaking, $\mathrm{Gr}_{n_0}(\mathfrak{h}) \times \mathcal{S}(\mathfrak{h})$ is an algebraic variety on which H acts algebraically (through the representation $\mathrm{Ad} : H \to \mathrm{GL}(\mathfrak{h})$). More precisely, $\mathrm{Gr}_{n_0}(\mathfrak{h}) \times \mathcal{S}(\mathfrak{h})$ is an open subset of a projective subvariety of $\mathbf{RP}^m$, for some integer m, and the H-action on it comes from that of a simple Lie subgroup of $\mathrm{GL}(m+1, \mathbb{R})$. Thus, our Gauss map has transformed our initial dynamical system into an algebraic one. Moreover, our pseudo-Riemannian metric defines a volume on M, giving volume 1 to every direct orthonormal frame (if M is not orientable, this makes only sense locally, but this still defines a smooth measure). As a consequence, H preserves a Borel measure μ on M, which is finite by compactness of M. Pushing forward the measure μ by $\alpha \times \beta$, we get a (nonzero) finite Borel measure ν on $\mathrm{Gr}_{n_0}(\mathfrak{h}) \times \mathcal{S}(\mathfrak{h})$, and this measure ν is H-invariant.

Now, algebraic actions preserving a finite measure are dynamically very poor, since from the measurable point of view, they factor through actions of compact groups. This is the content of the following statement, often called "Borel density theorem".

Theorem 5.3 (Borel density theorem). *Let $H \subset \mathrm{GL}(m+1, \mathbb{R})$ be an algebraic subgroup. If the action of H on $\mathbf{RP}^m$ preserves a finite Borel measure ν, then there exists a cocompact, normal, algebraic subgroup $H_0 \subset H$ which acts trivially on the support of ν.*

In our situation, Borel's density theorem 5.3 says that for μ-almost every point $x \in \Omega$, $(\sigma(x), \beta(x))$ is $\mathrm{Ad}\, H$-invariant. The $\mathrm{Ad}\, H$-invariance of the subspace $\sigma(x) = \mathfrak{h}_x$ means exactly that the Lie algebra $\mathfrak{h}_x$ is an ideal of $\mathfrak{h}$. By simpleness of $\mathfrak{h}$, we get $\mathfrak{h}_x = \{0\}$ or $\mathfrak{h}_x = \mathfrak{h}$. Points where $\mathfrak{h}_x = \mathfrak{h}$ have orbits of dimension 0, so the definition of Ω leads to $\mathfrak{h}_x = \{0\}$ μ–almost everywhere on Ω. Because μ is of full support, this implies $\mathfrak{h}_x = \{0\}$ on Ω. It follows that the dimension of the orbits is that of H.

We now use the fact that $\beta(x)$ is $\mathrm{Ad}\, H$-invariant for μ–almost every x of Ω, which implies that the kernel of $\beta(x)$ must be an ideal in $\mathfrak{h}$. We infer that $\beta(x)$ is either zero, or non-degenerate of type (p', q'), $p' \leq p, q' \leq q$.

1. If $\beta(x)$ is zero, the restriction of g to the orbit $H.x$ is zero as well. Because we already noticed the dimension of $H.x$ is that of H, we infer $\dim H \leq p$ (and thus $p \geq 3$). The second point of Theorem 5.1 follows because through the representation $\mathrm{ad} : \mathfrak{h} \to \mathrm{End}(\mathfrak{h})$, the Lie algebra $\mathfrak{h}$ embeds into $\mathfrak{sl}(d, \mathbb{R})$, where $d = \dim H$, and $\mathfrak{o}(p,q)$ contains a subalgebra which is conjugate in $\mathfrak{gl}(p+q, \mathbb{R})$ to:
$$\left\{ \begin{pmatrix} A & 0 & 0 \\ 0 & 0_{p+q-2d} & 0 \\ 0 & 0 & -{}^tA \end{pmatrix}, \; A \in \mathfrak{sl}(d, \mathbb{R}) \right\}.$$
2. If $\beta(x)$ is nonzero, the restriction of g to the orbit $H.x$ is non-degenerate of type (p', q'). Thus $p' \leq p$ and $q' \leq q$. The group $\mathrm{Ad}(H)$, whose Lie algebra is $\mathfrak{h}$ by simpleness, can be seen as a subgroup of $\mathrm{O}(p', q')$, and Theorem 5.1 follows.

5.4 Extension of Zimmer's result to Cartan geometries Zimmer's Theorem 5.1 does not apply to all $G-$structures. Indeed, the group G is required to be an algebraic subgroup of $\mathrm{SL}'(n,\mathbb{R})$ (the subgroup of linear transformations with determinant ± 1). This assumption is basically equivalent to the fact that the $G-$structure defines a natural (*i.e* invariant by automorphisms) smooth measure on the manifold M. That's why some natural geometric structures do not enter in the range of application of Theorem 5.1. Actually, the statement is even wrong for some of those structures. To check this, let us consider, for instance, the case of Riemannian conformal structures. To have a conformal class of Riemannian metrics on some $n-$dimensional manifold amounts to having an $\mathbb{R}_+^* \times \mathrm{O}(n)-$structure. Observe that $\mathbb{R}_+^* \times \mathrm{O}(n)$ is not included in $\mathrm{SL}'(n,\mathbb{R})$. Now, the Möbius group $\mathrm{PO}(1,n+1)$ acts conformally on the round sphere $\mathbf{S}^n$. On the other hand, one can show that any Lie algebra morphism

$$\sigma : \mathfrak{o}(1,n+1) \to \mathbb{R} \oplus \mathfrak{o}(n)$$

must be trivial, hence can never be an embedding, and the conclusions of Theorem 5.1 do not hold in this case.

In light of this example, it would be desirable to obtain statements in the spirit of Zimmer's embedding theorem, for structures, like conformal ones, which do not define natural invariant measures.

Some results in this direction were proved in [6]. They yield significant information about the automorphism groups of geometric structures which are not covered by Theorem 5.1, like conformal, or CR, or projective structures. The class of geometric structures covered by these results is no longer that of $G-$structures, but that of Cartan geometries. The Cartan geometries we will consider in the following will be modelled on homogeneous spaces $\mathbf{X} = G/P$ satisfying the two properties:

- The action of G on $\mathbf{X} = G/P$ has finite kernel.
- The image $\mathrm{Ad}_{\mathfrak{g}} P$ of P by the representation $\mathrm{Ad} : G \to \mathrm{GL}(\mathfrak{g})$ is almost algebraic, namely, it has finite index in its Zariski closure.

These restrictions are actually harmless since they are satisfied for most relevant examples of Cartan geometries.

The main result of [6] is probably too technical to be stated here. It says roughly that if a connected Lie group H acts on a compact manifold M preserving a Cartan geometry $\mathcal{S}$ modelled on $\mathbf{X} = G/P$, then the adjoint representation of most solvable Lie subgroups $S < H$ on $\mathfrak{h}$ is "contained" in the adjoint representation of P on $\mathfrak{g}$. Thus, the upshot is that for a Lie subgroup $H < \mathrm{Aut}(M,\mathcal{S})$, relevant algebraic information on H are controlled by the pair (G, P).

Some of these algebraic information consist of numerical invariants, whose definition we recall now. If $L \subset \mathrm{GL}(m,\mathbb{R})$ is a linear subgroup, one defines the *real rank of* L, denoted $\mathrm{rk}(L)$, as the maximal dimension of an abelian subgroup of L made of $\mathbb{R}-$split transformations. The *algebraic rank* $\mathrm{rk}^{\mathrm{alg}}(L)$ is the maximal real rank of the Zariski closure of an abelian subgroup of L made of $\mathbb{R}-$split transformations. One always has $\mathrm{rk}(L) \leq \mathrm{rk}^{\mathrm{alg}}(L)$, and the inequality can be strict (see examples in Section 5.5 below). The *nilpotency index* of L, denoted by $\mathrm{nilp}(L)$ is the maximal nilpotency index of a connected nilpotent Lie subgroup of L.

Theorem 5.4 ([6], Theorems 1.3 and 1.5). *Let $(M, \mathcal{S})$ be a Cartan geometry modelled on the homogeneous space $\mathbf{X} = G/P$. Let H be a connected Lie subgroup of* $\mathrm{Aut}(M, \mathcal{S})$. *Assume that the manifold M is compact. Then*

1. $$\mathrm{rk}^{\mathrm{alg}}(\mathrm{Ad}\, H) \leq \mathrm{rk}(\mathrm{Ad}_{\mathfrak{g}}\, P)$$
   and
   $$\mathrm{nilp}(\mathrm{Ad}\, H) \leq \mathrm{nilp}(\mathrm{Ad}_{\mathfrak{g}}\, P).$$
2. *If moreover $\mathbf{X} = G/P$ is a parabolic geometry, if $\mathcal{S}$ is regular, and if the equality*
   $$\mathrm{rk}(\mathrm{Ad}\, H) = \mathrm{rk}(\mathrm{Ad}_{\mathfrak{g}}\, P)$$
   holds, then $(M, \mathcal{S})$ is isomorphic, as a Cartan geometry, to a quotient $\Gamma\backslash\tilde{\mathbf{X}}$, for some discrete group $\Gamma \subset \tilde{G}$.

We refer to Section 2.1 for the definition of parabolic geometries. The regularity condition involves conditions on the curvature, and is harmless since it is part of the normalization made on the Cartan connection to ensure uniqueness when the equivalence problem is solved.

The second point of the theorem might be compared to Theorem 3.2. It is another nice illustration of the principle stated in the introduction, that rigid geometric structures with large automorphism group should be very peculiar. Here, the largeness of the automorphism group is expressed by the fact that $\mathrm{Aut}(M, \mathcal{S})$ has the maximal real rank allowed.

Remark 5.5. If H is not assumed to be connected, the inequality

$$\mathrm{rk}^{\mathrm{alg}}(\mathrm{Ad}\, H) \leq \mathrm{rk}(\mathrm{Ad}_{\mathfrak{g}}\, P)$$

is still true, provided the kernel of the morphism $\mathrm{Ad} : H \to \mathrm{GL}(\mathfrak{h})$ is amenable

5.5 Illustrations We are going to illustrate Theorem 5.4, by considering actions on several geometric structures. We will be interested in the following groups:

- For $n \geq 2$, we consider the subgroup of affine transformations of $\mathbb{R}^n$ given by $\Gamma_n = \mathrm{SL}(n, \mathbb{Z}) \ltimes \mathbb{R}^n$. The real rank $\mathrm{rk}(\mathrm{Ad}\, \Gamma_n)$ is zero, but its algebraic rank $\mathrm{rk}^{\mathrm{alg}}$ is $n - 1$.
- For $n \geq 2$, we introduce $R_n = L_n \ltimes \mathbb{R}^n$, the subgroup of affine transformations of $\mathbb{R}^n$, for which

$$L_n = \left\{ \begin{pmatrix} e^{t_1} & 0 & 0 \\ 0 & \ddots & 0 \\ 0 & 0 & e^{t_n} \end{pmatrix}, \ (t_1, \dots, t_n) \in \mathbb{R}^n \right\}.$$

The group R_n is a semi-direct product $\mathbb{R}^n \ltimes \mathbb{R}^n$. The real rank $\mathrm{rk}(\mathrm{Ad}(R_n))$ as well as the algebraic rank $\mathrm{rk}^{\mathrm{alg}}(\mathrm{Ad}(R_n))$ are equal to n.

- For every $n \geq 2$, the group U_n of unipotent upper-triangular matrices in $\mathrm{GL}(n, \mathbb{R})$ is nilpotent, with index of nilpotency equal to $n-1$. The index $\mathrm{nilp}(\mathrm{Ad}(U_n))$ is equal to $n-2$.

We can now state some consequences of the previous theorem.

1. Let (M, g) be a compact pseudo-Riemannian manifold of type (p, q), where $1 \leq p \leq q$. Pseudo-Riemannian structures of type (p, q) are Cartan geometries whose model space is $\mathbf{E}^{p,q} = G/P$ where $G = \mathrm{O}(p, q) \ltimes \mathbb{R}^{p+q}$ and $P = \mathrm{O}(p, q)$. The real rank of $\mathrm{Ad}_{\mathfrak{g}} P$ is p, and its index of nilpotency is $2p-1$. We knew thanks to Zimmer's Theorem 5.1 that $\mathrm{SL}(m, \mathbb{R})$ cannot act isometrically on (M, g) if $m \geq p+2$. Theorem 5.4 and Remark 5.5 yield the same conclusion for the group Γ_m. In the same way, there is no isometric action of R_m on (M, g) as soon as $m \geq p+1$, and the same is true for U_m if $m \geq 2p+2$.

2. When $p+q \geq 3$, the conformal class of the type (p, q) pseudo-Riemannian manifold (M, g) defines a unique normal Cartan geometry modelled on the space $\mathbf{Ein}^{p,q} = \mathrm{O}(p+1, q+1)/P$ (where P is the stabilizer of an isotropic line in $\mathrm{O}(p+1, q+1)$, see Section 2.1). One computes $\mathrm{rk}(\mathrm{Ad}_{\mathfrak{g}} P) = p+1$ and $\mathrm{nilp}(\mathrm{Ad}_{\mathfrak{g}} P) = 2p+1$.

   Hence, for instance, a Lie group H acting conformally on a compact Lorentzian manifold must satisfy $\mathrm{rk}(\mathrm{Ad}\, H) \leq 2$. We thus infer that $\mathrm{SL}(4, \mathbb{R})$, which has real rank 3, does not admit such a conformal action. Actually, this is also true for $\mathrm{SL}(3, \mathbb{R})$, even though this group has rank 2. Indeed, by the second point of Theorem 5.4, a conformal action of $\mathrm{SL}(3, \mathbb{R})$ on a compact Lorentzian manifold can only occur on a quotient $\Gamma \backslash \widetilde{\mathbf{Ein}}^{1,n-1}$, which is conformally flat. Hence, such an action would provide an embedding of Lie algebras $\mathfrak{sl}(3, \mathbb{R}) \to \mathfrak{o}(2, n)$, and it is rather easy to check that this is impossible.

   Using the bounds provided by Theorem 5.4 and Remark 5.5, we infer more generally that there does not exist any conformal action of Γ_m on a compact type (p, q) manifold (M, g) as soon as $m \geq p+3$. The same conclusion holds for R_m when $m \geq p+2$, and for U_m if $m \geq 2p+3$.

3. As a last example, let us consider a compact manifold M of dimension $n \geq 2$, endowed with a linear connection ∇. This connection defines a unique Cartan geometry modelled on the affine space $\mathbf{A}^n = G/P$, where $G = \mathrm{GL}(n, \mathbb{R}) \ltimes \mathbb{R}^n$ and $P = \mathrm{GL}(n, \mathbb{R})$. One checks that $\mathrm{rk}(\mathrm{Ad}\, P) = n$, and $\mathrm{nilp}(\mathrm{Ad}\, P) = n-1$. We infer that whenever $m \geq n+2$, neither Γ_m, nor U_m can act on M preserving ∇. The same conclusion holds for R_m if $m \geq n+1$.

Bibliography

[1] S. Adams and G. Stuck, The isometry group of a compact Lorentz manifold, I and II. *Invent. Math.* 129 (1997), 239–261 and 263–287.

[2] D. V. Alekseevskii, Groups of conformal transformations of Riemannian spaces. *Math. USSR. Sb.* 18 (1972), 285–301.

[3] D. V. Alekseevskii, Self-similar Lorentzian manifolds. *Ann. Global Anal. Geom.* 3 (1985), 59–84.

[4] D. Asimov, Finite groups as isometry groups. *Trans. Amer. Math. Soc.* 216 (1976), 388–390.

[5] Y. Benoist, P. Foulon, and F. Labourie, Flots d'Anosov à distributions stable et instable différentiables. *J. Amer. Math. Soc.* 5 (1992), 33–74.

[6] U. Bader, C. Frances, and K. Melnick, An embedding theorem for automorphism groups of Cartan geometries. *Geom. Funct. Anal.* 19 (2009), 333–355.

[7] U. Bader and A. Nevo, Conformal actions of simple Lie groups on compact pseudo-Riemannian manifolds. *J. Differ. Geom.* 60 (2002), 355–387.

[8] A. Čap and H. Schichl, Parabolic geometries and canonical Cartan connections. *Hokkaido Math. J.* 29 (2000), 453–505.

[9] A. Čap and J. Slovák, *Parabolic geometries. I. Background and general theory.* Mathematical Surveys and Monographs. 154, American Mathematical Society, Providence, RI, 2009.

[10] É. Cartan, Les espaces à connexion conforme. *Ann. Soc. pol. math.* 2 (1923), 171–221.

[11] É. Cartan, Sur les variétés à connexion projective. *Bull. Soc. Math. France* 52 (1924), 205–241.

[12] É. Cartan, Les espaces géneralisés. In *Notice sur les traveaux scientifique*. Oeuvres Complètes, Partie I, Vol. I (1952), 72–85.

[13] S. S. Chern and J. K. Moser, Real hypersurfaces in complex manifolds. *Acta Math.* 133 (1974), 219–271.

[14] G. D'Ambra, Isometry groups of Lorentz manifolds. *Invent. Math.* 92 (1988), 555–565.

[15] G. D'Ambra and M. Gromov, Lectures on transformation groups: geometry and dynamics. In *Surveys in differential geometry* (Cambridge, MA, 1990), Lehigh Univ., Bethlehem, PA, 1991, 19–111.

[16] S. Dumitrescu, Homogénéité locale pour les métriques riemanniennes holomorphes en dimension 3. *Ann. Instit. Fourier.* 57 (2007), 739–773.

[17] R. Feres, *Dynamical systems and semisimple groups: an introduction.* Cambridge Tracts in Mathematics. 126, Cambridge University Press, Cambridge, 1998.

[18] J. Ferrand, Transformations conformes et quasi-conformes des variétés riemanniennes compactes. *Mém. Acad. Royale Belgique.* 39 (1971), 1–44.

[19] J. Ferrand, Geometrical interpretations of scalar curvature and regularity of conformal homeomorphisms. In *Differential geometry and relativity.* Mathematical Phys. and Appl. Math., 3, Reidel, Dordrecht, 1976, 91–105.

[20] J. Ferrand, The action of conformal transformations on a Riemannian manifold. *Math. Ann.* 304 (1996), 277–291.

[21] J. Ferrand, Convergence and degeneracy of quasiconformal maps of Riemannian manifolds. *J. Anal. Math.* 69 (1996), 1–24.

[22] J. Ferrand, Histoire de la réductibilité du groupe conforme des variétés riemanniennes (1964–1994). In *Séminaire de Théorie Spectrale et Géométrie*, 17, Année 1998–1999, 9–25.

[23] D. Fisher, Groups acting on manifolds: around the Zimmer program. In *Geometry, Rigidity and Groups Actions*, Chicago Lectures in Mathematics, 2011.

[24] C. Frances, Sur le groupe d'automorphismes des géométries paraboliques de rang 1. *Ann. Sci. École Norm. Sup.* 40 (2007), 741–764.

[25] M. Gromov. Rigid transformations groups. In *Géométrie différentielle* (Paris, 1986). Travaux en Cours, 33 (1988), Hermann, Paris, 65–139.

[26] S. Kobayashi, *Transformation groups in differential geometry.* Reprint of the 1972 edition. *Classics in Mathematics*. Springer-Verlag, Berlin, 1995.

[27] N. H. Kuiper, On conformally-flat spaces in the large. *Ann. of Math.* 50 (1949), 916–924.

[28] F. Labourie, Large groups actions on manifolds. In *Proceedings of the International Congress of Mathematicians*, Vol. II (Berlin, 1998), 371–380.

[29] A. Lichnerowicz, Sur les transformations conformes d'une variété riemannienne compacte. *C. R. Acad. Sci. Paris.* 259 (1964), 697–700.

[30] K. Melnick, A Frobenius theorem for Cartan geometries, with applications. *Enseign. Math.* 57 (2011), 57–89.

[31] A. Medina and P. Revoy, Les groupes oscillateurs et leurs réseaux. *Manuscripta Math.* 52 (1985), 81–95.

[32] S. B Myers and N. E Steenrod, The group of isometries of a Riemannian manifold. *Ann. of Math.* 40 (1939), 400–416.

[33] M. Obata, The conjectures on conformal transformations of Riemannian manifolds. *J. Differential Geometry.* 6 (1971/72), 247–258.

[34] R. Quiroga-Barranco, Isometric actions of simple Lie groups on pseudo-Riemannian manifolds. *Ann. of Math.* 164 (2006), 941–969.

[35] R. S. Palais, A global formulation of the Lie theory of transformation groups. *Mem. Amer. Math. Soc.* 22 (1957).

[36] R. Schoen, On the conformal and CR automorphism groups. *Geom. Funct. Anal.* 5 (1995), 464–481.

[37] R. W. Sharpe, *Differential Geometry: Cartan's generalization of Klein's Erlangen Program.* New York, Springer, 1997.

[38] R. Saerens and W. Zame, The isometry groups of manifolds and the automorphism groups of domains. *Trans. Amer. Math. Soc.* 301 (1987), 413–429.

[39] N. Tanaka, On the equivalence problems associated with simple graded Lie algebras. *Hokkaido Math. J.* 8 (1979), 23–84.

[40] W. P. Thurston, *Three-dimensional geometry and topology.* Vol. 1. Edited by Silvio Levy. Princeton Mathematical Series, 35, Princeton University Press, Princeton, NJ, 1997.

[41] S. M. Webster On the transformation group of a real hypersurface. *Trans. Amer. Math. Soc.* 231 (1977), 179–190.

[42] D. Witte Morris and R. J. Zimmer, Ergodic theory, groups, and geometry. In *CBMS Regional Conference Series in Mathematics*. 109, American Mathematical Society, Providence, RI, 2008.

[43] B. Wong, Characterization of the unit ball in $\mathbf{C}^n$ by its automorphism group. *Invent. Math.* 41 (1977), 253–257.

[44] A. Zeghib, Killing fields in compact Lorentz 3-manifolds. *J. Differ. Geom.* 43 (1996), 859–894.

[45] A. Zeghib, Sur les actions affines des groupes discrets. *Ann. Inst. Fourier.* 47 (1997), 641–685.

[46] A. Zeghib, Sur les espaces-temps homogènes. In *The Epstein Birthday Schrift*, Geom. Topol. Monogr., vol. 1. Coventry, 1998, 551–576.

[47] A. Zeghib, The identity component of the isometry group of a compact Lorentz manifold. *Duke Math. J.* 92 (1998), 321–333.

[48] A. Zeghib, Isometry groups and geodesic foliations of Lorentz manifolds. I. Foundations of Lorentz dynamics. *Geom. Funct. Anal.* 9 (1999), 775–822.

[49] A. Zeghib, Isometry groups and geodesic foliations of Lorentz manifolds. II. Geometry of analytic Lorentz manifolds with large isometry groups. *Geom. Funct. Anal.* 9 (1999), 823–854.

[50] R. J. Zimmer, *Ergodic theory and semisimple groups*. Monographs in Mathematics, 81, Birkhäuser Verlag, Basel, 1984.

[51] R. J. Zimmer, Semisimple automorphism groups of G-structures. *J. Differ. Geom.* 19 (1984), 117–123.

[52] R. J. Zimmer, On the automorphism group of a compact Lorentz manifold and other geometric manifolds. *Invent. Math.* 83 (1986), 411–424.

[53] R. J. Zimmer, Ergodic theory and the automorphism group of a G-structure. In *Group representations, ergodic theory, operator algebras, and mathematical physics* (Berkeley, Calif., 1984), Math. Sci. Res. Inst. Publ., 6 (1987), Springer, New York, 247–278.

[54] R. J. Zimmer, Split rank and semisimple automorphism groups of G-structures. *J. Differ. Geom.* 26 (1987), 169–173.

Chapter 8

Transitional geometry

Norbert A'Campo and Athanase Papadopoulos

Contents

1 Introduction

There are well-known relations between the three constant-curvature geometries, and there are known instances in which a sequence of geometries of nonzero curvature converges to the Euclidean geometry. For instance, it is well known that the geometry of a sequence of spheres whose radii tend to infinity converges to the geometry of the Euclidean plane.[1] Likewise, a sequence of hyperbolic planes with curvature tending to 0 converges to the Euclidean plane. Lobachevsky, although he did not have the notion of curvature for his geometry, knew that hyperbolic geometry becomes Euclidean at the infinitesimal level, and he checked this fact whenever he found an occasion for doing so. He noticed that the first-order approximations of the non-Euclidean trigonometric formulae that he obtained are the usual formulae of Euclidean geometry, see e.g. his *Pangeometry* [15]. We can also mention Gauss, who wrote to his friend F. A. Taurinus on November 8, 1824: "The assumption that in a triangle the sum of three angles is less than 180° leads to a curious geometry, quite different from ours, but thoroughly consistent, which I have developed to my entire satisfaction, so that I can solve every problem in it with the exception of the determination of a constant, which cannot be designated *a priori*. The greater one takes this constant, the nearer

[1] The one-dimensional aspect was known since ancient times, even though the notion of convergence was not rigorously defined. For instance, Kepler already thought of a straight line as a circle whose center is at infinity, cf. [14], p. 290. Before Kepler, Nicholas of Cusa, in his *De docta ignorantia* (1440) described how a circle with increasing radius tends to a straight line.

one comes to Euclidean geometry, and when it is chosen infinitely large, the two coincide."[2]

In this paper, we relate the three geometries of constant curvature by constructing a continuous transition between them. This is done in the setting of geometry as a transformation group, as expressed in Klein's Erlangen program. We construct a fiber space $\mathcal{E} \to [-1,1]$, where the fiber above each point $t \in [-1,1]$ is a two-dimensional space of constant curvature which is positive for $t > 0$ and negative for $t < 0$ and such that when t tends to 0 from each side, the geometry converges to the geometry of the Euclidean plane. Making the last statement precise will consist in showing that segments, angles, figures and several properties and propositions of the two non-Euclidean geometries converge in some appropriate sense to those of the Euclidean one. We say that the geometries *transit*. At the level of the basic notions (points, lines, distances, angles, etc.) and of the geometrical properties of the figures (trigonometric formulae, area, etc.), we say that they *transit in a coherent way*. This is expressed by the existence of some analytic sections of the fiber space $\mathcal{E}$ which allows us to follow these notions and properties. The group-theoretic definitions of each of the primary notions of the geometries of constant curvature (points, lines, etc.) make them transit in a coherent way.

The idea of transition of geometries was already emitted by Klein in §15 of his paper [12] (see the comments in Chapter 5 of this volume [2]). We note however that the abstract theoretical setting of transformation groups was not really developed at the time Klein wrote the paper [12]. The notion of "classical group", which is probably the most convenient setting for the description of these groups, was developed later, in works of Weyl [22], Dieudonné [5] and others.

The construction we describe here is developed in Chapter 9 of our *Notes on hyperbolic geometry* [1]. We review the basic ideas and we add to them some new results and comments. Beyond the fact that it establishes relations between the three classical geometries, this theory highlights at the same time the important notions of *families* and of *deformations*.

2 The fiber space $\mathcal{E}$

We work in a fiber space over $[-1,1]$ where the fibers above each point are built out of the groups of transformations of the three geometries. Each fiber is a geometry of constant curvature considered as a homogeneous space in the sense of Lie group theory, that is, a space of cosets G/H_0 where G is a Lie group – the orthogonal group of some quadratic form – and H_0 the stabilizer of a point. This is an example of the classical concept of *Klein geometry*, see e.g. [7] in this volume.

When the parameter $t \in [-1,1]$ is negative, the homogeneous space is the hyperbolic plane of a certain constant negative curvature. When it is positive, the ho-

[2]The translation is from Greenberg [9]. Gauss's correspondence is included in Volume VIII of his *Collected Works* [8].

mogeneous space is the sphere (or the elliptic plane) of a certain constant positive curvature. When the parameter is zero, the homogeneous space is the Euclidean plane. We can define points, lines and other notions in the geometry using the underlying group. For instance, a *point* will be a maximal abelian compact subgroup of G (which can be thought of as the stabilizer group of the point in the ambient group). The *space* of the geometry is the set of points, and it is thus built out of the group. A *line* is a maximal set of maximal abelian compact subgroups having the property that the subgroup generated by any two elements in this set is abelian. Other basic elements of the geometry (angle, etc.) can be defined in a similar fashion. This is in the spirit of Klein's Erlangen program where a geometry consists of a group action. In this case, the underlying space of the geometry is even constructed from the abstract group. The definitions are made in such a way that points, distances, angles, geometrical figures and trigonometric formulae vary continuously from hyperbolic to spherical geometry, transiting through the Euclidean, and several phenomena can be explained in a coherent manner.

To be more precise, we consider the vector space $\mathbb{R}^3$ equipped with a basis, and we denote the coordinates of a point p by $(x(p), y(p), z(p))$ or more simply (x, y, z).

We recall that the isometry groups of the hyperbolic plane and of the sphere are respectively the orthogonal groups of the quadratic form

$$(x, y, z) \mapsto -x^2 - y^2 + z^2$$

and

$$(x, y, z) \mapsto x^2 + y^2 + z^2.$$

Introducing a nonzero real parameter t does not make a difference at the level of the axioms of the geometries (although it affects the curvature): for any $t < 0$ (respectively $t > 0$) the isometry group of the hyperbolic plane (respectively the sphere) is isomorphic to the orthogonal group of the quadratic form

$$q_t : (x, y, z) \mapsto tx^2 + ty^2 + z^2. \tag{8.1}$$

We wish to include the Euclidean plane in this picture.

The first guess to reach the Euclidean plane is to give the parameter t the value 0. This does not lead to the desired result. In fact, although the orthogonal groups of the quadratic forms q_t, for $t > 0$ (respectively for $t < 0$) are all isomorphic, they are not uniformly bounded in terms of t, and when $t \to 0$, their dimension blows up. Indeed, when $t \to 0$ from either side, the quadratic form q_t reduces to

$$(x, y, z) \mapsto z^2$$

whose orthogonal group is much larger than the isometry group of the Euclidean plane.

Thus, an adjustment is needed. For this, we shall introduce the notion of "coherent element". This will make the Euclidean plane automorphism group (and the Euclidean plane itself) appear in a continuous way between the hyperbolic and spherical automorphism group (respectively the hyperbolic plane and the sphere).

We now define the fiber space $\mathcal{E}$, equipped with its fibration

$$\mathcal{E} \to [-1, 1].$$

The fiber E_t above each $t \in [-1, 1]$ will be a space of constant positive curvature t^2 for $t > 0$ and of constant negative curvature $-t^2$ for $t < 0$, such that when t converges to 0 from either side, E_t converges to the Euclidean plane.

We explain this now. It will be useful to consider the elements of the space E_t as matrices, and when we do so we shall denote such an element by a capital letter A.

Let $J \subset \mathrm{GL}(3, \mathbb{R}) \times [-1, 1] \to [-1, 1]$ be the fibration whose fiber J_t above t is the subgroup of $\mathrm{GL}(3, \mathbb{R})$ consisting of the stabilizers of the form q_t. Finally, let $I \subset J$ be defined by taking fiberwise the connected component of the identity element in the orthogonal group of the quadratic form q_t, for each $t \in [-1, 1]$.

In other words, for every $t \neq 0$, I_t is the orientation-preserving component of the orthogonal group of the quadratic form q_t. It is a Lie group of dimension 3. However, as we already noticed, I_0 is a Lie group of dimension 6. It is the matrix group

$$I_0 = \{A = (A_{ij}) \in \mathrm{GL}(3, \mathbb{R}) \mid A_{3,1} = A_{3,2} = 0, A_{3,3} = 1, \det(A) > 0\}.$$

This is the group of matrices of the form

$$\begin{pmatrix} a & b & e \\ c & d & f \\ 0 & 0 & 1 \end{pmatrix}$$

with $ad - bc > 0$. The fact that the determinant is positive is a consequence of the orientation-preserving assumption. Thus, I_0 is the usual matrix representation group of the orientation-preserving group of affine transformations of $\mathbb{R}^2$. Equivalently, it is the group of matrices preserving the (x, y)-plane in $\mathbb{R}^3$. (One can deduce this fom the fact that the affine group is the subgroup of the group of projective transformations that preserve a hyperplane in projective space.) This shows again that I_0 is not isomorphic to the group of orientation-preserving Euclidean motions of the plane. We shall reduce the size of I_0, by restricting the type of matrices that we consider.

Definition 2.1 (Coherent element). An element A of I_0 is said to be *coherent* if for all i and j satisfying $1 \leq i, j \leq 3$, there exists an analytic function $A_{i,j}(t)$ such that for each $t \in [-1, 1]$ the matrix $A_t = \big(A_{i,j}(t)\big)$ belongs to I_t, and $A_0 = A$.

3 The space of coherent elements

We denote by $E \subset I \subset \mathrm{GL}(3, \mathbb{R}) \times [-1, 1]$ the set of coherent elements. For each $t \in [-1, 1]$ we denote by E_t the set of coherent elements that are above the point t.

From the definition, we have $E_t = I_t$ for all $t \neq 0$. We now study E_0.

The coherent elements of I_0 form a group and they have important features. We start with the following:

Proposition 3.1. *If $A \in I_0$ is coherent, then* $\det(A) = 1$.

Proof. For $t \neq 0$, we have $\det(A) = 1$ since A is an orthogonal matrix of a non-degenerate quadratic form. The result then follows from the definition of coherence and from the fact that the map $t \mapsto \det(A_t)$ is continuous. □

We also have the following:

Proposition 3.2. *An element $A \in I_0$ is coherent if and only if A is an orientation-preserving motion of the Euclidean plane.*

Proof. Let $A \in I_0$ be a coherent element. From the definition, we have $A = \lim_{t\to 0} A(t)$, with $A(t) \in I_t$, $t > 0$. Thus, the 9 sequences $A_{ij}(t)$, for $1 \leq i, j \leq 3$, converge and therefore they are bounded. For $t \neq 0$, since $A(t)$ preserves the quadratic form q_t, it also preserves the form $\frac{1}{t}q_t$. For $i = 1, 2, 3$, let $A_i(t)$ denote the i-th column of $A(t)$. We show that for $i, j < 3$, the $\frac{1}{t}q_t$-scalar product of $A_i(t), A_j(t)$ is the i, j-th coefficient of the Kronecker delta function.

We have

$$\begin{aligned}\frac{1}{t}q_t(A) &= \begin{pmatrix} A_{11}(t) & A_{21}(t) & A_{31}(t) \end{pmatrix} \begin{pmatrix} 1 & 0 & 0 \\ 0 & 1 & 0 \\ 0 & 0 & \frac{1}{t} \end{pmatrix} \begin{pmatrix} A_{11}(t) \\ A_{21}(t) \\ A_{31}(t) \end{pmatrix} \\ &= A_{11}(t)^2 + A_{21}(t)^2 + \frac{1}{t}A_{31}(t)^2.\end{aligned}$$

Since $A(t)$ preserve $\frac{1}{t}q_t$, we get $A_{11}(t)^2 + A_{21}(t)^2 + \frac{1}{t}A_{31}(t)^2 = 1$ for all $t \neq 0$. Since the sequences $A_{ij}(t)$ converge, we obtain $\lim_{t\to 0} A_{31}(t) = 0$, therefore the coherence property imposes $A_{31}(0) = 0$. Now we use the fact that the section $A_{31}(t)$ is differentiable (recall that by coherence, it is even analytic). This gives $A_{31}(t) = 0 + \lambda t + O(t^2)$, therefore $\frac{1}{t}A_{31}(t)^2 = \lambda^2 t + O(t^3)$, which implies $\lim_{t\to 0} \frac{1}{t}A_{31}(t)^2 = 0$. We get $A_{11}(0)^2 + A_{21}(0)^2 = 1$.

In the same way, we prove that $A_{12}(0)^2 + A_{22}(0)^2 = 1$ and $A_{11}(0)A_{12}(0) + A_{21}(0)A_{22}(0) = 0$.

It follows that the first principal 2-bloc of A is an orthogonal matrix, which implies that E_0 is a direct isometry group of the Euclidean plane. In other words, the 2×2 matrix

$$\begin{pmatrix} A_{11} & A_{12} \\ A_{21} & A_{22} \end{pmatrix}$$

preserves the standard quadratic form on $\mathbb{R}^2$. Thus, it is an element of the Euclidean rotation group SO(2). □

Note.— In the proof of the preceding proposition, we used the fact that the coherence property imposes that the sections A_{ij} are differentiable. There is another proof in [1] (Proposition 2.2) which only uses the existence of sections that are only continuous.

The set of coherent elements of I_0, being the group of orientation-preserving motions of the Euclidean plane, is the group we are looking for.

From now on, we use the notation I_0 for the subgroup of *coherent* elements instead of the previously defined group I_0. Thus, in the following, for each $t \in [-1, 1]$, any element of I_t is coherent.

The space of coherent elements (for variable t) is not a locally trivial fiber bundle (for instance, the topological type of the fiber I_t is not constant; it is compact for $t > 0$ and noncompact for $t < 0$). However this space has nice properties which follow from the fact that it has many continuous (and even analytic) sections.

4 The algebraic description of points and lines

For every $t \in [-1, 1]$, we define E_t as the set of maximal abelian compact subgroups of I_t. An element of E_t is a *point* of our geometry. Note that any maximal abelian compact subgroups of I_t is isomorphic to the circle group SO(2). With this in hand, a coherent family of points in the fiber space (E_t) is a coherent family of maximal abelian compact subgroups of I_t. This is a family depending analytically on the parameter t. We can also consider coherent families of pairs (respectively triples, etc.) of points. This defines the segments (respectively triangles, etc.) of our geometries. We can study the corresponding distance (respectively area, etc.) function of t defined by such a pair or triple, in algebraic terms.

For each $p \in E_t$, we denote by $K_p \subset I_t$ the maximal subgroup that defines p. Since K_p is a group isomorphic to the circle group SO(2), for each $p \in E_t$, there exists a unique order-two element $s_p \in K_p$. We shall make use of this element. In the circle group SO(2), this corresponds to the rotation of angle π. We call s_p the *reflection*, or *involution* in I_t, of center p. In this way, any point in E_t is represented by an involution. For each $t \neq 0$, an involution is a self-map of the space of our geometry (the sphere or the hyperbolic plane) that fixes the given point, whose square is the identity, and whose differential at the given point is $-\mathrm{Id}$.

This algebraic description of points in E_t as involutions has certain advantages. In particular, we can define compositions of involutions and we can use this operation to describe algebraically lines and other geometric objects in E_t.

A *line* in E_t is a maximal subset L of E_t satisfying the following property:

> (*) The subgroup of I_t generated by the all elements of the form $s_p s_{p'}$, for $p, p' \in L$, is abelian. In the the case $t = 0$, we ask furthermore that the group is proper. (For $t \neq 0$, this is automatic).

In other words, each time we take four points $s_{p_1}, s_{p_2}, s_{p_3}, s_{p_4}$ in E_t represented by involutions, then, $s_{p_1} s_{p_2}$ commutes with $s_{p_3} s_{p_4}$.

Given two distinct points in E_t, there is a unique line joining them; this is the maximal subset L of E_t containing them and satisfying property (*) above.

The group I_t acts by conjugation on E_t and by reflections along lines.

With these notions of points and lines defined group-theoretically, one can check that the postulates of the geometry can be expressed in group theoretic terms: for instance, any two distinct points belong to a line; two lines intersect in at most one point. (For this to hold, in the case $t > 0$, one has to be in the projective plane and not on the sphere.)

We now introduce circles. Let s be a point represented by a subgroup K_p and an involution s_p. Any element ρ in K_p acts on the space of points E_t by conjugation:

$$(\rho, K_p) \mapsto \rho K_p \rho^{-1}.$$

A *circle* of center p in our geometry is an orbit of such an action. For $t > 0$ and for every point p, among such circles, there is one and only one circle centered at p which is a line of our geometry.

A *triangle* in our space is determined by three distinct points with three lines joining them pairwise together with the choice of a connected component of the complement of these lines which contains the three given points on its boundary. The last condition is necessary in the case of the sphere, since, in general, three lines divide the sphere into eight connected components.

Let us consider in more detail the case $t > 0$. This case is particularly interesting, because we can formulate the elements of spherical polarity theory in this setting. We work in the projective plane (elliptic space) rather than the sphere. In this way, the set of points of the geometry E_t ($t > 0$) can be thought of as the set of unordered pairs of antipodal points on the sphere in the 3-dimensional Euclidean space $(\mathbb{R}^3, q_t)$.

To each line in E_t is associated a well-defined point called its *pole*. In algebraic terms (that is, in our description of points as involutions), the pole of a line is the unique involution s_p which, as an element of I_t, fixes globally the line and does not belong to it.

Conversely, to each point p, we can associate the line of which p is the pole. This line is called the *equator* of p. There are several equivalent algebraic characterizations of that line. For instance, it is the unique line L such that for any point q on L, the involution s_q fixes the point p. In other words, the equator of a point s_p is the set of points $q \neq p$ satisfying $s_q(p) = p$.[3]

This correspondence between points and lines is at the basis of duality theory.

If two points p and q in E_t are distinct, the product of the corresponding involutions s_p and s_q is a translation along the line joining these points. More concretely, $s_p s_q$ is a rotation along the line in 3-space which is perpendicular to the plane of the great circle determined by p and q. This line passes through the pole s_N of the great circle, and therefore the product $s_p s_q$ commutes with the pole s_N (seen as an involution).

Given $p \in E_t$, the equator of p is the set of all $q \in E_t$ whose image under the stabilizer of p in I_t is a straight line.

In spherical (or elliptic) geometry, a symmetry with respect to a point is also a symmetry with respect to a line. This can be seen using the above description

[3]This correspondence between poles and lines holds because we work in the elliptic plane, and not on the sphere. In the latter case, there would be two "poles", which are exchanged by the involution s_q associated to a point q on the line.

of points as involutions. From the definition, the involution corresponding to the pole of a line fixes the pole, but it also fixes pointwise the equator. The pole can be characterized in this setting as being the unique *isolated* fixed point of its involution. The set of other fixed points is the equator. Thus, to a line in elliptic geometry is naturally associated an involution.

We can use polarity to define perpendicularity between lines: Consider two lines L_1 and L_2 that intersect at a point p. Then L_1 and L_2 are *perpendicular* if the pole of L_1 belongs to L_2. This is equivalent to the fact that the polar dual to L_2 belongs to L_1. This also defines the notion of *right angle*.

In what follows, we shall measure lengths in the geometry E_t for $t > 0$. We know that in spherical (or elliptic) geometry, there is a natural length unit. The length unit in this space can be taken to be the diameter of a line (all lines in that geometry are homeomorphic to a circle), or as the diameter of the whole space (which is compact). We can also use the correspondence between lines and poles and define a normalized distance on E_t, by fixing once and for all the distance from a point to its equator. We normalize this distance by setting it equal to $\frac{\pi}{2\sqrt{t}}$. (Remember that each space E_t, for $t > 0$, is the sphere of constant curvature t^2.)

After defining the distance on E_t, we can check that the segment which joins a pole to a point on its equator is the shortest path to the equator.

5 Transition of points, lines, distances, curvature and triangles

Before we continue, let us summarize what we did. We chose a fixed basis for $\mathbb{R}^3$. We defined the space $\mathcal{E} = (E_t)_{t\in[-1,1]}$ of coherent elements as a subset of a space of matrices, equipped with a map onto $[-1, 1]$. A point $p = (x, y, z)$ in $\mathbb{R}^3$ equipped with a basis gives a group-theoretically defined point in E_t for every $t \in [-1, 1]$ with $t \neq 0$, namely, the stabilizer of the vector whose coordinates are (x, y, z) in the group of linear transformations of $\mathbb{R}^3$ preserving the form q_t. Using the coherence property, we extended to $t = 0$, and for any $t \in [-1, 1]$, we defined a point A_t in E_t as a stabilizer group under the action of the group I_t, namely, a maximal abelian subgroup K_t of I_t. The family A_t, $t \in [-1, 1]$ is a coherent family of points. This group-theoretic definition of points allows the continuous transition of points in the space $\mathcal{E}$. Likewise, the group-theoretic definition of lines makes this notion coherent transit from one geometry to another.

Now we introduce the notion of angle. We want a notion that transits between the various geometries. This can also be done group-theoretically in a coherent manner. There are various ways of doing it, and one way is the following. We start with the group-theoretic description of a point as a maximal compact abelian subgroup (the "stabilizer" of the point). Such a subgroup is isomorphic to a circle group. But the circle group is also the set of all oriented lines starting at the given point. It is equipped with its natural Haar measure, which we normalize so that the total measure of the circle is 2π. This gives a measure on the set of oriented lines. From these measures, we obtain the notion of angle at every point. It is also possible to give a coherent definition of an angle by using a formula, and we shall do this in §7.

Now we consider transition of distances.

To measure distances between points, we use a model for the space E_t for each $t \in [0, 1]$. We take the unit sphere S_t, that is, the space of vectors in $\mathbb{R}^3$ of norm one with respect to the quadratic form q_t. This is the subset defined by

$$S_t = \{(x, y, z) \mid tx^2 + ty^2 + z^2 = 1\} \subset \mathbb{R}^3. \tag{8.2}$$

We take the quotient of this set by the action of $\mathbb{Z}_2$ that sends a vector to its opposite. Note that the unit sphere S_t (before taking the quotient) is connected for $t > 0$, and it has two connected components for $t < 0$. After taking the quotient, all spaces become connected. The group of transformations of $\mathbb{R}^3$ which preserve the quadratic form q_t acts transitively on $S_t/\mathbb{Z}_2$ for each t. For $t = 0$, the unit sphere (before taking the quotient) is defined by the equation $z^2 = 1$, and it has two connected components, namely, the planes $z = 1$ and $z = -1$.

In this way, the elements of the geometries E_t can be seen either algebraically as cosets in a group (the isometry group modulo the stabilizer of a point) or, geometrically, as elements of the unit sphere quotiented by $\mathbb{Z}_2$.

Now we define distances between pairs of points in each E_t and then we study transition of distances.

For $t \neq 0$, we use the angular distance in E_t. To do so, for each $t \neq 0$, we let β_t be the bilinear form associated to the quadratic form q_t, that is,

$$\beta_t(x, y) = \frac{q_t(x + y) - q_t(x) - q_t(y)}{2}.$$

We set, for every x_t and y_t in E_t,

$$w_t(x_t, y_t) = \arccos \frac{\beta_t(x_t, y_t)}{\sqrt{q_t(x_t)}\sqrt{q_t(y_t)}} \tag{8.3}$$

We call $w_t(x_t, y_t)$ the *angular distance* between the points x_t and y_t. For each $t \neq 0$, the angular distance can be thought of as being defined on $S_t/\mathbb{Z}_2$, but also on the 2-dimensional projective plane, the quotient of $\mathbb{R}^3 \setminus \{0\}/\mathbb{Z}_2$ by the nonzero homotheties. As a distance on the space E_t, we shall take w_t multiplied by a constant c_t that we determine below. Note that although Equation (8.2) describes an ellipsoid in the Euclidean orthonormal coordinates (x, y, z), this surface, equipped with this angular metric induced from the quadratic form, is isometric to a round sphere.

For each $t \neq 0$, equipped with the normalized distance function $c_t w_t$, the point set E_t becomes a metric space. Its points and its geodesics coincide with the group-theoretically defined points and lines that we considered in §1. This can be deduced from the fact that the isometry group of the space acts transitively on points and on directions, and that both notions (the metric and the group-theoretic) are invariant by this action. There are other ways of seeing this fact.

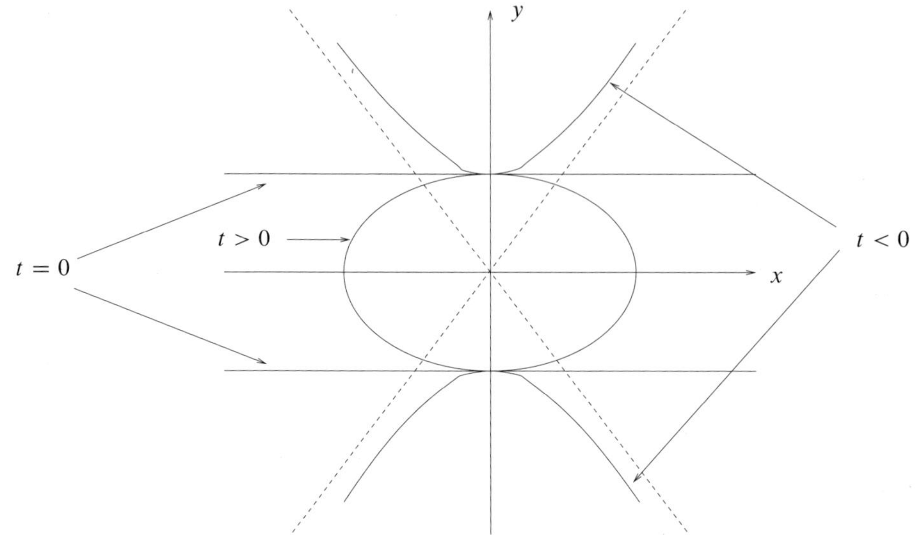

Figure 8.1.

We can draw a picture of S_t for each $t \in [0, 1]$. In dimension 2, this is represented in Figure 8.1. To get the 3-dimensional picture, this figure has to be rotated in space around the y-axis.

Any vector in $\mathbb{R}^3$ defines a point in each geometry E_t by taking the intersection of the ray containing it with S_t (or in the quotient of this intersection by $\mathbb{Z}_2$). This is represented in Figure 8.2, in which the vector is denoted by P, and where we see three intersection points with the level surfaces S_t: for $t > 0$, $t = 0$ and $t < 0$. Note that the two points with Cartesian coordinates $(0, 0, 1)$ and $(0, 0, -1)$ belong to all geometries (they belong to S_t for any $t \in [-1, 1]$). We shall use this in the following discussion about triangles, where some vertices will be taken to be at these points, and this will simplify the computations.

We now discuss transitions of distances.

We want the distance between two points in the geometry E_0 to be the limit as $t \to 0$ (from both sides, $t > 0$ and $t < 0$) of the distance in E_t between corresponding points (after normalization). The next proposition tells us how to choose the normalization factor c_t.

Let A and B be two points in $\mathbb{R}^3$. We have a natural way of considering each of these points to lie in E_t, for every $t \in [-1, 1]$. Let us denote by A_t, B_t the corresponding points.

Proposition 5.1. *The limit as $t \to 0$, $t > 0$ (respectively $t < 0$), of the distance from A_t to B_t in E_t normalized by the factor $1/\sqrt{t}$ is equal to the Euclidean distance between A_0 and B_0.*

Using this result, we shall define the distance between the points A and B in E_0 in such a way that the distances in Proposition 5.1 vary continuously for $t \in [-1, 1]$.

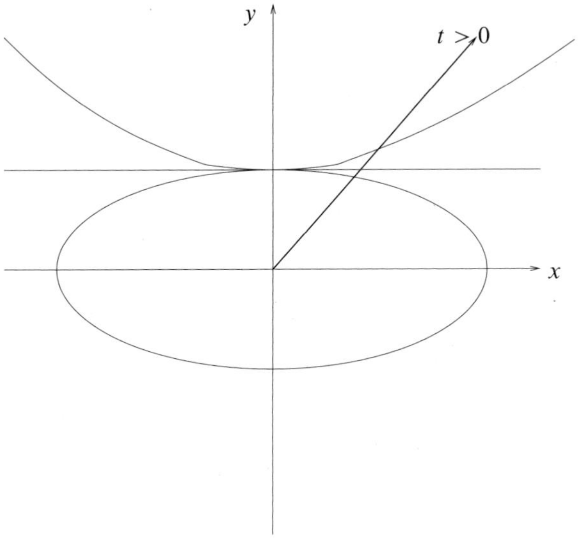

Figure 8.2.

Proof. For $t \in [-1,1]$, let $a = (0,0,1) \in \mathbb{R}^3$ and consider the coherent family of points $A_t \in E_t, t \in [-1,1]$ represented by a. For $x \in \mathbb{R}$, let $B_t^x \in E_t, t \in [-1,1]$ be the coherent family of points represented by the stabilizer of the vector $b^x = (x,0,1)$ of $\mathbb{R}^3$. The family $[A_t, B_t^x]$, $t \in [0,1]$ is a coherent family of segments in $\mathcal{E}$.

We start with $t > 0$. We compute the limit as $t \to 0,\ t > 0$, of the distance from A_t to B_t^x in E_t, normalized by dividing it by $\sqrt{t}$. We have

$$q_t(a) = 1,$$
$$q_t(b^x) = tx^2 + 1$$

and

$$\beta_t(a, b^x) = 1.$$

We use the angular distance from A_t to B_t^x, measured with $q_t, t > 0$. By Equation (8.3), we have

$$\begin{aligned} w_t(A_t, \beta_t^x) &= \arccos\left(\frac{\beta_t(a,b^x)}{\sqrt{q_t(a)}\sqrt{q_t(b^x)}}\right) \\ &= \arccos\left(\frac{1}{\sqrt{tx^2+1}}\right) \\ &= |\arctan\left(\sqrt{t}x\right)| \\ &= |\sqrt{t}x - \frac{1}{3}\sqrt{t^3}x^3 + \frac{1}{5}\sqrt{t^5}x^5 - \dots| \end{aligned}$$

Taking $c_t = 1/\sqrt{t}$ as a factor for distances in E_t, the distance $c_t w_t$ between the two points becomes

$$D_t(A_t, B_t^x) = \frac{1}{\sqrt{t}} \arccos\left(\frac{1}{\sqrt{tx^2+1}}\right) = |x - \frac{1}{3}tx^3 + \frac{1}{5}t^2x^5 - \dots|. \tag{8.4}$$

The limit as $t \to 0$, $t > 0$, of $D_t(A_t, B_t^x)$ is $|x|$, which is what the distance from A_0 to B_0^x in E_0 ought to be. □

Thus, we define the distance D_t on each E_t to be the angular distance w_t given in (8.3) divided by $\sqrt{t}$. In this way, the distance function transits.

The fact that points and distances transit between the various geometries implies that triangles also transit. We already noted that angles also transit. From this, we get several properties. For instance, since angle bisectors may be defined using equidistance, the following property transits between the various geometries:

Proposition 5.2. *The angle bisectors in a triangle intersect at a common point.*

The reader can search for other properties that transit between the various geometries.

Curvature is also a coherent notion. Indeed, curvature in spherical and in hyperbolic geometry is defined as the integral of the excess (respectively the defect) to two right angles of the angle sum in a triangle (and to a multiple of a right angle for more general polygons). This notion transits through Euclidean geometry, which is characterized by zero excess to two right angles. One can make this more precise, by choosing a triangle with fixed side lengths – for instance with three sides equal to a quarter of a circle – and making this triangle transit through the geometries.

6 Transition of trigonometric formulae

The trigonometric formulae, in a geometry, make relations between lengths and angles in triangles. They are at the bases of the geometry, since from these formulae, one can recover the geometric properties of the space. Let us recall the formal similarities between the trigonometric formulae in hyperbolic and in spherical geometry. One example is the famous "sine rule", which is stated as follows:

For any triangle ABC, with sides (or side lengths) a, b, c opposite to the vertices A, B, C, we have, in Euclidean geometry:

$$\frac{a}{\sin A} = \frac{b}{\sin B} = \frac{c}{\sin C},$$

in spherical geometry:

$$\frac{\sin a}{\sin A} = \frac{\sin b}{\sin B} = \frac{\sin c}{\sin C}$$

and in hyperbolic geometry:

$$\frac{\sinh a}{\sin A} = \frac{\sinh b}{\sin B} = \frac{\sinh c}{\sin C}.$$

Thus, to pass from the Euclidean to the spherical (respectively the hyperbolic), one replaces a side length by the sine of the side length (respectively the sinh of the length). Since using the sine rule one can prove many other trigonometric formulae, it is natural to expect that there are many occurrences of trigonometric formulae in non-Euclidean geometry where one replaces the circular functions of the side lengths (sin, cos, etc.) by the hyperbolic function of these side lengths, in order to get the hyperbolic geometry formulae from the spherical ones. Examples are contained in the following table. They concern a triangle ABC having a right angle at C:

Hyperbolic	Euclidean	Spherical
$\cosh c = \cosh a \cosh b$	$c^2 = a^2 + b^2$	$\cos c = \cos a \cos b$
$\sinh b = \sinh c \sin B$	$b = c \sin B$	$\sin b = \sin c \sin B$
$\tanh a = \tanh c \cos B$	$a = c \cos B$	$\tan a = \tan c \cos B$
$\cosh c = \cot A \cot B$	$1 = \cot A \cot B$	$\cos c = \cot A \cot B$
$\cos A = \cosh a \sin B$	$\cos A = \sin B$	$\cos A = \cos a \sin B$
$\tanh a = \sinh b \tan A$	$a = b \tan A$	$\tan a = \sin b \tan A$

In this table, the Euclidean formulae in the middle column are obtained by taking Taylor series expansions of any of the two corresponding non-Euclidean ones, for side lengths tending to 0. This is a consequence of the fact that the spherical and the hyperbolic geometries become Euclidean at the level of infinitesimal triangles.

The transitional geometry sheds a new light on the analogies between the trigonometric formulae.

First, we must recall that the above formulae are valid for spaces of constant curvatures -1, 0 and 1. In a geometry E_t, with $t \neq 0$, one has to introduce a parameter t in the above trigonometric formulae. For instance, for a right triangle ABC with right angle at C, we have, in spherical geometry ($t > 0$):

$$\cos \sqrt{t}c = \cos \sqrt{t}a \cos \sqrt{t}b$$

and in hyperbolic geometry ($t < 0$):

$$\cosh \sqrt{-t}a = \cosh \sqrt{-t}b \cosh \sqrt{-t}c.$$

Now we can study the transition of this formula, called the *Pythagorean theorem*.[4] We first study the transition of a triangle.

[4]By extension from the Euclidean case, a Pythagorean theorem, in a certain geometry, is a formula that makes relations between the edges and angles of a right triangle.

For $t \in [-1,1]$, let $a = (0,0,1) \in \mathbb{R}^3$ and consider the coherent family of points $A_t \in E_t, t \in [-1,1]$ represented by the vector a. Take a real number x. For any $t \in [0,1]$, let $B_t^x \in E_t$ and $C_t^x \in E_t$ be the two coherent families of points represented by the stabilizers in I_t of the vectors $b^x = (x,0,1)$ and $c^x = (0,x,1)$ of $\mathbb{R}^3$. The family $\Delta_t(x,y) = (A_t, B_t^x, C_t^y), t \in [0,1]$ is a coherent family of triangles. The triangle $\Delta_0(x,y)$ is a right triangle at A, with catheti ratio x/y. To see that each triangle $\Delta_t(x,y)$ is right, we note that if we change B into $-B$, we do not change the angle value. (Recall the definition of the angle measure as the Haar measure on SO(2).) Therefore there is a transformation of the space E_t that belongs to our isometry group and that sends this angle to its opposite. Therefore the angle is equal to its symmetric image, and therefore it is right. We saw (proof of Proposition 5.1) that in the geometry E_t, the distance between the two points A_t and B_t^x is given by

$$D_t(A_t, B_t^x) = \frac{1}{\sqrt{t}} \arccos\left(\frac{1}{\sqrt{tx^2+1}}\right). \tag{8.5}$$

Likewise, the distance between the two points A_t and C_t^y is equal to

$$D_t(A_t, C_t^y) = \frac{1}{\sqrt{t}} \arccos\left(\frac{1}{\sqrt{ty^2+1}}\right). \tag{8.6}$$

In the geometry E_0, the distance from A_0 to C_0^y is equal to $|y|$.
We need to know the distance in E_0 between the points B_0^x and C_0^y.
The angular distance from B_t^x to C_t^y, measured with $q_t, t > 0$, is

$$\begin{aligned} w_t(B_t^x, C_t^y) &= \arccos\left(\frac{\beta_t(b^x, c^y)}{\sqrt{q_t(b^x), q_t(c^y)}}\right) \\ &= \arccos\left(\frac{1}{\sqrt{tx^2+1}\sqrt{ty^2+1}}\right) \\ &= \arccos\left(\frac{1}{\sqrt{t^2x^4y^4+tx^2+ty^2+1}}\right). \end{aligned}$$

Up to higher order terms, this expression is equal to

$$\sqrt{t^2x^4y^4+tx^2+ty^2}.$$

We collect the information in the following lemma, using the fact that in the coherent geometry E_0, distances are obtained as a limit of normalized distances in E_t for $t > 0$ (or $t < 0$).

Lemma 6.1. *In the geometry E_t with $t > 0$, we have*

$$D_t(A_t, B_t^x) = \frac{1}{\sqrt{t}} \arccos\left(\frac{1}{\sqrt{tx^2+1}}\right),$$
$$D_t(A_t, C_t^y) = \frac{1}{\sqrt{t}} \arccos\left(\frac{1}{\sqrt{ty^2+1}}\right),$$

and

$$D_t(B_t^x, C_t^y) = \frac{1}{\sqrt{t}} \arccos\left(\frac{1}{\sqrt{tx^2+1}\sqrt{ty^2+1}}\right).$$

In the coherent geometry E_0, we have

$$\lim_{t\to 0} D_t(A_t, B_t^x) = D_0(A_0, B_0^x) = |x|,$$
$$\lim_{t\to 0} D_t(A_t, C_t^x) = D_0(A_0, C_0^y) = |y|$$

and

$$\lim_{t\to 0} D_t(B_t, C_t^x) = D_0(B_0^x, C_0^y) = \sqrt{x^2+y^2}.$$

Using this lemma, we now show that the Euclidean Pythagorean formula in E_0 is a limit of the Pythagorean formula in E_t for $t > 0$.

Let x and y be real numbers and let $t > 0$. In the triangle $\Delta_t(x, y)$, the lengths of the three sides (measured in E_t) are the angular distances $\sqrt{t} D_t(A_t, B_t^x)$, $\sqrt{t} D_t(A_t, C_t^y)$ and $\sqrt{t} D_t(Cy_t, B_t^x)$. The first natural attempt is to write the Pythagorean theorem in spherical geometry, for the triangle with right angle at A_t:

$$\begin{aligned}\cos\left(\sqrt{t} D_t(A_t, B_t^x)\right) \cos\left(\sqrt{t} D_t(A_t, C_t^y)\right) &= \frac{1}{\sqrt{tx^2+1}\sqrt{ty^2+1}} \\ &= \cos\left(\sqrt{t} D_t(C_t^y, B_t^x)\right).\end{aligned}$$

Taking the limit, as $t \to 0$ gives

$$1 = 1 \times 1,$$

which is correct but which is not the Euclidean Pythagorean theorem. We obtain a more useful result by taking another limit.

We transform the spherical Pythagorean theorem into the following one:

$$\begin{aligned}&\sqrt{t}\left(1 - \cos(\sqrt{t} D_t(A_t, B_t^x)) \cos(\sqrt{t} D_t(A_t, C_t^x))\right) = \\ &\sqrt{t}\left(1 - \cos(\sqrt{t} D_t(C_t^y, B_t^x))\right).\end{aligned}$$

Writing this equation at the first order using the expansion

$$\frac{1}{\sqrt{t}} \arccos\left(\frac{1}{\sqrt{tx^2+1}}\right) = \left| x - \frac{1}{3}tx^3 + \frac{1}{5}t^2x^5 - \dots \right|$$

and taking the limit as $t \to 0$, with $t > 0$ (use Equation 8.5) gives

$$D_0(A_0, B_0^x)^2 + D_0(A_0, C_0^y)^2 = D_0(B_0^x, C_0^y)^2,$$

which is the Pythagorean theorem in the geometry E_0. This is indeed the familiar Pythagorean theorem in Euclidean geometry.

It is possible to do the same calculation in the hyperbolic case ($t < 0$), where the hyperbolic Pythagorean theorem is:

Proposition 6.1.

$$\cosh\left(\sqrt{-t}\, D_t(A_t, B_t^x)\right) \cosh\left(\sqrt{-t}\, D_t(A_t, C_t^y)\right) = \cosh\left(\sqrt{-t}\, D_t(C_t^y, B_t^x)\right).$$

7 Transition of angles and of area

Using the Pythagorean theorem in spherical and hyperbolic geometry, it is possible to prove the other formulae in the above table, in particular the formulae

$$\cos A_t = \frac{\tan\sqrt{a_t}}{\tan\sqrt{b_t}}$$

in spherical geometry and

$$\cos A_t = \frac{\tanh\sqrt{a_t}}{\tanh\sqrt{b_t}}$$

in hyperbolic geometry.

At the same time, we can study convergence of angles. For this, take a triangle ABC with right angle at C and let α be the angle at A. For each t, the (lengths of the) sides opposite to A, B, C are respectively a_t, b_t, c_t in the geometry E_t. Take the side lengths a and b to be constant, $a_t = a$ and $b_t = b$. For each t, we have, in the geometry E_t,

$$\cos\alpha_t = \frac{\tan\sqrt{t}a}{\tan\sqrt{t}b}.$$

As $t \to 0$, we have $\cos\alpha_t \to a/b$. Thus, the angle α_t in the triangle $A_tB_tC_t$ converges indeed to the Euclidean angle.

Note that this can also be used to define the notion of angle in each geometry. The reader can check that this definition amounts to the one we gave in §5.

Finally we consider transition of area.

There are several works dealing with non-Euclidean area. For instance, in the paper [6], Euler obtained the following formula for the area Δ of a spherical triangle of side lengths a, b, c:

$$\cos\frac{1}{2}\Delta = \frac{1+\cos a+\cos b+\cos c}{4\cos\frac{1}{2}a\cos\frac{1}{2}b\cos\frac{1}{2}c}. \tag{8.7}$$

This formula should be compared to the Heron formula in Euclidean geometry that gives the area Δ of a triangle in terms of its side length. In the geometry E_t ($t>0$), Euler's formula becomes

$$\cos\frac{1}{2}\Delta = \frac{1+\cos\sqrt{t}a+\cos\sqrt{t}b+\cos\sqrt{t}c}{4\cos\frac{\sqrt{t}}{2}a\cos\frac{\sqrt{t}}{2}b\cos\frac{\sqrt{t}}{2}c}.$$

Setting $s=\sqrt{t}$ to simplify notation, we write this formula as

$$\cos\frac{1}{2}\Delta = \frac{1+\cos sa+\cos sb+\cos sc}{4\cos\frac{s}{2}a\cos\frac{s}{2}b\cos\frac{s}{2}c}. \tag{8.8}$$

We would like to see that Euler's formula (8.8) transits and leads to the area formula in Euclidean geometry for $t=0$. It suffices to deal with right triangles, and we therefore assume that the angle at c is right. We use the Pythagorean theorem

$$\cos\sqrt{t}c = \cos\sqrt{t}a\cos\sqrt{t}b.$$

We take the Taylor expansion in t in Formula (8.8).

Using the formula for the cosine of the double and taking the square, we obtain the formula

$$\left(\cos\frac{1}{2}\Delta\right)^2 = \frac{(1+\cos sa+\cos sb+\cos sa\times\cos sb)^2}{2(1+\cos sa)(1+\cos sb)(1+\cos sa\times\cos sb)}.$$

The degree-8 Taylor expansion of $1-(\cos\frac{1}{2}\Delta)^2$ is

$$\frac{1}{16}a^2b^2s^4+\left(\frac{1}{96}a^2b^4+\frac{1}{96}a^4b^2\right)s^6+O(s^8).$$

The first term in this expansion is $\frac{1}{16}a^2b^2s^4$, that is, $\frac{1}{16}a^2b^2t^2$, which is the square of the expression $\frac{1}{4}abt$. Recall now that for the transition of the distance function (§5), we had to normalize the length function in E_t, dividing it by $\sqrt{t}$. It is natural then, for the transition of the area function, to divide it by t. Thus, the result that we obtain is $\frac{1}{2}ab$, which is indeed the area of a Euclidean triangle with base a and altitude b.

Finally, we note that there has been a recent activity on transition of geometries in dimension 3, namely, on moving continuously between the eight Thurston geometries, and also on varying continuously between Riemannian and Lorentzian geometries on orbifolds. We mention the works of Porti [17], [19], Porti and Weiss [18],

Cooper, Hodgson and Kerckhoff [3], Kerckhoff and Storm [11] and Dancieger [4]. The ideas and the methods are different from those of the present paper.[5]

As a conclusion, we propose the following problem.

Problem 7.1. Extend this theory of transition of geometries to a parameter space which, instead of being the interval $[-1, 1]$, is a neighborhood of the origin in the complex plane $\mathbb{C}$.

This is in accordance with a well-established tradition to trying to extend to the complex world. We recall in this respect the following words of Painlevé [16] "Entre deux vérités du domaine réel, le chemin le plus facile et le plus court passe bien souvent par le domaine complexe." (Between two truths of the real domain, the easiest and shortest path quite often passes through the complex domain.)[6] We also recall Riemann's words, from his *Inaugural dissertation* [20], concerning the introduction of complex numbers: "If one applies these laws of dependence in an extended context, by giving the related variables complex values, there emerges a regulatrity and harmony which would otherwise have remained concealed."

Acknowledgements. The authors were supported by the French ANR project FINSLER, by the GEAR network of the National Science Foundation (GEometric structures And Representation varieties) and by the Erwin Schrödinger Institute (Vienna). The second author was also supported by a TUBITAK 2221 grant for a visit to Galatasaray University (Istanbul).

Bibliography

[1] N. A'Campo and A. Papadopoulos, Notes on hyperbolic geometry. In *Strasbourg Master class on Geometry*, IRMA Lectures in Mathematics and Theoretical Physics, vol. 18, Zürich, European Mathematical Society (EMS). 1–182, 2012.

[2] N. A'Campo and A. Papadopoulos, On Klein's *So-called Non-Euclidean geometry*. In *Sophus Lie and Felix Klein: The Erlangen program and its impact in mathematics and physics*, ed. L. Ji and A. Papadopoulos, Zürich, European Mathematical Society (EMS), 2015, 91–136.

[3] D. Cooper, C. Hodgson and S. Kerckhoff, *Three-dimensional orbifolds and cone-manifolds. With a postface by S. Kojima.* MSJ Memoirs, 5. Mathematical Society of Japan, Tokyo, 2000.

[4] J. Dancieger, A geometric transition from hyperbolic to anti de Sitter geometry. Preprint, 2013.

[5] J. Dieudonné, *La géométrie des groupes classiques*. Ergebnisse der Mathematik und ihrer Grenzgebiete (N. F.), Heft 5, Berlin, New York: Springer-Verlag, 1955.

[5] Added on proofs: J. McCleary pointed out the following interesting paper: J. McCleary, Trigonometries. *Amer. Math. Monthly* 109 (2002), 623–638.

[6] Painlevé's sentence has been taken over by Hadamard in [10]: "It has been written that the shortest and best way between two truths of the real domain often passes through the imaginary one."

[6] L. Euler, Variae speculationes super area triangulorum sphaericorum, *Nova Acta academiae scientarum imperialis Petropolitinae* 10, 1797, 47–62; *Opera omnia* Series 1, vol. XXIX, 253–266.

[7] C. Frances, Transformation groups in non-Riemannian geometry. In *Sophus Lie and Felix Klein: The Erlangen program and its impact in mathematics and physics*, ed. L. Ji and A. Papadopoulos, Zürich, European Mathematical Society (EMS), 2015, 191–216.

[8] C. F. Gauss, *Collected Works*, Vol. VIII. Königliche Gesellshaft der Wissenschaften, Göttingen, 1900.

[9] M. J. Greenberg, *Euclidean and Non-Euclidean Geometries*. 4th edition, New York, N. Y., W. H. Freeman, 2008 (first edition 1980).

[10] J. Hadamard, *An Essay on the Psychology of Invention in the Mathematical Field*. Princeton University Press, Princeton, 1945, reprint Dover, 1954 and in *The Mathematician's Mind*, 1996, Princeton University Press, Princeton, 123.

[11] S. P. Kerckhoff and P. A. Storm, From the hyperbolic 24-cell to the cuboctahedron. Preprint, 2008.

[12] F. Klein, Über die sogenannte Nicht-Euklidische Geometrie. *Mathematische Annalen* IV (1871), 573–625. English translation (On the co-called non-Euclidean geometry) by J. Stillwell in [21]. French translation (Sur la Géométrie dite non euclidienne) by L. Laugel, Mém. de la Fac. des Sciences de Toulouse, t. XI (1898), p. 334–335.

[13] F. Klein, Über die sogenannte Nicht-Euklidische Geometrie (Zweiter Aufsatz). *Mathematische Annalen* VI (1873), 112–145.

[14] M. Kline, *Mathematical thought from ancient to modern times*. Oxford University Press, 1972.

[15] N. I. Lobachevsky, *Pangeometry*. New edition with an English translation and a commentary by A. Papadopoulos, Heritage of European Mathematics, Vol. 4, European Mathematical Society, Zürich, 2010.

[16] P. Painlevé, *Analyse des travaux scientifiques*. Gauthier-Villars, 1900. Reproduced in Œuvres de Paul Painlevé, Éditions du CNRS, Paris, 1972–1975, vol. 1, 72–73.

[17] J. Porti, Regenerating hyperbolic cone structures from Nil. *Geometry and Topology*, Vol. 6 (2002), 815–852.

[18] J. Porti and H. Weiss, Deforming Euclidean cone 3-manifolds. *Geometry and Topology* 11 (2007) 1507–1538.

[19] J. Porti, Regenerating hyperbolic cone 3-manifolds from dimension 2. Preprint, 2010.

[20] B. Riemann, Grundlagen für eine allgemeine Theorie der Functionen einer veränderlichen complexen Grösse, (Göttingen, 1851). Published in his Collected works: *Gesammelte mathematische Werke, wissenschaftlicher Nachlass und Nachträge*. Collected papers. Nach der Ausgabe von Heinrich Weber und Richard Dedekind neu hrsgg. von Raghavan Narasimhan. (Collected mathematical works, scientific posthumous works and supplements. Collected papers. According to the edition by H. Weber and R. Dedekind newly edited by R. Narasimhan). Teubner Verlagsgesellschaft, Leipzig, 1862. Reprint: Springer-Verlag, Berlin (1990), 3–48.

[21] J. Stillwell, *Sources of Hyperbolic geometry*. History of Mathematics Vol. 10, AMS-LMS, 1996.

[22] H. Weyl, *The Classical Groups: Their Invariants and Representations*. Princeton University Press, Princeton, N. J., 1939.

Chapter 9

On the projective geometry of constant curvature spaces

Athanase Papadopoulos and Sumio Yamada

Contents

1 Cross Ratio

We consider the projective geometry of a Riemannian manifold (M, g) as the geometry of the space of lines/unparametrized geodesics of the manifold. For each $n \geq 2$, the Euclidean space $\mathbb{R}^n$, the sphere S^n and the hyperbolic space $\mathbb{H}^n$ provide a collection of Riemannian manifolds of constant sectional curvatures 0, 1 and -1 that are distinct from each other in the Riemannian sense, but one can consider the projective geometry on each of these spaces and ask whether these spaces are distinct in this setting. The most important qualitative feature of projective geometry is the notion of cross ratio, and in this geometry the Riemannian distance and angle invariance under isometric transformations of the underlying Riemannian manifold are replaced by a weaker invariance of the cross ratio under projective transformations. The goal of this chapter is to highlight the relations among the projective geometries of these three space forms through the projective models of these spaces in the Minkowski space $\mathbb{R}^{n,1}$, by using a cross ratio notion which is proper to each of the three geometries.

Felix Klein was the first to give a formula for distances in hyperbolic metric. This formula uses the disc Euclidean model, which Beltrami had already presented as a model for hyperbolic geometry, noticing that the Euclidean lines in this model satisfy all the axioms of hyperbolic geometry. Klein's formula (see [2]) uses the cross-ratio, which is a fundamental object in projective geometry. This study is also in the spirit of Klein's Erlangen program, in which Klein proposed to study Euclidean, hyperbolic and spherical geometry in the unifying setting of projective geometry.

We start by recalling some classical facts.

Consider a geodesic line in Euclidean space and let A_1, A_2, A_3, A_4 be four ordered pairwise distinct points on that line. Their cross ratio $[A_1, A_2, A_3, A_4]$ is defined by the formula:

$$[A_1, A_2, A_3, A_4]_e := \frac{A_2A_4}{A_3A_4} \cdot \frac{A_3A_1}{A_2A_1}$$

where for $1 \leq i, j \leq 4$, A_iA_j denotes the Euclidean distance.

Now consider four ordered distinct lines l_1, l_2, l_3, l_4 in the Euclidean plane $\mathbb{R}^2$ that are concurrent at a point A and let l be a line that intersects these four lines at points A_1, A_2, A_3, A_4 respectively. Then the cross ratio $[A_1, A_2, A_3, A_4]$ does not depend on the choice of the line l. This property expresses the fact that the cross ratio is a *projectivity invariant.*

As a matter of fact, Menelaus of Alexandria (2nd century A.D.) knew the above property not only for the Euclidean plane, but also on the sphere, where the lines are the spherical geodesics, that is, the great circles of the sphere, and he used it in his work; see [7].

Once there is a parallel between the Euclidean geometry and the spherical geometry, it is natural to expect a corresponding statement for hyperbolic geometry. We now define the cross ratio for the non-Euclidean geometries.

Definition 1.1 (Non-Euclidean cross ratio). Consider a geodesic line in the hyperbolic n-space or on the n-sphere respectively and let A_1, A_2, A_3, A_4 be four ordered pairwise distinct points on that line. We define the cross ratio $[A_1, A_2, A_3, A_4]$, in the hyperbolic case, by:

$$[A_1, A_2, A_3, A_4]_h := \frac{\sinh A_2A_4}{\sinh A_3A_4} \cdot \frac{\sinh A_3A_1}{\sinh A_2A_1},$$

and in the spherical case, by:

$$[A_1, A_2, A_3, A_4]_s := \frac{\sin A_2A_4}{\sin A_3A_4} \cdot \frac{\sin A_3A_1}{\sin A_2A_1},$$

where A_iA_j stands for the distance between the pair of points A_i and A_j. (For this, we shall assume that in the case of the sphere the four points lie in a hemisphere; instead, we could work in the elliptic space, that is, the quotient of the sphere by its canonical involution.)

We denote by U^n the open upper hemisphere of S^n equipped with the induced metric and we let X and X' belong to the set $\{\mathbb{R}^n, \mathbb{H}^n, U^n\}$. We shall say that a map $X \to X'$ is a *perspectivity*, or a *perspective-preserving transformation* if it preserves lines and if it preserves the cross ratio of quadruples of points on lines. (We note that these terms are classical in the Euclidean setting, see e.g. Hadamard [4] or Busemann [3]. In the Euclidean world, such maps arise indeed in perspective drawing.) In what follows, using well-known projective models in $\mathbb{R}^{n+1}$ of the hyperbolic space $\mathbb{H}^n$ and of the sphere S^n, we define natural homeomorphisms between $\mathbb{R}^n$, $\mathbb{H}^n$ and the open upper hemisphere of S^n which are perspective-preserving transformations. The proofs are elementary and are based on first principles of geometry.

The existence and the invariance properties of this cross ratio is used to develop an analogue of the classical Funk and Hilbert geometries on convex subsets of hyperbolic space and on the sphere, see [6].

2 Projective Geometry

The word "projective" is often used as a property of the incidence of lines (or of geodesics) in the underlying space. On the other hand, the n-dimensional sphere and the n-dimensional hyperbolic space are realized projectively in $n+1$-dimensional Euclidean space as sets of unit length vectors. Namely the sphere S^n is the set of unit vectors in $\mathbb{R}^{n+1}$ with respect to the Euclidean norm

$$x_1^2 + \cdots + x_n^2 + x_{n+1}^2 = 1$$

and the hyperbolic space $\mathbb{H}^n$ is one of the two components the set of "vectors of imaginary norm i" with $x_{n+1} > 0$ in $\mathbb{R}^{n+1}$ with respect to the Minkowski norm

$$x_1^2 + \cdots + x_n^2 - x_{n+1}^2 = -1.$$

One reason for which these models of the two constant curvature spaces are called *projective* is that the geodesics in the curved spaces are realized as the intersection of the unit spheres with the two-dimensional subspaces of $\mathbb{R}^{n+1}$ through the origin of this space. We also note that each two-dimensional linear subspace intersects the hyperplane $\{x_{n+1} = 1\}$ in a Euclidean geodesic. Conversely, each two-dimensional linear subspace of $\mathbb{R}^{n+1}$ represents a geodesic in each of the three geometries, consequently establishing the correspondence between the three incidence geometries.

We prove the following two theorems:

Theorem 2.1 (Spherical Case). *Let P_s be the projection map from the origin of $\mathbb{R}^{n+1}$ sending the open upper hemisphere U^n of S^n onto the affine hyperplane $\{x_{n+1} = 1\} \subset \mathbb{R}^{n+1}$. Then the projection map P_s is a perspectivity. In particular it preserves the values of cross ratio; namely for a set of four ordered pairwise distinct points A_1, A_2, A_3, A_4 aligned on a great circle in the upper hemisphere, we have*

$$[P_s(A_1), P_2(A_2), P_s(A_3), P_s(A_4)]_e = [A_1, A_2, A_3, A_4]_s.$$

Proof. Let u, v be the two points on the hyperplane $\{x_{n+1} = 1\}$, let $P_s^{-1}(u) =: [u]$, $P_s^{-1}(v) =: [v]$ be the points in U, and let $d([u],[v])$ be the spherical distance between them. Finally let $\|x\|$ be the Euclidean norm of the vector $x \in \mathbb{R}^{n+1}$. Assume that u, v and the origin $(0,\ldots,0,1)$ of the hyperplane $\{x_{n+1} = 1\}$ are collinear.

We claim that

$$\sin d([u],[v]) = \frac{\|u - v\|}{\|u\|\|v\|}.$$

This follows from the following trigonometric relations:

$$\begin{aligned}
\sin d([u],[v]) &= \sin\Big[\cos^{-1}\Big(\frac{u}{\|u\|}\cdot\frac{v}{\|v\|}\Big)\Big]\\
&= \sqrt{1-\cos^2\Big[\cos^{-1}\Big(\frac{u}{\|u\|}\cdot\frac{v}{\|v\|}\Big)\Big]}\\
&= \sqrt{1-\Big(\frac{u}{\|u\|}\cdot\frac{v}{\|v\|}\Big)^2} = \frac{1}{\|u\|\|v\|}\sqrt{\|u\|^2\|v\|^2-(u\cdot v)^2}\\
&= \frac{1}{\|u\|\|v\|}\times(\text{the area of parallelogram spanned by } u \text{ and } v)\\
&= \frac{\|u-v\|}{\|u\|\|v\|}.
\end{aligned}$$

(The last equality holds since the line through the origin of $\mathbb{R}^{n+1}$ and the origin of the plane $\{x_{n+1}=1\}$ is perpendicular to the line through u and v.)

Now consider a set of four ordered pairwise distinct points A_1, A_2, A_3, A_4 aligned on a great circle in the upper hemisphere, whose images $P_s(A_1)$, $P_s(A_2)$, $P_s(A_3)$, $P_s(A_4)$ by P_s are aligned on a line ℓ_0 through the origin $(0,\dots,0,1)$ of the hyperplane $\{x_{n+1}=1\}$. Their spherical cross ratio

$$[A_1, A_2, A_3, A_4]_s$$

is equal to the Euclidean cross ratio of their images,

$$[P_s(A_1), P_s(A_2), P_s(A_3), P_s(A_4)]_e$$

since we have:

$$\begin{aligned}
\frac{\sin d([A_2],[A_4])}{\sin d([A_3],[A_4])}\cdot\frac{\sin d([A_3],[A_1])}{\sin d([A_2],[A_1])} &= \frac{\dfrac{\|A_2-A_4\|}{\|A_2\|\|A_4\|}}{\dfrac{\|A_3-A_4\|}{\|A_3\|\|A_4\|}}\cdot\frac{\dfrac{\|A_3-A_1\|}{\|A_3\|\|A_1\|}}{\dfrac{\|A_2-A_1\|}{\|A_2\|\|A_1\|}}\\
&= \frac{\|A_2-A_4\|}{\|A_3-A_4\|}\cdot\frac{\|A_3-A_1\|}{\|A_2-A_1\|}.
\end{aligned}$$

For the general case, suppose that the line ℓ_0, on which the set of four points $P_s(A_1), P_s(A_2), P_s(A_3), P_s(A_4)$ are aligned, does not go through the origin of $(0,\dots,0,1)$ of the hyperplane $\{x_{n+1}=1\}$. Now consider a line ℓ through the origin, which is $\Pi\cap\{x_{n+1}=1\}$ for some two-dimensional subspace $\Pi\subset\mathbb{R}^{n+1}$. There is some two-dimensional subspace $\Pi_0\subset\mathbb{R}^{n+1}$ such that $\ell_0=\Pi_0\cap\{x_{n+1}=1\}$. As the group $\mathrm{SO}(n+1)$ acts transitively on the space of lines in $\{x_{n+1}=1\}$, there exists an element $L\in\mathrm{SO}(n+1)$ satisfying $L(\Pi_0)=\Pi$, which induces a projective transformation $\tilde{L}$ of $\{x_{n+1}=1\}$ sending ℓ_0 to ℓ, in particular preserving the value of the cross ratio

$$[P_s(A_2), P_s(A_3), P_s(A_4), P_s(A_1)]_e = [L \circ P_s(A_2), L \circ P_s(A_3), L \circ P_s(A_4), L \circ P_s(A_1)]_e.$$

On the other hand, the elements of $SO(n+1)$ are isometries on the unit sphere in $\mathbb{R}^{n+1}$. Hence the spherical cross ratio is preserved by the action of L;

$$[L(A_1), L(A_2), L(A_3), L(A_4)]_s = [A_1, A_2, A_3, A_4]_s.$$

Note that for a point $x \in U \in \mathbb{R}^{n+1}$, the transformations L and P_s commute in the following sense:

$$P_s \circ L(x) = \tilde{L} \circ P_s(x).$$

Hence we have $\tilde{L} \circ P_s(A_i) = P_s \circ L(A_i)$ for $i = 1, 2, 3, 4$. As the four points $\{P_s \circ L(A_i)\}$ are on the line ℓ going through the origin, from the special case considered at the beginning of the proof, we have

$$[P_s \circ L(A_1), P_s \circ L(A_2), P_s \circ L(A_3), P_s \circ L(A_4)]_e = [L(A_1), L(A_2), L(A_3), L(A_4)]_s.$$

Putting the equalities together, we conclude the equality

$$[P_s(A_1), P_s(A_2), P_s(A_3), P_s(A_4)]_e = [A_1, A_2, A_3, A_4]_s. \qquad \square$$

Theorem 2.2 (Hyperbolic Case). *Let P_h be the projection map of the upper sheet of the hyperboloid $\mathbb{H}^n \subset \mathbb{R}^{n+1}$ from the origin of $\mathbb{R}^{n+1}$ onto the unit disc of the hyperplane $\{x_{n+1} = 1\} \subset \mathbb{R}^{n+1}$. Then the projection map P_h is a perspectivity. In particular it preserves the values of cross ratio; namely for a set of four ordered pairwise distinct points A_1, A_2, A_3, A_4 aligned on a geodesic in the upper sheet of the hyperbolid, we have*

$$[P_h(A_1), P_h(A_2), P_h(A_3), P_h(A_4)]_e = [A_1, A_2, A_3, A_4]_h.$$

Proof. We follow the reasoning we did in the spherical case, replacing the sphere of unit radius in $\mathbb{R}^{n+1}$ by the upper sheet of "the sphere of radius i", namely the hyperboloid in $\mathbb{R}^{n,1}$. Let u, v be the two points on the hyperplane $\{x_{n+1} = 1\}$ and $P_h^{-1}(u) =: [u]$, $P_h^{-1}(v) =: [v]$ be the points in the hyperboloid, that is, the time-like vectors of unit (Minkowski) norm. We first consider a special case where u, v and the origin $(0, \dots, 0, 1)$ of the hyperplane $\{x_{n+1} = 1\}$ are collinear.

Denote by $d([u], [v])$ the hyperbolic distance between the points and $\|x\|$ be the Minkowski norm of the vector $x \in \mathbb{R}^{n,1}$. We will show that

$$\sinh d([u], [v]) = -\frac{\|u - v\|}{\|u\| \|v\|}.$$

Note that the number on the right hand side is positive, for $\|u\|, \|v\|$ are positive imaginary numbers, and $u - v$ is a purely space-like vector, on which the Minkowski norm of $\mathbb{R}^{n,1}$ and the Euclidean norm of $\mathbb{R}^n$ coincide.

We now use the hyperbolic trigonometry relations:

$$\begin{aligned}
\sinh d([u],[v]) &= \sinh\Big[\cosh^{-1}\Big(\frac{u}{\|u\|}\cdot\frac{v}{\|v\|}\Big)\Big] \\
&= \sqrt{\cosh^2\Big[\cosh^{-1}\Big(\frac{u}{\|u\|}\cdot\frac{v}{\|v\|}\Big)\Big]-1} \\
&= \sqrt{\Big(\frac{u}{\|u\|}\cdot\frac{v}{\|v\|}\Big)^2-1} = \frac{\sqrt{-1}}{\|u\|\|v\|}\sqrt{\|u\|^2\|v\|^2-(u\cdot v)^2} \\
&= \frac{\sqrt{-1}}{\|u\|\|v\|}\sqrt{-1}\times(\text{the area of parallelogram spanned by } u \text{ and } v) \\
&= -\frac{\|u-v\|}{\|u\|\|v\|}.
\end{aligned}$$

The formula

$$\sqrt{\|u\|^2\|v\|^2-(u\cdot v)^2} = \sqrt{-1}\times(\text{the area of parallelogram spanned by } u \text{ and } v)$$

is in Thurston's notes (Section 2.6 [8]).

Now the proof of the theorem, in the case where the four points A_1, A_2, A_3, A_4 on the hyperboloid are such that their images $P_h(A_1), P_h(A_2), P_h(A_3), P_h(A_4)$ by P_h are on a line ℓ_0 through the origin $(0,\dots,0,1)$ of the hyperplane $\{x_{n+1} = 1\}$ follows by the same argument as in the spherical case, namely, the hyperbolic cross ratio $[A_1, A_2, A_3, A_4]_h$ is equal to the Euclidean cross ratio $[P(A_1), P(A_2), P(A_3), P(A_4)]_e$.

Now we consider the general case, namely the line ℓ_0, on which the set of four points $P_h(A_1), P_h(A_2), P_h(A_3), P_h(A_4)$ are aligned, does not go through the origin $(0,\dots,0,1)$ of the hyperplane $\{x_{n+1} = 1\}$. Consider a line ℓ through the origin of this plane. It can be written as $\Pi \cap \{x_{n+1} = 1\}$ for some two-dimensional subspace $\Pi \subset \mathbb{R}^{n+1}$. There is also some two-dimensional subspace $\Pi_0 \subset \mathbb{R}^{n+1}$ such that $\ell_0 = \Pi_0 \cap \{x_{n+1} = 1\}$. Since the group $\mathrm{SO}(n,1)$ acts transitively on the space of lines intersecting the unit ball $\{\sum_{i=1}^n x_i^2 < 1\}$ in $\{x_{n+1} = 1\}$, there exists an element $L \in \mathrm{SO}(n,1)$ satisfying $L(\Pi_0) = \Pi$, which induces a projective transformation $\tilde{L}$ of $\{x_{n+1} = 1\}$ sending ℓ_0 to ℓ, in particular preserving the cross ratios:

$$\begin{aligned}
&[P_h(A_1), P_h(A_2), P_h(A_3), P_h(A_4)]_e \\
&= [\tilde{L}\circ P_h(A_1), \tilde{L}\circ P_h(A_2), \tilde{L}\circ P_h(A_3), \tilde{L}\circ P_h(A_4)]_e.
\end{aligned}$$

On the other hand, the elements of $\mathrm{SO}(n,1)$ are isometries on the hyperboloid $\mathbb{H}^n$. Hence the hyperbolic cross ratio is preserved by the action of L;

$$[L(A_1), L(A_2), L(A_3), L(A_4)]_h = [A_1, A_2, A_3, A_4]_h.$$

The transformations L and P_h commute in the sense that for any point $x \in \mathbb{H}^n \subset \mathbb{R}^{n+1}$

$$P_h \circ L(x) = \tilde{L}\circ P_h(x).$$

Hence we have $\tilde{L} \circ P_h(A_i) = P_h \circ L(A_i)$ for $i = 1, 2, 3, 4$. From the special case considered at the beginning of the proof, we have

$$[P_h \circ L(A_1), P_h \circ L(A_2), P_h \circ L(A_3), P_h \circ L(A_4)]_e = [L(A_1), L(A_2), L(A_3), L(A_4)]_h$$

Putting these equalities together, we conclude the equality

$$[P_h(A_1), P_h(A_2), P_h(A_3), P_h(A_4)]_e = [A_1, A_2, A_3, A_4]_s \qquad \square$$

Corollary 2.3. *The spherical and hyperbolic cross ratios are projectivity invariants.*

This follows from the fact that the projection map P_s and P_h are both perspective-preserving transformations. The classical proofs of this result in the cases of Euclidean and spherical geometry relies on the fact that the cross ratio is completely determined by the angles among the lines/geodesic l_i's at the vertex A. This proof can easily be done using the Sine Rule, see [7] for the case of spherical geometry. For this and for other non-Euclidean trigonometric formulae in hyperbolic trigonometry we refer the reader to [1] where the proofs are given in a model-free setting. In such a setting the proofs in the hyperbolic and the spherical cases can be adapted from each other with very little changes.

3 Generalized Beltrami–Klein models of $\mathbb{H}^n$

Given a bounded open convex set Ω in a Euclidean space, D. Hilbert in ([5] 1895) proposed a natural metric $H(x, y)$, now called the *Hilbert metric*, defined for $x \neq y$ in Ω as the logarithm of the cross ratio of the quadruple $(x, y, b(x, y), b(y, x))$, where $b(x, y)$ is the point where the ray $R(x, y)$ from x through y hits the boundary $\partial\Omega$ of Ω. This defines a metric on Ω, which is Finslerian and projective. We refer to the article [6] for a parallel treatment of the subject in hyperbolic and in spherical geometry.

A special case of the Hilbert metric gives the Beltrami–Klein model of hyperbolic space, where the underlying convex set Ω is the unit ball in $\mathbb{R}^n$. In fact, this special case was Hilbert's primary motivation to define the so-called Hilbert metric on an arbitrary bounded open convex set Ω in a Euclidean space. Actually, for the Beltrami–Klein model, the size of the ball is irrelevant, as the Hilbert metric is invariant under homothety of the underlying Euclidean space, so that for each Hilbert metric $H_\rho(x, y)$ defined on the ball $B_\rho(0)$ of radius $\rho > 0$ centered at the origin, (B_ρ, H_ρ) is isometric to the hyperbolic space $\mathbb{H}^n$.

Immediate corollaries of Theorems 2.1 and 2.2 are that there are new models of the hyperbolic space, which we call generalized Beltrami–Klein models. To describe them, we first set some notation: let B^h_ρ and B^s_ρ be the geodesic balls of $\mathbb{H}^n$ and S^n respectively, both centered at a fixed reference point which is identified with the point $(0, \cdots, 0, 1)$ of $\mathbb{R}^{n+1}$ via the projective models. Then we define the spherical Hilbert metric as follows:

Definition 3.1. For a pair of points x and y in B^s_ρ with $0 < \rho \leq \pi/2$, the Hilbert distance from x to y is defined by

$$H^s_\rho(x,y) = \begin{cases} \log \dfrac{\sin d(x,b(x,y))}{\sin d(y,b(x,y))} \cdot \dfrac{\sin d(y,b(y,x))}{\sin d(x,b(y,x))} & \text{if } x \neq y, \\ 0 & \text{otherwise} \end{cases}$$

where $b(x,y)$ is the point where the geodesic ray $R(x,y)$ from x through y hits the boundary ∂B^s_ρ of B^s_ρ.

We also define hyperbolic Hilbert metric:

Definition 3.2. For a pair of points x and y in B^h_ρ with $\rho > 0$, the Hilbert distance from x to y is defined by

$$H^h_\rho(x,y) = \begin{cases} \log \dfrac{\sinh d(x,b(x,y))}{\sinh d(y,b(x,y))} \cdot \dfrac{\sinh d(y,b(y,x))}{\sinh d(x,b(y,x))} & \text{if } x \neq y, \\ 0 & \text{otherwise} \end{cases}$$

where $b(x,y)$ is the point where the geodesic ray $R(x,y)$ from x through y hits the boundary ∂B^h_ρ of B^h_ρ.

Through the projective maps P_s and P_h, each geodesic ball in the curved spaces corresponds to a Euclidean ball. From Theorems 2.1 and 2.2, the Hilbert distance, which is the logarithm of the spherical/hyperbolic cross ratio of the quadruple $(x, y, b(x,y), b(y,x))$ defined on the geodesic ball, is preserved by projective maps; that is, P_h and P_s are isometries of the Hilbert metrics. Hence it follows that in the spherical case, we have the following:

Corollary 3.3. *The geodesic ball B^s_ρ in the unit sphere S^n with $0 < \rho < \pi/2$ with its spherical Hilbert metric H^s_ρ is isometric to the hyperbolic space $\mathbb{H}^n$.*

And in the hyperbolic case, we have:

Corollary 3.4. *The geodesic ball B^h_ρ in the hyperbolic space $\mathbb{H}^n$ with its hyperbolic Hilbert metric H^h_ρ is isometric to the hyperbolic space $\mathbb{H}^n$.*

We can consider the spaces (B^s_ρ, H^s_ρ) and (B^h_ρ, H^h_ρ) as *generalized Beltrami–Klein models of the hyperbolic space.*

Acknowledgements. The first author is supported by the French ANR project FINSLER and the GEAR network of the National Science Foundation (GEometric structures And Representation varieties). The second author is supported in part by JSPS Grant-in-aid for Scientific Research No.24340009. Part of this work was done during a stay of the two authors at the Erwin Schrödinger Institute (Vienna).

The authors would like to thank Norbert A'Campo for sharing his enthusiasm and ideas.

Bibliography

[1] N. A'Campo and A. Papadopoulos, Notes on hyperbolic geometry. In *Strasbourg Master class on Geometry*, IRMA Lectures in Mathematics and Theoretical Physics, vol. 18, Zürich, European Mathematical Society (EMS). 1–182, 2012.

[2] N. A'Campo and A. Papadopoulos, On Klein's *So-called Non-Euclidean geometry*. In *Sophus Lie and Felix Klein: The Erlangen program and its impact in mathematics and physics*. Ed. L. Ji and A. Papadopoulos, Zürich, European Mathematical Society (EMS), 2015, 91–136.

[3] H. Busemann and P. J. Kelly, *Projective geometry and projective metrics*. Academic Press, 1953.

[4] J. Hadamard, *Leçons de géométrie élémentaire*. Librairie Armand Colin, Paris, 1898 (several later editions).

[5] D. Hilbert. Über die gerade Linie als kürzeste Verbindung zweier Punkte. *Math. Ann.* 46 (1895) 91–96.

[6] A. Papadopoulos and S. Yamada. The Funk and Hilbert geometries for spaces of constant curvature. *Monatsh. Math.* 172, 1 (2013), 97–120.

[7] R. Rashed and A. Papadopoulos, On Menelaus' *Spherics* III.5 in Arabic mathematics, I: Ibn 'Irāq. *Arabic Science and Philosophy* (Cambridge University Press), vol. 24 (2014), 1–68.

[8] W. Thurston. The Geometry and Topology of Three-Manifolds. MSRI, available at library.msri.org/books/gt3m/.

Chapter 10

The Erlangen program and discrete differential geometry

Yuri B. Suris

Contents

1 Introduction

The field of discrete differential geometry is presently emerging on the border between differential and discrete geometry. Whereas classical differential geometry investigates smooth geometric shapes, discrete differential geometry studies geometric shapes with a finite number of elements and aims to develop discrete equivalents of the geometric notions and methods of classical differential geometry. The latter appears then as a limit of refinements of the discretization.

One may suggest many different reasonable discretizations with the same smooth limit. Which one is the best? From our point of view, the best discretization is the one which preserves the fundamental properties of the smooth theory. Such a discretization often clarifies the structures of the smooth theory and it possesses important connections with other fields of mathematics (projective geometry, integrable systems, algebraic geometry, complex analysis, etc.).

In particular, in this chapter, we will be mainly interested in relations with integrable systems. It is well known that differential equations describing interesting special classes of surfaces and parametrizations are integrable, and, conversely, many interesting integrable systems admit a differential-geometric interpretation. The area of integrable differential geometry was founded in the fundamental treatises by Bianchi [4], Darboux [14, 15], and Eisenhart [20]. A progress in understanding the unifying fundamental structure that the classical differential geometers were looking for, and simultaneously in understanding of the very nature of integrability, came from the efforts to discretize these theories. It turns out that many sophisticated properties of differential-geometric objects find their simple explanation within discrete differential geometry. The early period of this development is documented in the work of Sauer [23, 24]. The modern period began with the work by Bobenko and Pinkall [8, 9] and by Doliwa and Santini [18, 13]. An extensive exposition is given in the textbook by Bobenko and Suris [12], on which this chapter is mainly based. We refer the reader to this book for all the proofs omitted here, as well as for the detailed bibliographical remarks.

Discrete differential geometry related to integrable systems deals with multidimensional discrete nets, i.e., maps from the regular cubic lattice $\mathbb{Z}^m$ into $\mathbb{R}^N$ specified by certain geometric properties. We will be most interested in this chapter in the case $N = 3$. In this setting discrete surfaces appear as two-dimensional layers of multidimensional discrete nets, and their transformations correspond to shifts in the transversal lattice directions. A characteristic feature of the theory is that all lattice directions are on equal footing with respect to the defining geometric properties. Discrete surfaces and their transformations become indistinguishable. We associate such a situation with the *multidimensional consistency*, and this is one of our fundamental discretization principles. The multidimensional consistency, and therefore the existence and construction of multidimensional nets, relies just on certain incidence theorems of elementary geometry.

Conceptually, one can think of passing to a continuum limit by refining the mesh size in some of the lattice directions. In these directions, the net converges to smooth surfaces whereas those directions that remain discrete correspond to transformations of the surfaces (see Figure 10.1). The smooth theory comes as a corollary of a more fundamental discrete master theory. The true roots of classical surface theory are found, quite unexpectedly, in various incidence theorems of elementary geometry. This phenomenon, which has been shown for many classes of surfaces and coordinate systems [10, 12], is currently getting accepted as one of the fundamental features of classical integrable differential geometry.

Note that finding simple discrete explanations for complicated differential geometric theories is not the only outcome of this development. Having identified the roots of integrable differential geometry in the multidimensional consistency of

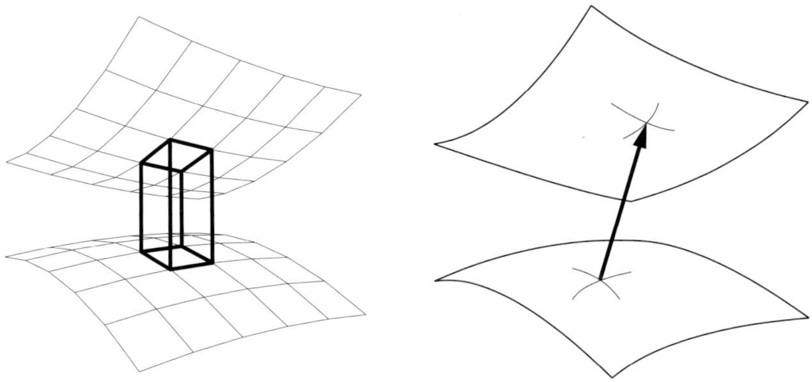

Figure 10.1. From the discrete master theory to the classical theory: surfaces and their transformations appear by refining two of three net directions.

discrete nets, we are led to a new (geometric) understanding of integrability itself [11, 2, 12].

The simplest and at the same time the basic example of consistent multidimensional nets are multidimensional Q-nets [18], or discrete conjugate nets [23, 24], which are characterized by planarity of all quadrilaterals. The planarity property is preserved by projective transformations and thus Q-nets are subject of projective geometry (like conjugate nets, which are smooth counterparts of Q-nets).

Here we come to the next basic discretization principle. According to Klein's Erlangen program, geometries are classified by their transformation groups. Classical examples are projective, affine, Euclidean, spherical, hyperbolic geometries, and the sphere geometries of Lie, Möebius, and Laguerre. We postulate that the transformation group as the most fundamental feature should be preserved by discretization. This can be seen as a sort of *discrete Erlangen program.*

Thus we come to the following fundamental *Discretization Principles*:

1. *Transformation group principle:* smooth geometric objects and their discretizations belong to the same geometry, i.e. they are invariant with respect to the same transformation group.
2. *Multidimensional consistency principle:* discretizations of surfaces, coordinate systems, and other smooth parametrized objects can be extended to multidimensional consistent nets.

Let us explain why such different imperatives as the transformation group principle and the consistency principle can be simultaneously imposed for the discretization of classical geometries. The transformation groups of various geometries are subgroups of the projective transformation group. Classically, such a subgroup is described as consisting of projective transformations which preserve some distinguished quadric called the absolute, see [1] in this volume. A remarkable result by Doliwa [16] is that multidimensional Q-nets can be restricted to an arbitrary quadric. This is the reason why the discretization principles work for the classical geometries.

2 Multidimensional consistency as a discretization principle

2.1 Q-nets We use the following standard notation: for a function f on $\mathbb{Z}^m$ we write

$$\tau_i f(u) = f(u + e_i),$$

where e_i is the unit vector of the i-th coordinate direction, $1 \le i \le m$. We also use the shortcut notation f_i for $\tau_i f$, f_{ij} for $\tau_i \tau_j f$, etc.

The most general known discrete 3D systems possessing the property of 4D consistency are nets consisting of planar quadrilaterals, or Q-nets. Two-dimensional Q-nets were introduced by Sauer [23, 24], and the multi-dimensional generalization has been given by Doliwa and Santini [18]. Our presentation in this section follows the latter. The fundamental importance of multi-dimensional consistency of discrete systems as their integrability has been put forward in [11, 2, 12].

Definition 2.1. (Q-net) A map $f : \mathbb{Z}^m \to \mathbb{RP}^N$ is called an m-dimensional Q-net (quadrilateral net, or discrete conjugate net) in $\mathbb{RP}^N$ ($N \ge 3$), if all its elementary quadrilaterals (f, f_i, f_{ij}, f_j) (at any $u \in \mathbb{Z}^m$ and for all pairs $1 \le i \ne j \le m$) are planar.

Given three points f, f_1, f_2 in $\mathbb{RP}^N$, one can take any point of the plane through these three points as the fourth vertex f_{12} of an elementary quadrilateral (f, f_1, f_{12}, f_2) of a Q-net. Correspondingly, given any two discrete curves $f : \mathbb{Z} \times \{0\} \to \mathbb{RP}^N$ and $f : \{0\} \times \mathbb{Z} \to \mathbb{RP}^N$ with a common point $f(0,0)$, one can construct infinitely many Q-surfaces $f : \mathbb{Z}^2 \to \mathbb{RP}^N$ with these curves as coordinate ones: the construction goes inductively: at each step one has the freedom of choosing a point in a plane (two real parameters).

On the other hand, constructing elementary hexahedra of Q-nets corresponding to elementary 3D cubes of the lattice $\mathbb{Z}^m$ admits a well-posed initial value problem with a unique solution; therefore one says that Q-nets are described by a *discrete 3D system*:

Theorem 2.2. (Elementary hexahedron of a Q-net) *Given seven points f, f_i and f_{ij} ($1 \le i < j \le 3$) in $\mathbb{RP}^N$, such that each of the three quadrilaterals (f, f_i, f_{ij}, f_j) is planar (i.e., f_{ij} lies in the plane Π_{ij} through f, f_i, f_j), define three planes $\tau_k \Pi_{ij}$ as those passing through the point triples f_k, f_{ik}, f_{jk}, respectively. Then these three planes intersect generically at one point:*

$$f_{123} = \tau_1 \Pi_{23} \cap \tau_2 \Pi_{13} \cap \tau_3 \Pi_{12}\,.$$

The elementary construction step from Theorem 2.2 is symbolically represented in Figure 10.2, which is the picture we have in mind when thinking about discrete three-dimensional systems with dependent variables (fields) attached to the vertices of a regular cubic lattice.

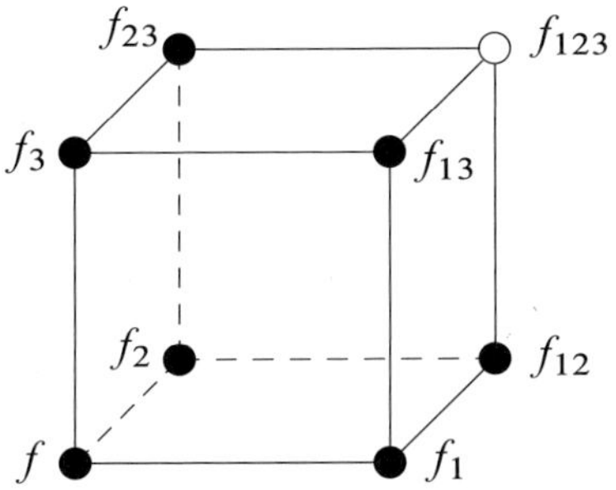

Figure 10.2. A 3D system on an elementary cube: the field at the white vertex is determined by the seven fields at the black vertices (initial data)

As follows from Theorem 2.2, a three-dimensional Q-net $f : \mathbb{Z}^3 \to \mathbb{RP}^N$ is completely determined by its three coordinate surfaces

$$f : \mathbb{Z}^2 \times \{0\} \to \mathbb{RP}^N, \quad f : \mathbb{Z} \times \{0\} \times \mathbb{Z} \to \mathbb{RP}^N, \quad f : \{0\} \times \mathbb{Z}^2 \to \mathbb{RP}^N.$$

Turning to an elementary cube of dimension $m \geq 4$, we see that one can prescribe all points f, f_i and f_{ij} for all $1 \leq i < j \leq m$. Indeed, these data are clearly independent, and one can construct all other vertices of an elementary cube starting from these data, *provided one does not encounter contradictions*. To see a possible source of contradiction, consider in detail first the case $m = 4$. From f, f_i and f_{ij} $(1 \leq i < j \leq 4)$, one determines all f_{ijk} uniquely. After that, one has, in principle, four different ways to determine f_{1234}, from four 3D cubic faces adjacent to this point; see Figure 10.3. Absence of contradictions means that these four values for f_{1234} automatically coincide. We call this property the 4D consistency.

Definition 2.3. (4D consistency) A 3D system is called 4D consistent, if it can be imposed on all three-dimensional faces of an elementary cube of $\mathbb{Z}^4$ (see Figure 10.3).

Remarkably, the construction of Q-nets based on the planarity of all elementary quadrilaterals enjoys this property.

Theorem 2.4. (Q-nets are 4D consistent) *The 3D system governing Q-nets is 4D-consistent.*

The m-dimensional consistency of a 3D system for $m > 4$ is defined analogously to the $m = 4$ case. Remarkably and quite generally, the 4-dimensional consistency already implies m-dimensional consistency for all $m > 4$.

Theorem 2.5. (4D consistency yields consistency in all higher dimensions) *Any 4D consistent discrete 3D system is also m-dimensionally consistent for any $m > 4$.*

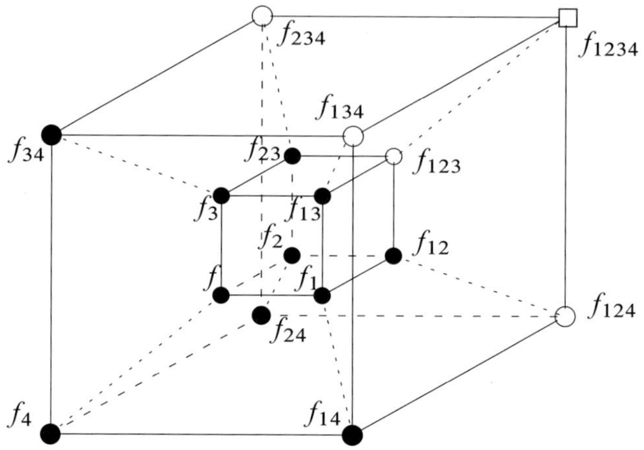

Figure 10.3. 4D consistency of 3D systems: the fields at the black vertices (initial data) determine, in virtue of the 3D system, the fields f_{ijk} at the white vertices. Then the 3D system gives four a priori different values for f_{1234}. The system is 4D consistent, if these four values coincide for any initial data

Theorems 2.4, 2.5 yield that Q-nets are m-dimensionally consistent for any $m \geq 4$. This fact, in turn, yields the existence of transformations of Q-nets with remarkable permutability properties. Referring for the details to [19, 12], we mention here only the definition.

Definition 2.6. (F-transformation of Q-nets) Two m-dimensional Q-nets $f, f^+ : \mathbb{Z}^m \to \mathbb{RP}^N$ are called F-transforms (fundamental transforms) of one another, if all quadrilaterals (f, f_i, f_i^+, f^+) (at any $u \in \mathbb{Z}^m$ and for all $1 \leq i \leq m$) are planar, i.e., if the net $F : \mathbb{Z}^m \times \{0, 1\} \to \mathbb{RP}^N$ defined by $F(u, 0) = f(u)$ and $F(u, 1) = f^+(u)$ is a two-layer $(m + 1)$-dimensional Q-net.

It follows from Theorem 2.2 that, given a Q-net f, its F-transform f^+ is uniquely defined as soon as its points along the coordinate axes are suitably prescribed.

2.2 Discrete line congruence Another important geometrical objects described by a discrete 3D system which is 4D consistent, are *discrete line congruences*. Their theory has been developed by Doliwa, Santini and Mañas [19].

Let $\mathcal{L}^N$ be the space of lines in $\mathbb{RP}^N$; it can be identified with the Grassmannian $\mathrm{Gr}(N + 1, 2)$ of two-dimensional vector subspaces of $\mathbb{R}^{N+1}$.

Definition 2.7. (Discrete line congruence) A map $\ell : \mathbb{Z}^m \to \mathcal{L}^N$ is called an m-dimensional discrete line congruence in $\mathbb{RP}^N$, $(N \geq 3)$, if any two neighboring lines ℓ, ℓ_i (at any $u \in \mathbb{Z}^m$ and for any $1 \leq i \leq m$) intersect (are co-planar).

For instance, lines $\ell = (ff^+)$ connecting corresponding points of two Q-nets $f, f^+ : \mathbb{Z}^m \to \mathbb{RP}^N$ in the relation of F-transformation clearly build a discrete line congruence.

A discrete line congruence is called *generic*, if for any $u \in \mathbb{Z}^m$ and for any $1 \le i \ne j \ne k \ne i \le m$, the four lines ℓ, ℓ_i, ℓ_j and ℓ_k span a four-dimensional space (i.e., a space of a maximal possible dimension). This yields, in particular, that for any $u \in \mathbb{Z}^m$ and for any $1 \le i \ne j \le m$, the three lines ℓ, ℓ_i and ℓ_j span a three-dimensional space.

The construction of line congruences is similar to that of Q-nets. Given three lines ℓ, ℓ_1, ℓ_2 of a congruence, one has a two-parameter family of lines admissible as the fourth one ℓ_{12}: connect by a line any point of ℓ_1 with any point of ℓ_2. Thus, given any two sequences of lines $\ell : \mathbb{Z} \times \{0\} \to \mathcal{L}^N$ and $\ell : \{0\} \times \mathbb{Z} \to \mathcal{L}^N$ with a common line $\ell(0,0)$ such that any two neighboring lines are co-planar, one can extend them to a two-dimensional line congruence $f : \mathbb{Z}^2 \to \mathcal{L}^N$ in an infinite number of ways: at each step of the inductive procedure, one has a freedom of choosing a line from a two-parameter family.

Non-degenerate line congruences are described by a *discrete 3D system*, which is, moreover, multidimensionally consistent:

Theorem 2.8. (Elementary hexahedron of a discrete line congruence) *Given seven lines ℓ, ℓ_i and ℓ_{ij} $(1 \le i < j \le 3)$ in $\mathbb{RP}^N$ such that ℓ intersects each of ℓ_i, the space V_{123} spanned by ℓ, ℓ_1, ℓ_2, ℓ_3 has dimension four, and each ℓ_i intersects both ℓ_{ij} and ℓ_{ik}, there is generically a unique line ℓ_{123} that intersects all three ℓ_{ij}.*

Theorem 2.9. (Discrete line congruences are 4D consistent) *The 3D system governing discrete line congruences is 4D-consistent.*

Like in the case of Q-nets, this theorem yields the existence of transformations of discrete line congruences with remarkable permutability properties.

Definition 2.10. (F-transformation of line congruences) Two m-dimensional line congruences $\ell, \ell^+ : \mathbb{Z}^m \to \mathcal{L}^N$ are called F-transforms of one another if the corresponding lines ℓ and ℓ^+ intersect (at any $u \in \mathbb{Z}^m$), i.e., if the map $L : \mathbb{Z}^m \times \{0,1\} \to \mathcal{L}^N$ defined by $L(u,0) = \ell(u)$ and $L(u,1) = \ell^+(u)$ is a two-layer $(m+1)$-dimensional line congruence.

It follows from Theorem 2.8 that given a line congruence ℓ, its F-transform ℓ^+ is uniquely defined as soon as its lines along the coordinate axes are suitably prescribed.

According to Definition 2.7, any two neighboring lines $\ell = \ell(u)$ and $\ell_i = \ell(u + e_i)$ of a line congruence intersect at exactly one point $f = \ell \cap \ell_i \in \mathbb{RP}^N$ which is thus combinatorially associated with the edge $(u, u + e_i)$ of the lattice $\mathbb{Z}^m$: $f = f(u, u+e_i)$. It is, however, sometimes more convenient to use the notation $f(u, u + e_i) = f^{(i)}(u)$ for this point, thus associating it to the vertex u of the lattice (and, of course, to the coordinate direction i). See Figure 10.4.

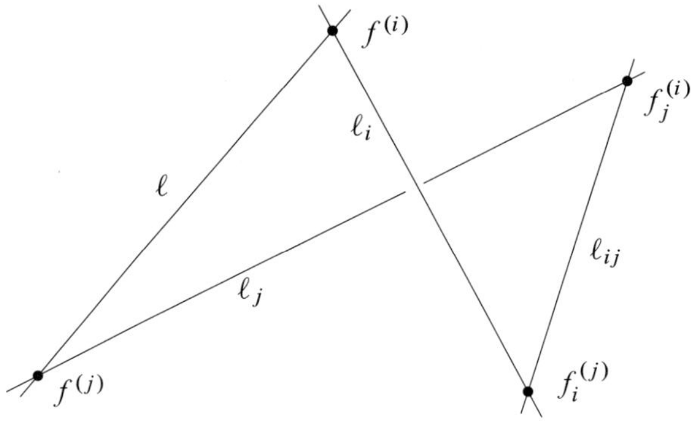

Figure 10.4. Four lines of a discrete line congruence

Definition 2.11. (Focal net) For a discrete line congruence $\ell : \mathbb{Z}^m \to \mathcal{L}^N$, the map $f^{(i)} : \mathbb{Z}^m \to \mathbb{RP}^N$ defined by $f^{(i)}(u) = \ell(u) \cap \ell(u + e_i)$ is called its i-th focal net (see Figure 10.5).

Theorem 2.12. *For a non-degenerate discrete line congruence $\ell : \mathbb{Z}^m \to \mathcal{L}^N$, all its focal nets $f^{(k)} : \mathbb{Z}^m \to \mathbb{RP}^N$, $1 \le k \le m$, are Q-nets.*

Corollary 2.13. (Focal net of F-transformation of a line congruence) *Given two generic line congruences $\ell, \ell^+ : \mathbb{Z}^m \to \mathcal{L}^N$ in the relation of F-transformation, the intersection points $f = \ell \cap \ell^+$ form a Q-net $f : \mathbb{Z}^m \to \mathbb{RP}^N$.*

2.3 Q-nets in quadrics We consider an important admissible reduction of Q-nets: they can be restricted to an arbitrary quadric in $\mathbb{RP}^N$. In the smooth differential geometry, i.e., for conjugate nets, this is due to Darboux [14]. In the discrete differential geometry this result has been found by Doliwa [16].

A deep reason for this result is the following fundamental fact well known in classical projective geometry (see, e.g., [6]):

Theorem 2.14. (Associated point) *Given eight distinct points which are the intersection set of three quadrics in $\mathbb{RP}^3$, all quadrics through any seven of these points must pass through the eighth point. Such sets of points are called associated.*

Theorem 2.15. (Elementary hexahedron of a Q-net in a quadric) *If seven points f, f_i, and f_{ij} $(1 \le i < j \le 3)$ of an elementary hexahedron of a Q-net $f : \mathbb{Z}^m \to \mathbb{RP}^N$ belong to a quadric $\mathcal{Q} \subset \mathbb{RP}^N$, then so does the eighth point f_{123}.*

As a global corollary of this local statement, we immediately obtain the following:

Theorem 2.16. (Reduction of Q-nets to a quadric) *If the coordinate surfaces of a Q-net $f : \mathbb{Z}^m \to \mathbb{RP}^N$ belong to a quadric $\mathcal{Q}$, then so does the entire f.*

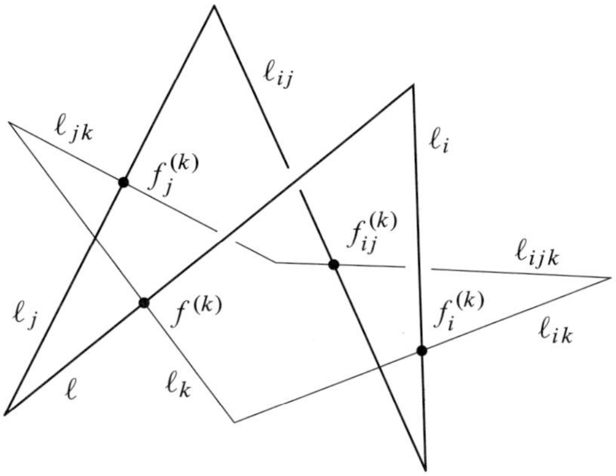

Figure 10.5. Elementary (ij) quadrilateral of the k-th focal net is planar

Another version of Theorem 2.15 can be formulated as follows. It is based on an obvious fact that for a nonisotropic line ℓ with a nonempty intersection with $\mathcal{Q}$, this intersection consists generically of *two* points (because it is governed by a quadratic equation).

Theorem 2.17. (Elementary Ribaucour transformation in a quadric) *Let (f, f_1, f_{12}, f_2) be a planar quadrilateral in a quadric $\mathcal{Q}$. Let $\ell, \ell_1, \ell_2, \ell_{12}$ be non-isotropic lines in $\mathbb{RP}^N$ containing the corresponding points f, f_1, f_2, f_{12} and such that every two neighboring lines intersect. Denote the second intersection points of the lines with $\mathcal{Q}$ by f^+, f_1^+, f_{12}^+, f_2^+, respectively. Then the quadrilateral $(f^+, f_1^+, f_{12}^+, f_2^+)$ is also planar.*

Again, a global version of this local statement is immediate:

Theorem 2.18. (Ribaucour transformation of a Q-net in a quadric) *Consider a quadric $\mathcal{Q} \subset \mathbb{RP}^N$ and a Q-net $f : \mathbb{Z}^m \to \mathcal{Q}$. Let a discrete congruence of nonisotropic lines $\ell : \mathbb{Z}^m \to \mathcal{L}^N$ be given such that $f(u) \in \ell(u)$ for all $u \in \mathbb{Z}^m$. Denote by $f^+(u)$ the second intersection point of $\ell(u)$ with $\mathcal{Q}$, so that $\ell(u) \cap \mathcal{Q} = \{f(u), f^+(u)\}$. Then $f^+ : \mathbb{Z}^m \to \mathcal{Q}$ is also a Q-net, called a Ribaucour transformation of f.*

Similarly, the 3D system describing line congruences admits a reduction to an arbitrary quadric. Suppose that the quadratic form in the space $\mathbb{R}^{N+1}$ of homogeneous coordinates, generating a quadric $\mathcal{Q} \subset \mathbb{RP}^N$, has signature containing at least two positive and two negative entries. In this case, the quadric $\mathcal{Q}$ carries isotropic lines $\ell \subset \mathcal{Q}$; actually, one can draw at least two such lines through any point of $\mathcal{Q}$. We denote the set of isotropic lines on $\mathcal{Q}$ by $\mathcal{L}_{\mathcal{Q}}^N$.

Theorem 2.19. (Elementary hexahedron of an isotropic line congruence) *Given seven isotropic lines ℓ, ℓ_i, $\ell_{ij} \in \mathcal{L}_Q^N$ $(1 \le i < j \le 3)$ such that ℓ intersects each of ℓ_i, the space V_{123} spanned by ℓ, ℓ_1, ℓ_2, ℓ_3 has dimension four, and each ℓ_i intersects both ℓ_{ij} and ℓ_{ik}, generically there is a unique isotropic line $\ell_{123} \in \mathcal{L}_Q^N$ that intersects all three ℓ_{ij}.*

An important interplay between these two 3D systems is given in the following result.

Theorem 2.20. (Extending a Q-net in a quadric to an isotropic line congruence) *Given a Q-net $f : \mathbb{Z}^m \to \mathcal{Q}$, there exist discrete congruences of isotropic lines $\ell : \mathbb{Z}^m \to \mathcal{L}_Q^N$ such that, for every $u \in \mathbb{Z}^m$, we have $f(u) \in \ell(u)$. Such a congruence is uniquely determined by prescribing an isotropic line $\ell(0)$ through the point $f(0)$.*

3 Example 1: Plücker line geometry and asymptotic nets

3.1 Plücker line geometry In the standard way, projective subspaces of $\mathbb{RP}^3$ are projectivizations of vector subspaces of $V = \mathbb{R}^4$. In particular, let $x, y \in \mathbb{RP}^3$ be any two different points, and let $\hat{x}, \hat{y} \in V$ be their arbitrary representatives in the space of homogeneous coordinates. Then the line $g = (xy) \subset \mathbb{RP}^3$ is the projectivization of the two-dimensional vector subspace $\mathrm{span}(\hat{x}, \hat{y}) \subset V$.

After Grassmann and Plücker, the latter subspace can be identified with (a projectivization of) the decomposable bivector

$$\hat{g} = \hat{x} \wedge \hat{y} \in \Lambda^2 V.$$

Denote the homogeneous coordinates of a point $x \in \mathbb{RP}^3$ by

$$\hat{x} = (x^0, x^1, x^2, x^3) \in \mathbb{R}^4.$$

Choose a basis of $\Lambda^2 V$ consisting of $\mathbf{e}_i \wedge \mathbf{e}_j$ with $0 \le i < j \le 3$. A coordinate representation of the bivector $\hat{g}$ in this basis is

$$\hat{g} = \sum_{(ij)} g^{ij}\, \mathbf{e}_i \wedge \mathbf{e}_j, \qquad g^{ij} = x^i y^j - x^j y^i.$$

The numbers $(g^{01}, g^{02}, g^{03}, g^{12}, g^{13}, g^{23})$ are called *Plücker coordinates* of the line g. They are defined projectively (up to a common factor). Not every bivector represents a line in $\mathbb{RP}^3$, since not every bivector is decomposable. An obvious necessary condition for a non-zero $\hat{g} \in \Lambda^2 V$ to be decomposable is

$$\hat{g} \wedge \hat{g} = 0.$$

It can be shown that this condition is also sufficient. In the Plücker coordinates, this condition can be written as

$$g^{01}g^{23} - g^{02}g^{13} + g^{03}g^{12} = 0.$$

Summarizing, we have the following description of $\mathcal{L}^3$, the set of lines in $\mathbb{RP}^3$, within Plücker line geometry. The six-dimensional vector space $\Lambda^2 V$ with the basis $\mathbf{e}_j \wedge \mathbf{e}_k$ is supplied with a nondegenerate scalar product defined by the following list of nonvanishing scalar products of the basis vectors:

$$\langle \mathbf{e}_0 \wedge \mathbf{e}_1, \mathbf{e}_2 \wedge \mathbf{e}_3 \rangle = -\langle \mathbf{e}_0 \wedge \mathbf{e}_2, \mathbf{e}_1 \wedge \mathbf{e}_3 \rangle = \langle \mathbf{e}_0 \wedge \mathbf{e}_3, \mathbf{e}_1 \wedge \mathbf{e}_2 \rangle = 1.$$

It is not difficult to verify that the signature of this scalar product is $(3,3)$, so that we can write $\Lambda^2 V \simeq \mathbb{R}^{3,3}$. We set

$$\mathbb{L}^{3,3} = \{\hat{g} \in \Lambda^2 V : \langle \hat{g}, \hat{g} \rangle = 0\}.$$

The points of the *Plücker quadric* $\mathbb{P}(\mathbb{L}^{3,3})$ are in a one-to-one correspondence with elements of $\mathcal{L}^3$.

Fundamental features of this model include:

- Two lines g, h in $\mathbb{RP}^3$ intersect if and only if their representatives in $\Lambda^2 V$ are polar of each other:

$$\langle \hat{g}, \hat{h} \rangle = g^{01}h^{23} - g^{02}h^{13} + g^{03}h^{12} + g^{23}h^{01} - g^{13}h^{02} + g^{12}h^{03} = 0.$$

  In this case the line $\ell \subset \mathbb{P}(\Lambda^2 V)$ through $[\hat{g}]$ and $[\hat{h}]$ is isotropic: $\ell \subset \mathbb{P}(\mathbb{L}^{3,3})$.
- Any *isotropic line* $\ell \subset \mathbb{P}(\mathbb{L}^{3,3})$ corresponds to a one-parameter family of lines in $\mathbb{RP}^3$ through a common point, which lie in one plane. Such a family of lines is naturally interpreted as a *contact element* (a point and a plane through this point) within the line geometry.

3.2 Asymptotic nets

Definition 3.1. (Discrete A-net) A map $f : \mathbb{Z}^m \to \mathbb{R}^3$ is called an m-dimensional discrete A-net (discrete asymptotic net) if for every $u \in \mathbb{Z}^m$ all the points $f(u \pm e_i)$, $i \in \{1, \dots, m\}$, lie in some plane $\mathcal{P}(u)$ through $f(u)$.

Two-dimensional A-nets were introduced by Sauer [23, 24] as a discrete analogue of surfaces parametrized along asymptotic lines. In this case, the plane $\mathcal{P}(u)$ is a discrete analogue of the *tangent plane* to the surface f at the point $f(u)$. A discrete A-net f is called *non-degenerate*, if all its elementary quadrilaterals $(f, \tau_i f, \tau_i \tau_j f, \tau_j f)$ are non-planar. In principle, it would be possible to consider discrete A-nets in $\mathbb{R}^N$ with $N > 3$, however this would not lead to an essential generalization, as any non-degenerate A-net $\mathbb{R}^N$ can be shown to lie in a three-dimensional affine subspace of

$\mathbb{R}^N$. Note that Definition 3.1 belongs to projective geometry and could equally well be formulated for the ambient space $\mathbb{RP}^3$.

Multidimensional A-nets are due to Doliwa [17]. The following result is based on the famous theorem of Möbius about pairs of mutually inscribed tetrahedra (see [6], [12]).

Theorem 3.2. (Consistency of discrete A-nets) *Discrete A-nets are multidimensionally consistent.*

As usual, this leads to a natural class of transformations with permutability properties, which can be seen, in the discrete context, as adding an additional lattice dimension to a given A-net.

Definition 3.3. (Discrete Weingarten transformation) A pair of discrete A-nets $f, f^+ : \mathbb{Z}^m \to \mathbb{R}^3$ is related by a Weingarten transformation if, for every $u \in \mathbb{Z}^m$, the line $(f(u)f^+(u))$ lies in both tangent planes to f and f^+ at the points $f(u)$ and $f^+(u)$, respectively. The net f^+ is called a Weingarten transform of the net f.

We now want to demonstrate how these notions can be reduced to the previously discussed notions of Q-nets and discrete line congruences in the context of Plücker line geometry.

It is not difficult to realize that Definition 3.1 of a discrete asymptotic net allows for the following reformulation:

Definition 3.4. (Discrete A-net, Euclidean model of line geometry) A map

$$(f, \mathcal{P}) : \mathbb{Z}^m \to \{\text{contact elements in } \mathbb{R}^3\}$$

is called an A-net if each pair of neighboring contact elements $(f, \mathcal{P})$, $(f_i, \mathcal{P}_i)$ has a line in common, that is, if the line (ff_i) belongs to both planes $\mathcal{P}, \mathcal{P}_i$. Such nets of contact elements are called *principal contact element nets.*

Thus, one views an A-net as a surface consisting not just of points, but rather of contact elements (points and tangent planes through these points). This can be immediately translated into the language of the projective model of Plücker line geometry, where contact elements are represented by the set $\mathcal{L}_0^{3,3}$ of isotropic lines in the Plücker quadric $\mathbb{P}(\mathbb{L}^{3,3}) \subset \mathbb{P}(\mathbb{R}^{3,3})$.

Definition 3.5. (Discrete A-net, projective model of line geometry) A map $\ell : \mathbb{Z}^m \to \mathcal{L}_0^{3,3}$ is called an A-net if it is a discrete congruence of isotropic lines in $\mathbb{P}(\mathbb{R}^{3,3})$, that is, if every two neighboring lines intersect:

$$\ell(u) \cap \ell(u + e_i) = \hat{l}^{(i)}(u) \in \mathbb{P}(\mathbb{L}^{3,3}), \quad \forall u \in \mathbb{Z}^m, \ \forall i \in \{1, 2, \ldots, m\}.$$

The elements of the focal nets $\hat{l}^{(i)} : \mathbb{Z}^m \to \mathbb{P}(\mathbb{L}^{3,3})$ represent the lines (ff_i) of the A-net in $\mathbb{R}^3$. We proceed with a re-formulation of Definition 3.3:

Definition 3.6. (Discrete W-congruence, Euclidean model of line geometry) Two discrete A-nets

$$(f, \mathcal{P}),\ (f^+, \mathcal{P}^+) : \mathbb{Z}^m \to \{\text{contact elements in } \mathbb{R}^3\}$$

are called Weingarten transforms of each other if for every pair of corresponding contact elements $(f, \mathcal{P})$, $(f^+, \mathcal{P}^+)$ the line $l = (ff^+)$ belongs to both tangent planes $\mathcal{P}$, $\mathcal{P}^+$. The connecting lines $l : \mathbb{Z}^m \to \{\text{lines in } \mathbb{R}^3\}$ of a Weingarten pair are said to constitute a discrete W-congruence.

In the language of the projective model, this becomes:

Definition 3.7. (Discrete W-congruence, projective model of line geometry) Two discrete A-nets $\ell,\ \ell^+ : \mathbb{Z}^m \to \mathcal{L}_0^{3,3}$ are called Weingarten transforms of each other if these discrete congruences of isotropic lines are related by an F-transformation, that is, if every two corresponding lines intersect:

$$\ell(u) \cap \ell^+(u) = \hat{l}(u) \in \mathbb{P}(\mathbb{L}^{3,3}), \quad \forall u \in \mathbb{Z}^m.$$

The intersection points $\hat{l} : \mathbb{Z}^m \to \mathbb{P}(\mathbb{L}^{3,3})$ represent the lines of a discrete W-congruence.

In the situation of Definition 3.6, both A-nets $(f, \mathcal{P})$ and $(f^+, \mathcal{P}^+)$ are said to be focal nets of the W-congruence $l = (ff^+)$. More generally, a discrete A-net $(f, \mathcal{P})$ is called *focal* for a discrete W-congruence l if each line l belongs to the corresponding contact element $(f, \mathcal{P})$, that is, $f \in l$ and $l \subset \mathcal{P}$. It is important to note a terminological confusion which is unfortunately unavoidable for historical reasons: a discrete W-congruence is *not* a discrete line congruence in the sense of Definition 2.7, and a focal A-net of a discrete W-congruence is *not* a focal net in the sense of Definition 2.11.

A characterization of discrete W-congruences which does not refer to their focal A-nets follows immediately from Theorem 2.12:

Corollary 3.8. (W-congruences are Q-nets in the Plücker quadric) *A generic W-congruence of lines is represented by a Q-net in the Plücker quadric* $\mathbb{P}(\mathbb{L}^{3,3})$.

In particular, four vectors $(\hat{l}, \hat{l}_i, \hat{l}_{ij}, \hat{l}_j)$ in $\mathbb{R}^{3,3}$ representing the four lines of an elementary quadrilateral of a generic W-congruence are linearly dependent. This means that the four lines (l, l_i, l_{ij}, l_j) in $\mathbb{R}^3$ belong to a regulus (a hyperboloidic family of lines).

As a corollary of Theorem 2.20, the following statement holds.

Theorem 3.9. (Focal A-nets of a discrete W-congruence) *Given a generic discrete W-congruence*

$$l : \mathbb{Z}^m \to \{\text{lines in } \mathbb{R}^3\},$$

there exists a two-parameter family of discrete A-nets

$$(f, \mathcal{P}) : \mathbb{Z}^m \to \{\text{contact elements in } \mathbb{R}^3\}$$

such that, for every $u \in \mathbb{Z}^m$, the line l belongs to the contact element $(f, \mathcal{P})$, that is, passes through the point f and lies in the plane $\mathcal{P}$. Such a discrete A-net is uniquely determined by prescribing a contact element $(f, \mathcal{P})(0)$ containing the line $l(0)$.

4 Example 2: Lie sphere geometry and principal curvature nets

4.1 Lie geometry A classical source on *Lie geometry* is Blaschke's book [5]. The following geometric objects in the Euclidean space $\mathbb{R}^3$ are elements of Lie geometry:

- *Oriented spheres.* A sphere in $\mathbb{R}^3$ with center $c \in \mathbb{R}^3$ and radius $r > 0$ is described by the equation $S = \{x \in \mathbb{R}^3 : |x - c|^2 = r^2\}$. It divides $\mathbb{R}^3$ in two parts, inner and outer. If one denotes one of two parts of $\mathbb{R}^3$ as positive, one comes to the notion of an oriented sphere. Thus, there are two oriented spheres $S^\pm$ for any S. One can take the orientation of a sphere into account by assigning a *signed radius* $\pm r$ to it. For instance, one can assign positive radii $r > 0$ to spheres with inward field of unit normals and negative radii $r < 0$ to spheres with outward field of unit normals.
- *Oriented planes.* A plane in $\mathbb{R}^3$ is given by the equation $P = \{x \in \mathbb{R}^3 : \langle v, x \rangle = d\}$, with a unit normal $v \in \mathbb{S}^2$ and $d \in \mathbb{R}$. Clearly, the pairs (v, d) and $(-v, -d)$ represent one and the same plane. It divides $\mathbb{R}^3$ in two halfspaces. Denoting one of two halfspaces as positive, one arrives at the notion of an oriented plane. Thus, there are two oriented planes $P^\pm$ for any P. One can take the orientation of a hyperplane into account by assigning the pair (v, d) to the plane with the unit normal v pointing into the positive halfspace.
- *Points.* One considers points $x \in \mathbb{R}^3$ as spheres of vanishing radius.
- *Infinity.* One compactifies the space $\mathbb{R}^3$ by adding the point at infinity ∞, with the understanding that a basis of open neighborhoods of ∞ is given, e.g., by the outer parts of the spheres $|x|^2 = r^2$. Topologically the so-defined compactification is equivalent to a sphere $\mathbb{S}^3$.
- *Contact elements.* A contact element of a surface in $\mathbb{R}^3$ is a pair consisting of a point $x \in \mathbb{R}^3$ and an (oriented) plane P through x; alternatively, one can use a normal vector v to P at x. In the framework of Lie geometry, a contact element can be identified with a set (a pencil) of all spheres S through x which are in an oriented contact with P (and with one another), thus sharing the normal vector v at x, see Figure 10.6.

All these elements are modelled in Lie geometry as points, resp. lines, in the 5-dimensional projective space $\mathbb{P}(\mathbb{R}^{4,2})$ with the space of homogeneous coordinates

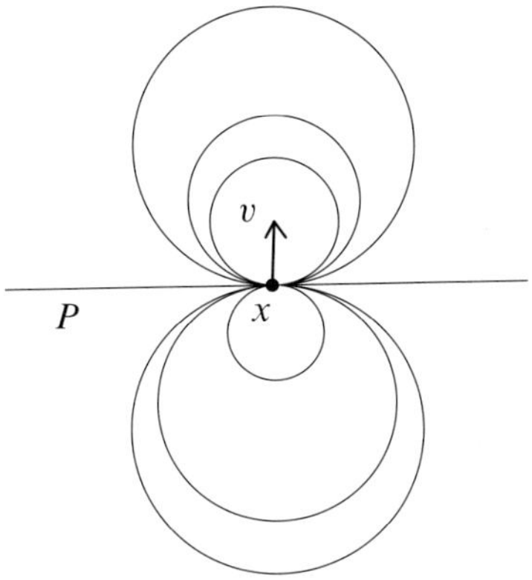

Figure 10.6. Contact element

$\mathbb{R}^{4,2}$. The latter is the space spanned by 6 linearly independent vectors $\mathbf{e}_1, \ldots, \mathbf{e}_6$ and equipped with the pseudo-Euclidean scalar product

$$\langle \mathbf{e}_i, \mathbf{e}_j \rangle = \begin{cases} 1, & i = j \in \{1, \ldots, 4\}, \\ -1, & i = j \in \{5, 6\}, \\ 0, & i \neq j. \end{cases}$$

It is convenient to introduce two isotropic vectors

$$\mathbf{e}_0 = \tfrac{1}{2}(\mathbf{e}_5 - \mathbf{e}_4), \quad \mathbf{e}_\infty = \tfrac{1}{2}(\mathbf{e}_5 + \mathbf{e}_4),$$

for which

$$\langle \mathbf{e}_0, \mathbf{e}_0 \rangle = \langle \mathbf{e}_\infty, \mathbf{e}_\infty \rangle = 0, \quad \langle \mathbf{e}_0, \mathbf{e}_\infty \rangle = -\tfrac{1}{2}.$$

The models of the above elements in the space $\mathbb{R}^{4,2}$ of homogeneous coordinates are as follows:

– *Oriented sphere with center $c \in \mathbb{R}^3$ and signed radius $r \in \mathbb{R}$:*

$$\hat{s} = c + \mathbf{e}_0 + (|c|^2 - r^2)\mathbf{e}_\infty + r\mathbf{e}_6.$$

– *Oriented plane $\langle v, x \rangle = d$ with $v \in \mathbb{S}^2$ and $d \in \mathbb{R}$:*

$$\hat{p} = v + 0 \cdot \mathbf{e}_0 + 2d\,\mathbf{e}_\infty + \mathbf{e}_6.$$

– *Point $x \in \mathbb{R}^3$:*

$$\hat{x} = x + \mathbf{e}_0 + |x|^2\mathbf{e}_\infty + 0 \cdot \mathbf{e}_6.$$

– *Infinity ∞:*

$$\hat{\infty} = \mathbf{e}_\infty.$$

– *Contact element (x, P):*

$$\operatorname{span}(\hat{x}, \hat{p}).$$

In the projective space $\mathbb{P}(\mathbb{R}^{4,2})$, the first four types of elements are represented by the points which are equivalence classes of the above vectors with respect to the relation $\xi \sim \eta \Leftrightarrow \xi = \lambda\eta$ with $\lambda \in \mathbb{R}^*$ for $\xi, \eta \in \mathbb{R}^{4,2}$. A contact element is represented by the line in $\mathbb{P}(\mathbb{R}^{4,2})$ through the points with representatives $\hat{x}$ and $\hat{p}$. We mention several fundamentally important features of this model:

1. All the above elements belong to the *Lie quadric* $\mathbb{P}(\mathbb{L}^{4,2})$, where
$$\mathbb{L}^{4,2} = \big\{\xi \in \mathbb{R}^{4,2} : \langle \xi, \xi \rangle = 0\big\}.$$
Moreover, points of $\mathbb{P}(\mathbb{L}^{4,2})$ are in a one-to-one correspondence with oriented spheres in $\mathbb{R}^3$, including the degenerate cases: proper spheres correspond to points of $\mathbb{P}(\mathbb{L}^{4,2})$ with both $\mathbf{e}_0$- and $\mathbf{e}_6$-components non-vanishing, planes correspond to points of $\mathbb{P}(\mathbb{L}^{4,2})$ with vanishing $\mathbf{e}_0$-component, points correspond to points of $\mathbb{P}(\mathbb{L}^{4,2})$ with vanishing $\mathbf{e}_6$-component, and infinity corresponds to the only point of $\mathbb{P}(\mathbb{L}^{4,2})$ with both $\mathbf{e}_0$- and $\mathbf{e}_6$-components vanishing.
2. Two oriented spheres S_1, S_2 are in an oriented contact (i.e., are tangent to each other with the unit normals at tangency pointing in the same direction) if and only if $|c_1 - c_2|^2 = (r_1 - r_2)^2$, and this is equivalent to
$$\langle \hat{s}_1, \hat{s}_2 \rangle = 0.$$
3. An oriented sphere $S = \{x \in \mathbb{R}^3 : |x - c|^2 = r^2\}$ is in an oriented contact with an oriented plane $P = \{x \in \mathbb{R}^3 : \langle v, x \rangle = d\}$ if and only if $\langle c, v \rangle - r - d = 0$, and the latter equation is equivalent to
$$\langle \hat{s}, \hat{p} \rangle = 0.$$
4. A point x can be considered as a sphere of radius $r = 0$ (in this case both oriented spheres coincide). Incidence relation $x \in S$ with a sphere S (resp. $x \in P$ with a plane P) can be interpreted as a particular case of an oriented contact of a sphere of radius $r = 0$ with S (resp. with P), and it takes place if and only if
$$\langle \hat{x}, \hat{s} \rangle = 0, \quad \text{resp.} \quad \langle \hat{x}, \hat{p} \rangle = 0.$$
5. For any plane P, we have $\langle \hat{\infty}, \hat{p} \rangle = 0$. One can interpret planes as spheres (of an infinite radius) through ∞. Analogously, the infinity ∞ can be considered as a limiting position of any sequence of points x with $|x| \to \infty$.
6. Any two spheres S_1, S_2 in an oriented contact determine a contact element (their point of contact and their common tangent hyperplane). If $\hat{s}_1$, $\hat{s}_2$ in $\mathbb{R}^{4,2}$ are representatives of S_1, S_2, then the line in $\mathbb{P}(\mathbb{R}^{4,2})$ through the corresponding points in $\mathbb{P}(\mathbb{L}^{4,2})$ is *isotropic*, i.e., lies entirely on the Lie quadric $\mathbb{P}(\mathbb{L}^{4,2})$. Such a line contains exactly one point whose representative $\hat{x}$ has vanishing $\mathbf{e}_6$-component (and corresponds to x, the common point of contact of all spheres), and, if $x \neq \infty$, exactly one point whose representative $\hat{p}$ has vanishing $\mathbf{e}_0$-component (and corresponds to P, the common tangent plane of all spheres). In the case where an isotropic line contains $\hat{\infty}$, all its points represent parallel planes, which constitute a contact element through ∞.

Thus, if one considers planes as spheres of infinite radii, and points as spheres of vanishing radii, then one can conclude that:

1. oriented spheres are in a one-to-one correspondence with points of the Lie quadric $\mathbb{P}(\mathbb{L}^{4,2})$ in the projective space $\mathbb{P}(\mathbb{R}^{4,2})$;
2. oriented contact of two oriented spheres corresponds to orthogonality of (any) representatives of the corresponding points in $\mathbb{P}(\mathbb{R}^{4,2})$.
3. contact elements of surfaces are in a one-to-one correspondence with isotropic lines in $\mathbb{P}(\mathbb{R}^{4,2})$. We will denote the set of all such lines by $\mathcal{L}_0^{4,2}$.

According to Klein's Erlangen program, Lie geometry is the study of properties of transformations which map oriented spheres (including points and planes) to oriented spheres and, moreover, preserve the oriented contact of sphere pairs. In the projective model described above, Lie geometry is the study of projective transformations of $\mathbb{P}(\mathbb{R}^{4,2})$ which leave $\mathbb{P}(\mathbb{L}^{4,2})$ invariant, and, moreover, preserve orthogonality of points of $\mathbb{P}(\mathbb{L}^{4,2})$ (which is understood as orthogonality of their lifts to $\mathbb{L}^{4,2} \subset \mathbb{R}^{4,2}$; clearly, this relation does not depend on the choice of lifts). Such transformations are called *Lie sphere transformations*.

Since (non-)vanishing of the $\mathbf{e}_0$- or of the $\mathbf{e}_6$-component of a point in $\mathbb{P}(\mathbb{L}^{4,2})$ is not invariant under a general Lie sphere transformation, there is no distinction between oriented spheres, oriented planes and points in Lie sphere geometry.

4.2 Curvature line parametrized surfaces in Lie geometry In Lie sphere geometry, where the notions of points and planes make no sense, a surface is naturally viewed as built of its contact elements. These contact elements are interpreted as points of the surface and tangent planes (or, equivalently, normals) at these points. This can be discretized in a natural way: a discrete surface is a map

$$(x, P) : \mathbb{Z}^2 \to \{\text{contact elements of surfaces in } \mathbb{R}^3\},$$

or, in the projective model of Lie geometry, a map

$$\ell : \mathbb{Z}^2 \to \mathcal{L}_0^{4,2}, \tag{10.1}$$

where, recall, $\mathcal{L}_0^{4,2}$ denotes the set of isotropic lines in $\mathbb{P}(\mathbb{R}^{4,2})$.

Lie sphere geometry is a natural framework to define an extremely important parametrization of surfaces, namely the parametrization along curvature lines. To see this, consider an infinitesimal neighborhood U of a point x of an oriented smooth surface in $\mathbb{R}^3$, and the pencil of spheres $S(r)$ of the signed radii r, touching the surface at x, see Figure 10.7. The signed radius r is assumed positive if $S(r)$ lies on the same side of the tangent plane as the normal n, and negative otherwise; $S(\infty)$ is the tangent plane. For small $r_0 > 0$ the spheres $S(r_0)$ and $S(-r_0)$ intersect U in x only. The set of the touching spheres with this property (intersecting U in x only) has two connected components: M_+ containing $S(r_0)$ and M_- containing $S(-r_0)$ for small $r_0 > 0$. The boundary values

$$k_1 = \inf\left\{\frac{1}{r} : S(r) \in M_+\right\}, \qquad k_2 = \sup\left\{\frac{1}{r} : S(r) \in M_-\right\}$$

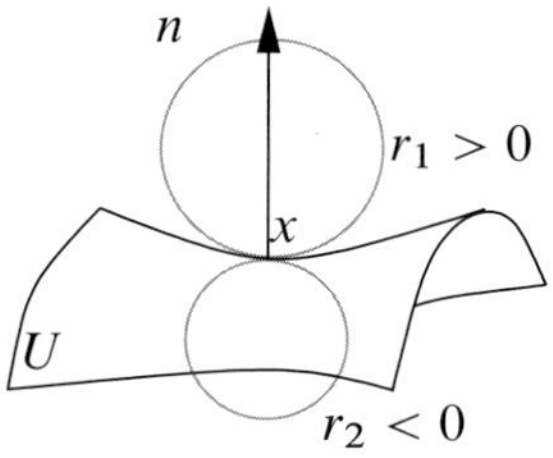

Figure 10.7. Principal directions through touching spheres.

are the principal curvatures of the surface in x. The directions in which $S(r_1)$ and $S(r_2)$ touch U are the principal directions. Clearly, all ingredients of this description are Möbius-invariant. Under a normal shift by distance d the centers of the principal curvature spheres are preserved and their radii are shifted by d. This implies that the principal directions and thus the curvature lines are preserved under normal shifts, as well. Thus, these notions are Lie-invariant.

A Lie-geometric nature of the curvature line parametrization yields that it has a Lie-invariant description. Such a description can be found in Blaschke's book [5]. A surface in Lie geometry, as already said, is considered as consisting of contact elements. Two infinitesimally close contact elements (sphere pencils) belong to the same curvature line if and only if they have a sphere in common, which is the principal curvature sphere.

The following definition is a literal discretization of this Lie-geometric description of curvature line parametrized surfaces.

Definition 4.1. (Principal contact element nets, Euclidean model) A map

$$(x, P) : \mathbb{Z}^2 \to \{\text{contact elements of surfaces in } \mathbb{R}^3\}$$

is called a principal contact element net, if any two neighboring contact elements (x, P), (x_i, P_i) have a sphere $S^{(i)}$ in common, that is, a sphere touching both planes P, P_i at the corresponding points x, x_i.

Thus, the normals to the neighboring planes P, P_i at the corresponding points x, x_i intersect at a point $c^{(i)}$ (the center of the sphere $S^{(i)}$), and the distances from $c^{(i)}$ to x and to x_i are equal, see Figure 10.8. The spheres $S^{(i)}$, attached to the edges of $\mathbb{Z}^2$ parallel to the i-th coordinate axis, will be called *principal curvature spheres* of the discrete surface.

A direct translation of Definition 4.1 into the projective model looks as follows:

Definition 4.2. (Principal contact element nets, projective model) A map $\ell : \mathbb{Z}^2 \to \mathcal{L}_0^{4,2}$ is called a principal contact element net, if it is a discrete congruence of isotropic lines in $\mathbb{P}(\mathbb{R}^{4,2})$, that is, any two neighboring lines intersect:

$$\ell(u) \cap \ell(u + e_i) = \hat{s}^{(i)}(u) \in \mathbb{P}(\mathbb{L}^{4,2}), \quad \forall u \in \mathbb{Z}^2, \quad \forall i = 1, 2.$$

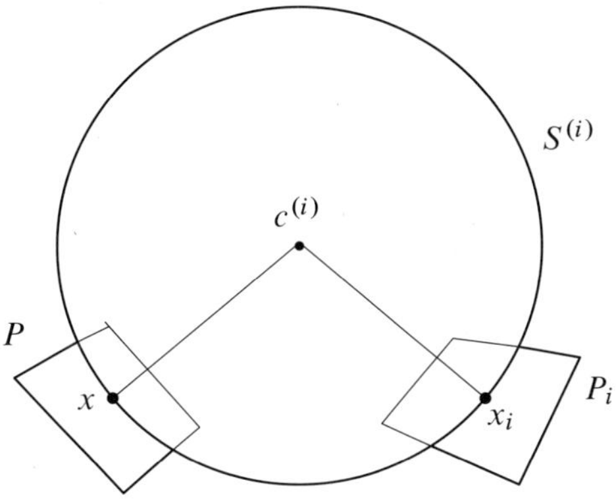

Figure 10.8. Principal curvature sphere

A comparison of the latter definition with Definition 3.5 shows that the only difference between the principal contact element nets and discrete asymptotic nets is the signature of the basic quadric of the projective model of the corresponding geometry. This is an instance of the famous Lie correspondence between spheres and lines in $\mathbb{R}^3$.

In the projective model, the representatives of the principal curvature spheres $S^{(i)}$ of the i-th coordinate direction build the corresponding focal net of the line congruence ℓ;

$$\hat{s}^{(i)} : \mathbb{Z}^2 \to \mathbb{P}(\mathbb{L}^{4,2}), \quad i = 1, 2, \tag{10.2}$$

cf. Definition 2.11. According to Theorem 2.12, both focal nets are Q-nets in $\mathbb{P}(\mathbb{R}^{4,2})$. This motivates the following definition.

Definition 4.3. (Discrete R-congruence of spheres) A map

$$S : \mathbb{Z}^m \to \{\text{oriented spheres in } \mathbb{R}^3\}$$

is called a discrete R-congruence (Ribaucour congruence) of spheres, if the corresponding map

$$\hat{s} : \mathbb{Z}^m \to \mathbb{P}(\mathbb{L}^{4,2})$$

is a Q-net in $\mathbb{P}(\mathbb{R}^{4,2})$.

Corollary 4.4. (Curvature spheres build an R-congruence) *For a discrete contact element net, the principal curvature spheres of the i-th coordinate direction (i=1,2) build a two-dimensional discrete R-congruence.*

Turning to transformations of principal contact element nets, we introduce the following definition.

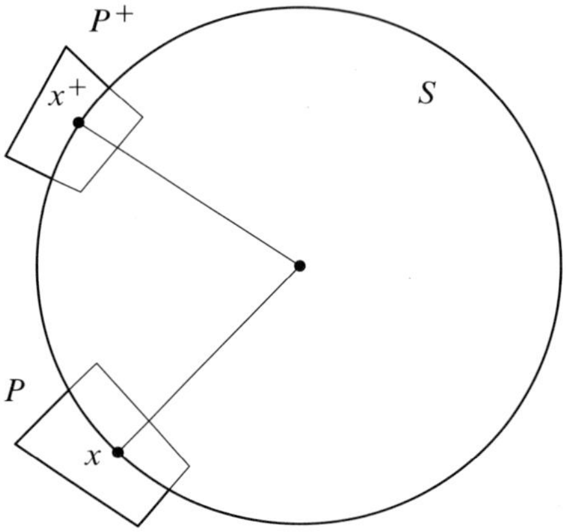

Figure 10.9. Ribaucour transformation

Definition 4.5. (Ribaucour transformation, projective model) Two principal contact element nets $\ell, \ell^+ : \mathbb{Z}^2 \to \mathcal{L}_0^{4,2}$ are called Ribaucour transforms of one another, if these discrete congruences of isotropic lines are in the relation of F-transformation, that is, if any pair of the corresponding lines intersect:

$$\ell(u) \cap \ell^+(u) = \hat{s}(u) \in \mathbb{P}(\mathbb{L}^{4,2}), \quad \forall u \in \mathbb{Z}^2. \tag{10.3}$$

Its direct translation into the geometric language gives:

Definition 4.6. (Ribaucour transformation, Euclidean model) Two principal contact element nets

$$(x, P), (x^+, P^+) : \mathbb{Z}^2 \to \{\text{contact elements of surfaces in } \mathbb{R}^3\}$$

are called Ribaucour transforms of one another, if any two corresponding contact elements (x, P) and (x^+, P^+) have a sphere S in common, that is, a sphere which touches both planes P, P^+ at the corresponding points x, x^+ (see Figure 10.9).

Spheres S of a Ribaucour transformation are attached to the vertices u of the lattice $\mathbb{Z}^2$, or, better, to the "vertical" edges connecting the vertices $(u, 0)$ and $(u, 1)$ of the lattice $\mathbb{Z}^2 \times \{0, 1\}$. In the projective model, their representatives

$$\hat{s} : \mathbb{Z}^2 \to \mathbb{P}(\mathbb{L}^{4,2}) \tag{10.4}$$

build the focal net of the three-dimensional line congruence for the third coordinate direction. From Theorem 2.12 there follows:

Corollary 4.7. (Spheres of a Ribaucour transformation build an R-congruence) *The spheres of a generic Ribaucour transformation build a discrete R-congruence.*

5 Example 3: Möbius geometry and circular nets

5.1 Möbius geometry Blaschke's book [5] serves also as a classical source on *Möbius geometry.*

Möbius geometry is a subgeometry of Lie geometry, with points distinguishable among all spheres as those of radius zero. Thus, Möbius geometry studies properties of spheres invariant under the subgroup of Lie sphere transformations preserving the set of points. In the projective model, points of $\mathbb{R}^3$ are distinguished as points of $\mathbb{P}(\mathbb{L}^{4,2})$ with vanishing $\mathbf{e}_6$-component. (Of course, one could replace here $\mathbf{e}_6$ by any time-like vector.) Thus, Möbius geometry studies the subgroup of Lie sphere transformations preserving the subset of $\mathbb{P}(\mathbb{L}^{N+1,2})$ with vanishing $\mathbf{e}_6$-component. The following geometric objects in $\mathbb{R}^6$ are elements of Möbius geometry.

- *(Non-oriented) spheres* $S = \{x \in \mathbb{R}^3 : |x-c|^2 = r^2\}$ with centers $c \in \mathbb{R}^3$ and radii $r > 0$.
- *(Non-oriented) planes* $P = \{x \in \mathbb{R}^3 : \langle v, x\rangle = d\}$, with unit normals $v \in \mathbb{S}^2$ and $d \in \mathbb{R}$.
- *Points* $x \in \mathbb{R}^3$.
- *Infinity* ∞ which compactifies $\mathbb{R}^3$ into $\mathbb{S}^3$.

In modelling these elements, one can use the Lie-geometric description and just omit the $\mathbf{e}_6$-component. The resulting objects are points of the four-dimensional projective space $\mathbb{P}(\mathbb{R}^{4,1})$ with the space of homogeneous coordinates $\mathbb{R}^{4,1}$ (classically called pentaspherical coordinates). The latter is the space spanned by five linearly independent vectors $\mathbf{e}_1, \dots, \mathbf{e}_5$ and equipped with the Minkowski scalar product

$$\langle \mathbf{e}_i, \mathbf{e}_j\rangle = \begin{cases} 1, & i = j \in \{1,\dots,4\}, \\ -1, & i = j = 5, \\ 0, & i \neq j. \end{cases}$$

We continue to use the notation

$$\mathbf{e}_0 = \frac{1}{2}(\mathbf{e}_5 - \mathbf{e}_4), \quad \mathbf{e}_\infty = \frac{1}{2}(\mathbf{e}_5 + \mathbf{e}_4)$$

in the context of Möbius geometry. The above elements are modelled in the space $\mathbb{R}^{4,1}$ of homogeneous coordinates as follows:

- *Sphere with center* $c \in \mathbb{R}^3$ *and radius* $r > 0$*:*

$$\hat{s} = c + \mathbf{e}_0 + (|c|^2 - r^2)\mathbf{e}_\infty.$$

- *Plane* $\langle v, x\rangle = d$ *with* $v \in \mathbb{S}^2$ *and* $d \in \mathbb{R}$*:*

$$\hat{p} = v + 0 \cdot \mathbf{e}_0 + 2d\mathbf{e}_\infty.$$

– *Point* $x \in \mathbb{R}^3$*:*

$$\hat{x} = x + \mathbf{e}_0 + |x|^2 \mathbf{e}_\infty.$$

– *Infinity* ∞*:*

$$\hat{\infty} = \mathbf{e}_\infty.$$

In the projective space $\mathbb{P}(\mathbb{R}^{4,1})$ these elements are represented by points which are equivalence classes of the listed vectors with respect to the usual relation $\xi \sim \eta \Leftrightarrow \xi = \lambda\eta$ with $\lambda \in \mathbb{R}^*$ for $\xi, \eta \in \mathbb{R}^{4,1}$. Fundamental features of these identifications are:

1. The infinity $\hat{\infty}$ can be considered as a limit of any sequence of $\hat{x}$ for $x \in \mathbb{R}^3$ with $|x| \to \infty$. Elements $x \in \mathbb{R}^3 \cup \{\infty\}$ are in a one-to-one correspondence with points of the projectivized *light cone* $\mathbb{P}(\mathbb{L}^{4,1})$, where
$$\mathbb{L}^{4,1} = \{\xi \in \mathbb{R}^{4,1} : \langle \xi, \xi \rangle = 0\}.$$
Points $x \in \mathbb{R}^3$ correspond to points of $\mathbb{P}(\mathbb{L}^{4,1})$ with a non-vanishing $\mathbf{e}_0$-component, while ∞ corresponds to the only point of $\mathbb{P}(\mathbb{L}^{4,1})$ with vanishing $\mathbf{e}_0$-component.
2. Spheres $\hat{s}$ and planes $\hat{p}$ belong to $\mathbb{P}(\mathbb{R}^{4,1}_{\text{out}})$, where
$$\mathbb{R}^{4,1}_{\text{out}} = \{\xi \in \mathbb{R}^{4,1} : \langle \xi, \xi \rangle > 0\}$$
is the set of space-like vectors of the Minkowski space $\mathbb{R}^{4,1}$. Planes can be interpreted as spheres (of infinite radius) through ∞.
3. Two spheres S_1, S_2 with centers c_1, c_2 and radii r_1, r_2 intersect orthogonally, if and only if $|c_1 - c_2|^2 = r_1^2 + r_2^2$, which is equivalent to
$$\langle \hat{s}_1, \hat{s}_2 \rangle = 0.$$
Similarly, a sphere S intersects orthogonally a plane P if and only if its center lies in P: $\langle c, v \rangle - d = 0$, which is equivalent to
$$\langle \hat{s}, \hat{p} \rangle = 0.$$
4. A point x can be considered as a limiting case of a sphere with radius $r = 0$. An incidence relation $x \in S$ with a sphere S (resp. $x \in P$ with a plane P) can be interpreted as a particular case of an orthogonal intersection of a sphere of radius $r = 0$ with S (resp. with P), and it takes place if and only if
$$\langle \hat{x}, \hat{s} \rangle = 0, \quad \text{resp.} \quad \langle \hat{x}, \hat{p} \rangle = 0.$$

Note that a sphere S can also be interpreted as the set of points $x \in S$. Correspondingly, it admits, along with the representation $\hat{s}$, the dual representation as a transversal intersection of $\mathbb{P}(\mathbb{L}^{4,1})$ with the projective 3-space $\mathbb{P}(\hat{s}^\perp)$, polar to the point $\hat{s}$ with respect to $\mathbb{P}(\mathbb{L}^{4,1})$; here, of course, $\hat{s}^\perp = \{\hat{x} \in \mathbb{R}^{4,1} : \langle \hat{s}, \hat{x} \rangle = 0\}$. This can be generalized to model circles.

- *Circles.* A circle is a (generic) intersection of two spheres S_1, S_2. The intersection of two spheres represented by $\hat{s}_1, \hat{s}_2 \in \mathbb{R}^{4,1}_{\mathrm{out}}$ is generic if the two-dimensional linear subspace of $\mathbb{R}^{4,1}$ spanned by the $\hat{s}_1, \hat{s}_2$ is space-like:

$$\Sigma = \operatorname{span}(\hat{s}_1, \hat{s}_2) \subset \mathbb{R}^{4,1}_{\mathrm{out}}.$$

As a set of points, this circle is represented as $\mathbb{P}(\mathbb{L}^{4,1} \cap \Sigma^{\perp})$, where

$$\Sigma^{\perp} = \bigcap_{i=1}^{2} \hat{s}_i^{\perp} = \Big\{\hat{x} \in \mathbb{R}^{4,1} : \langle \hat{s}_1, \hat{x}\rangle = \langle \hat{s}_2, \hat{x}\rangle = 0\Big\}$$

is a three-dimensional linear subspace of $\mathbb{R}^{4,1}$ of signature $(2,1)$.

Dually, through any three points $x_1, x_2, x_3 \in \mathbb{R}^3$ in general position one can draw a unique circle. It corresponds to the three-dimensional linear subspace

$$\Sigma^{\perp} = \operatorname{span}(\hat{x}_1, \hat{x}_2, \hat{x}_3),$$

of signature $(2,1)$, with three linearly independent isotropic vectors $\hat{x}_1, \hat{x}_2, \hat{x}_3 \in \mathbb{L}^{4,1}$. In the polar formulation, this circle corresponds to the two-dimensional space-like linear subspace

$$\Sigma = \bigcap_{i=1}^{3} \hat{x}_i^{\perp} = \Big\{\hat{s} \in \mathbb{R}^{4,1} : \langle \hat{s}, \hat{x}_1\rangle = \langle \hat{s}, \hat{x}_2\rangle = \langle \hat{s}, \hat{x}_3\rangle = 0\Big\}.$$

Möbius geometry is the study of properties of (non)-oriented spheres invariant with respect to projective transformations of $\mathbb{P}(\mathbb{R}^{4,1})$ which map points to points, i.e., which leave $\mathbb{P}(\mathbb{L}^{4,1})$ invariant. Such transformations are called *Möbius transformations.*

Since the (non-)vanishing of the $\mathbf{e}_\infty$-component of a point in $\mathbb{P}(\mathbb{R}^{N+1,1})$ is not invariant under a general Möbius transformation, there is no distinction in Möbius geometry between hyperspheres and hyperplanes.

5.2 Circular nets *Caution:* in this section, the notation $\hat{x}$ refers to the Möbius-geometric representatives in $\mathbb{L}^{4,1}$, and not to the Lie-geometric ones in $\mathbb{L}^{4,2}$. The former are obtained from the latter one by omitting the (vanishing) $\mathbf{e}_6$-component.

In *Möbius geometry*, a surface is viewed simply as built of points. A discrete surface is a map

$$x : \mathbb{Z}^2 \to \mathbb{R}^3,$$

or, in the projective model, a map

$$\hat{x} : \mathbb{Z}^2 \to \mathbb{P}(\mathbb{L}^{4,1}).$$

We now introduce the notion of circular nets [7, 13, 21] whose $m = 2$ particular case can be considered as a Möbius-geometric discretization of curvature line parametrized surfaces. These nets can be defined in two different ways.

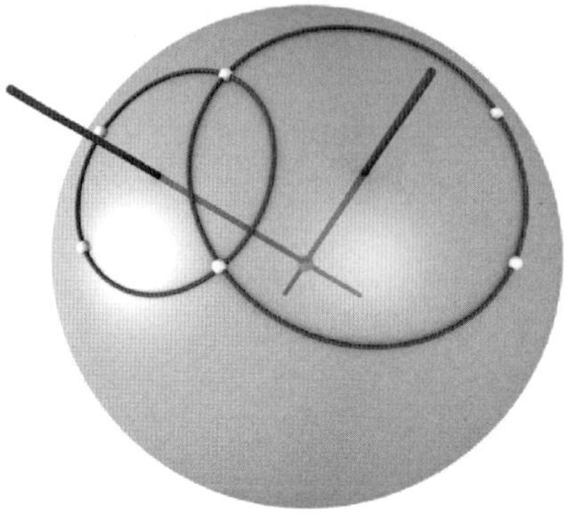

Figure 10.10. Normals of two neighboring quadrilaterals of a circular net intersect.

Definition 5.1. (Circular net, Euclidean model) A net $x : \mathbb{Z}^m \to \mathbb{R}^3$ is called circular, if the vertices of any elementary quadrilateral (x, x_i, x_{ij}, x_j) (at any $u \in \mathbb{Z}^m$ and for all pairs $1 \le i \ne j \le m$) lie on a circle (in particular, are co-planar).

Definition 5.2. (Circular net, projective model) A net $x : \mathbb{Z}^m \to \mathbb{R}^3$ is called circular, if the corresponding $\hat{x} : \mathbb{Z}^m \to \mathbb{P}(\mathbb{L}^{4,1})$ is a Q-net in $\mathbb{P}(\mathbb{R}^{4,1})$.

To see that Definitions 5.1 and 5.2 are equivalent, we observe that the linear subspace of $\mathbb{R}^{4,1}$ spanned by the isotropic vectors $\hat{x}, \hat{x}_i, \hat{x}_j, \hat{x}_{ij}$ is three-dimensional. Its orthogonal complement is therefore two-dimensional and lies in $\mathbb{R}^{4,1}_{\text{out}}$. Therefore, it represents a circle through the points x, x_i, x_j, x_{ij}.

Two-dimensional circular nets ($m = 2$) are discrete analogues of the curvature lines parametrized surfaces. It is natural to regard the lines passing through the centers of the circles orthogonally to their respective planes as the normals to the discrete circular surface. These normals behave in a way characteristic for the normals to a smooth surface along curvature lines; namely, for any two neighboring quadrilaterals of a circular net, the discrete normals intersect. Indeed, both normals lie in the bisecting orthogonal plane of the common edge. The intersection point is the center of a sphere containing both circles; see Figure 10.10.

On the other hand, circular nets are described by a 3D system, as follows from the next result.

Theorem 5.3. (Elementary hexahedron of a circular net) *Given seven points x, x_i, and x_{ij} $(1 \le i < j \le 3)$ in $\mathbb{R}^3$, such that each of the three quadruples (x, x_i, x_j, x_{ij}) lies on a circle C_{ij}, define three new circles $\tau_i C_{jk}$ as those passing through the triples (x_i, x_{ij}, x_{ik}), respectively. Then these new circles intersect at one point, see Figure* 10.11*:*

$$x_{123} = \tau_1 C_{23} \cap \tau_2 C_{31} \cap \tau_3 C_{12}\,.$$

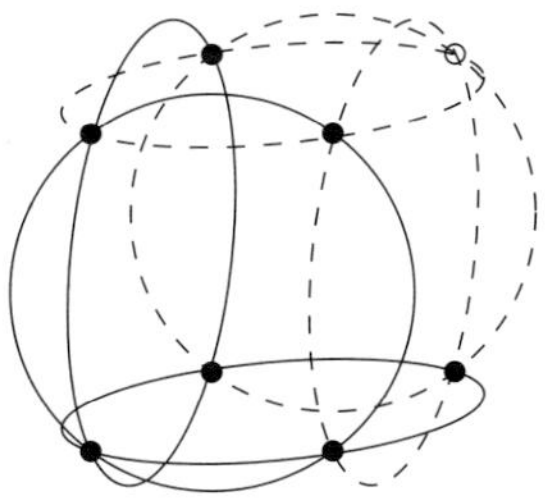

Figure 10.11. Construction of an elementary hexahedron of a circular net via the Miquel theorem

This is a particular case of Theorem 2.15, applied to the quadric $\mathbb{P}(\mathbb{L}^{4,1})$ underlying the Möbius geometry. This shows also that the 3D system describing circular nets is multidimensionally consistent. In particular, we have:

Theorem 5.4. (Circular reduction of Q-nets) *If the coordinate surfaces of a Q-net $f : \mathbb{Z}^m \to \mathbb{R}^N$ are circular surfaces, then f is a circular net.*

An important conceptual remark consists in the possibility of an elementary geometric proof of Theorem 5.3, based on the classical Miquel theorem, see Figure 10.11.

As usual, the multidimensional consistency yields the possibility of introducing transformations of circular nets with permutability properties, which is nothing but adding an additional dimension to the system under consideration. One arrives at the Möbius-geometric version of the discrete Ribaucour transformations:

Definition 5.5. (Discrete Ribaucour transformation) Two m-dimensional circular nets $f, f^+ : \mathbb{Z}^m \to \mathbb{R}^N$ are related by a Ribaucour transformation if the four points f, $\tau_i f$, f^+ and $\tau_i f^+$ are concircular at any $u \in \mathbb{Z}^m$ and for any $1 \le i \le m$. The net f^+ is called a Ribaucour transform of the net f.

6 Example 4: Laguerre geometry and conical nets

6.1 Laguerre geometry Blaschke's book [5] serves as the indispensable classical source also in the case of *Laguerre geometry*.

Laguerre geometry is a subgeometry of Lie geometry, with planes distinguishable among all spheres, as spheres through ∞. Thus, Laguerre geometry studies properties of spheres invariant under the subgroup of Lie sphere transformations which preserve the set of planes. The following objects in $\mathbb{R}^3$ are elements of the Laguerre geometry.

- *(Oriented) spheres* $S = \{x \in \mathbb{R}^3 : |x-c|^2 = r^2\}$ with centers $c \in \mathbb{R}^3$ and signed radii $r \in \mathbb{R}$, can be put into correspondence with 4-tuples (c, r).
- *Points* $x \in \mathbb{R}^3$ are considered as spheres of radius zero, and are put into correspondence with 4-tuples $(x, 0)$.
- *(Oriented) planes* $P = \{x \in \mathbb{R}^3 : \langle v, x\rangle = d\}$, with unit normals $v \in \mathbb{S}^2$ and $d \in \mathbb{R}$, can be put into correspondence with 4-tuples (v, d).

In the projective model of Lie geometry, planes are distinguished as elements of $\mathbb{P}(\mathbb{L}^{4,2})$ with vanishing $\mathbf{e}_0$-component. (Of course, one could replace here $\mathbf{e}_0$ by any isotropic vector.) Thus, Laguerre geometry studies the subgroup of Lie sphere transformations preserving the subset of $\mathbb{P}(\mathbb{L}^{4,2})$ with vanishing $\mathbf{e}_0$-component.

The scene of the *Blaschke cylinder model* of Laguerre geometry consists of *two* four-dimensional projective spaces, whose spaces of homogeneous coordinates, $\mathbb{R}^{3,1,1}$ and $(\mathbb{R}^{3,1,1})^*$, are dual to one another and arise from $\mathbb{R}^{4,2}$ by "forgetting" the $\mathbf{e}_0$-components. Thus, $\mathbb{R}^{3,1,1}$ is spanned by five linearly independent vectors $\mathbf{e}_1$, $\mathbf{e}_2$, $\mathbf{e}_3$, $\mathbf{e}_6$, $\mathbf{e}_\infty$, and is equipped with a degenerate bilinear form of signature $(3, 1, 1)$ in which the above vectors are pairwise orthogonal, the first three being space-like: $\langle \mathbf{e}_i, \mathbf{e}_i\rangle = 1$ for $1 \le i \le 3$, while the last two are time-like and isotropic, respectively: $\langle \mathbf{e}_6, \mathbf{e}_6\rangle = -1$ and $\langle \mathbf{e}_\infty, \mathbf{e}_\infty\rangle = 0$. Similarly, $(\mathbb{R}^{3,1,1})^*$ is assumed to have an orthogonal basis consisting of $\mathbf{e}_1$, $\mathbf{e}_2$, $\mathbf{e}_3$, $\mathbf{e}_6$, $\mathbf{e}_0$, again with an isotropic last vector: $\langle \mathbf{e}_0, \mathbf{e}_0\rangle = 0$. Note that one and the same symbol $\langle \cdot, \cdot\rangle$ is used to denote two degenerate bilinear forms in our two spaces. We will overload this symbol even more and use it also for the (non-degenerate) pairing between these two spaces, which is established by setting $\langle \mathbf{e}_0, \mathbf{e}_\infty\rangle = -\frac{1}{2}$, additionally to the above relations. (Note that a degenerate bilinear form cannot be used to identify a vector space with its dual.)

We have

- *A plane* $P = (v, d)$ *is modelled as a point in the space* $\mathbb{P}(\mathbb{R}^{3,1,1})$ *with a representative*
$$\hat{p} = v + 2d\mathbf{e}_\infty + \mathbf{e}_6.$$
- *A sphere* $S = (c, r)$ *is modelled as a point in the space* $\mathbb{P}\big((\mathbb{R}^{3,1,1})^*\big)$ *with a representative*
$$\hat{s} = c + \mathbf{e}_0 + r\mathbf{e}_6.$$

The preferred space is $\mathbb{P}(\mathbb{R}^{3,1,1})$ whose points model planes $P \subset \mathbb{R}^N$. A sphere $S \subset \mathbb{R}^N$ is then modelled as a hyperplane $\{\xi \in \mathbb{P}(\mathbb{R}^{N,1,1}) : \langle \hat{s}, \xi\rangle = 0\}$ in the space $\mathbb{P}(\mathbb{R}^{N,1,1})$. The following are basic features of this model:

1. Oriented planes $P \subset \mathbb{R}^N$ are in a one-to-one correspondence with points $\hat{p}$ of the quadric $\mathbb{P}(\mathbb{L}^{3,1,1})$, where
$$\mathbb{L}^{3,1,1} = \big\{\xi \in \mathbb{R}^{N,1,1} : \langle \xi, \xi\rangle = 0\big\}.$$
2. Two oriented planes $P_1, P_2 \subset \mathbb{R}^3$ are in an oriented contact (parallel), if and only if their representatives $\hat{p}_1$, $\hat{p}_2$ differ by a vector parallel to $\mathbf{e}_\infty$.

3. An oriented sphere $S \subset \mathbb{R}^3$ is in an oriented contact with an oriented hyperplane $P \subset \mathbb{R}^N$, if and only if $\hat{p} \in \hat{s}$, that is, if $\langle \hat{p}, \hat{s} \rangle = 0$. Thus, a sphere S is interpreted as its set of tangent planes.

The quadric $\mathbb{P}(\mathbb{L}^{3,1,1})$ is diffeomorphic to the *Blaschke cylinder*

$$\mathcal{Z} = \{(v, d) \in \mathbb{R}^4 : |v| = 1\} = \mathbb{S}^2 \times \mathbb{R} \subset \mathbb{R}^4.$$

Two points of this cylinder represent parallel planes if they lie on one straight line generator of $\mathcal{Z}$ parallel to its axis. In the ambient space $\mathbb{R}^4$ of the Blaschke cylinder, oriented spheres $S \subset \mathbb{R}^N$ are in one-to-one correspondence with planes non-parallel to the axis of $\mathcal{Z}$:

$$S \sim \{(v, d) \in \mathbb{R}^4 : \langle c, v \rangle - d - r = 0\}.$$

An intersection of such a hyperplane with $\mathcal{Z}$ consists of points in $\mathcal{Z}$ which represent tangent planes to $S \subset \mathbb{R}^N$.

6.2 Conical nets *Caution:* in this section, the notation $\hat{p}$ refers to the Laguerre-geometric representatives in $\mathbb{L}^{3,1,1}$, and not to the Lie-geometric ones in $\mathbb{L}^{4,2}$. The former are obtained from the latter by omitting the (vanishing) $\mathbf{e}_0$-component.

In Laguerre geometry, one thinks of a surface as of the envelope of its tangent planes. Correspondingly, a general discrete surface in the Laguerre geometry is just a net of oriented planes,

$$P : \mathbb{Z}^m \to \{\text{oriented planes in } \mathbb{R}^3\}.$$

A discrete analogue of curvature line parametrized surfaces is given by the $m = 2$ case of the following definition [22].

Definition 6.1. (Conical net, Euclidean model) A net

$$P : \mathbb{Z}^m \to \{\text{oriented planes in } \mathbb{R}^3\}$$

is called conical, if at any $u \in \mathbb{Z}^m$ and for all pairs $1 \le i \ne j \le m$ the four planes P, P_i, P_{ij}, P_j are in oriented contact with a cone of revolution (in particular, intersect at the tip of the cone), see Figure 10.12.

One can think of two-dimensional conical nets as another (besides circular nets) discretization of curvature line parametrized surfaces. The axes of the cones of revolution mentioned in Definition 6.1 are thought of as discrete normals to the surface, assigned to the elementary squares of the lattice $\mathbb{Z}^2$. For any two neighboring cones, there is a unique sphere touching both of them. The center of this sphere is the intersection point of the axes of the cones; see Figure 10.13. Indeed, two neighboring quadruples of planes share two planes, and the plane bisecting the dihedral angle between these two contains the axes of both cones, which therefore intersect (the orientations of the planes fix one of the two dihedral angles).

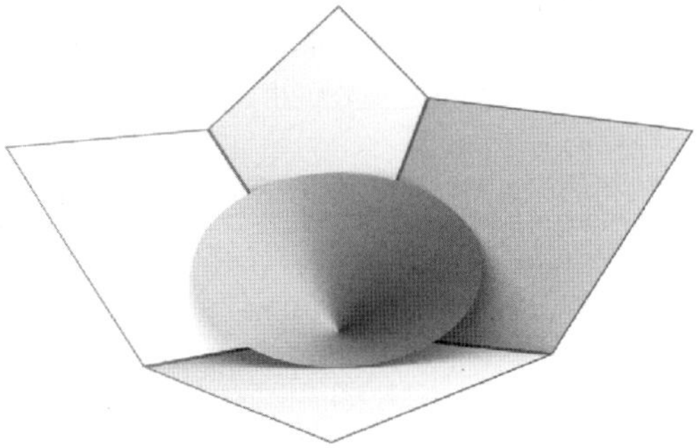

Figure 10.12. Four planes of a conical net.

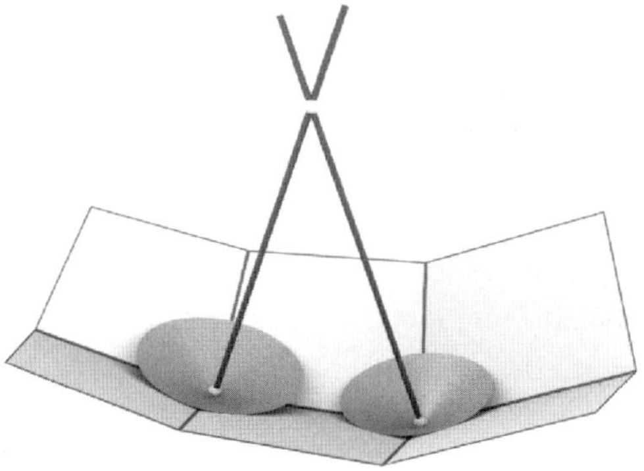

Figure 10.13. Axes of two neighboring cones of a conical net intersect.

A simple geometric criterion for a net P of planes to be conical can be given in terms of the *Gauss map* of P. Recall that an oriented plane P in $\mathbb{R}^3$ can be described by a pair $(v, d) \in \mathbb{S}^2 \times \mathbb{R}$, where

$$P = \{x \in \mathbb{R}^3 : \langle v, x \rangle = d\},$$

so that $v \in \mathbb{S}^2$ is the unit normal vector to P, and d is the distance of P to the origin. The net

$$v : \mathbb{Z}^m \to \mathbb{S}^2,$$

comprised by the (directed) unit normal vectors v to the planes P, is called the Gauss map.

Theorem 6.2. (Conical nets have circular Gauss maps) *A net of planes* $P : \mathbb{Z}^m \to$ $\{$oriented planes in $\mathbb{R}^3\}$ *is conical if and only if any four neighboring planes have a point in common, and the elementary quadrilaterals of its Gauss map* $v : \mathbb{Z}^m \to \mathbb{S}^2$ *are planar, that is, if the Gauss map is a circular net in* $\mathbb{S}^2$.

Indeed, the angles between all four unit vectors v, v_i, v_j, v_{ij} and the axis of the cone are equal; therefore their tips are at an equal (spherical) distance from the point of $\mathbb{S}^2$ representing the cone axis direction. Thus, the quadrilateral (v, v_i, v_{ij}, v_j) in $\mathbb{S}^2$ is circular, with the spherical center of the circle given by the direction of the axis of the tangent cone.

It is important to observe that Definition 6.1 actually belongs to Laguerre geometry. This means that the property of touching a common cone of revolution for given planes is invariant under Laguerre transformations, in particular, under shifting all the planes by the same distance in their corresponding normal directions (*normal shift*). Recall that a plane $P = \{x \in \mathbb{R}^3 : \langle v, x\rangle = d\}$ is represented in the projective model of Laguerre geometry (Blaschke cylinder model) by the point in $\mathbb{P}(\mathbb{L}^{3,1,1}) \subset \mathbb{P}(\mathbb{R}^{3,1,1})$ with representative $\hat{p} = v + 2d\mathbf{e}_\infty + \mathbf{e}_6$ in the space of homogeneous coordinates.

Theorem 6.3. (Conical net, Laguerre-geometric characterization) *A net*

$$P : \mathbb{Z}^m \to \{\text{oriented planes in } \mathbb{R}^3\}$$

is conical if and only if the corresponding points $\hat{p} : \mathbb{Z}^m \to \mathbb{P}(\mathbb{L}^{3,1,1})$ *form a Q-net in* $\mathbb{P}(\mathbb{R}^{3,1,1})$.

Thus, conical nets constitute a further example of a multidimensional Q-net restricted to a quadric (the absolute quadric in the projective model of Laguerre geometry).

7 Discrete curvature line parametrization in various sphere geometries

We have seen that the description of a discrete surface in Lie sphere geometry (through contact elements, i.e., points and tangent planes) contains more information than description of a discrete surface in Möbius geometry (through points only) or in Laguerre geometry (through tangent planes only). Actually, the former merges the two latter ones.

To clarify the relations between these various descriptions, we consider the geometry of an elementary quadrilateral of contact elements of a principal contact element net, consisting of $\ell \sim (x, P)$, $\ell_1 \sim (x_1, P_1)$, $\ell_2 \sim (x_2, P_2)$, and $\ell_{12} \sim (x_{12}, P_{12})$.

We leave aside a degenerate *umbilic* situation, when all four lines have a common point and span a four-dimensional space. Geometrically, this means that one is dealing with four contact elements of a sphere $S \subset \mathbb{R}^3$. In this situation, one cannot draw any further conclusion about the four points x, x_1, x_2, x_{12} on the sphere S: they can be arbitrary. Thus, we assume that the principal contact element nets under consideration are generic in the sense that they do not contain umbilic quadruples.

Theorem 7.1. (Points of principal contact element nets form circular nets) *For a principal contact element net*

$$(x, P) : \mathbb{Z}^2 \to \{\text{contact elements of surfaces in } \mathbb{R}^3\},$$

its points $x : \mathbb{Z}^2 \to \mathbb{R}^3$ *form a circular net.*

Indeed, in the non-umbilic situation, the space spanned by the four lines $\ell, \ell_1, \ell_2, \ell_{12}$ is three-dimensional. The four elements $\hat{x}, \hat{x}_1, \hat{x}_2, \hat{x}_{12} \in \mathbb{P}(\mathbb{L}^{4,2})$ corresponding to the points $x, x_1, x_2, x_{12} \in \mathbb{R}^3$ are obtained as the intersection of the four isotropic lines $\ell, \ell_1, \ell_2, \ell_{12}$ with the projective hyperplane $\mathbb{P}(\mathbf{e}_6^\perp)$ in $\mathbb{P}(\mathbb{R}^{4,2})$. Therefore, the four elements $\hat{x}, \hat{x}_1, \hat{x}_2, \hat{x}_{12}$ lie in a plane. Omitting the inessential (vanishing) $\mathbf{e}_6$-component, we arrive at a planar quadrilateral in the Möbius sphere $\mathbb{P}(\mathbb{L}^{4,1})$.

Theorem 7.2. (Tangent planes of principal contact element nets form conical nets) *For a principal contact element net*

$$(x, P) : \mathbb{Z}^2 \to \{\text{contact elements of surfaces in } \mathbb{R}^3\},$$

its tangent planes $P : \mathbb{Z}^2 \to \{\text{oriented planes in } \mathbb{R}^3\}$ *form a conical net.*

Indeed, the four elements $\hat{p}, \hat{p}_1, \hat{p}_2, \hat{p}_{12} \in \mathbb{P}(\mathbb{L}^{4,2})$ corresponding to the planes $P, P_1, P_2, P_{12} \in \mathbb{R}^3$ are obtained as the intersection of the four isotropic lines $\ell, \ell_1, \ell_2, \ell_{12}$ with the projective hyperplane $\mathbb{P}(\mathbf{e}_\infty^\perp)$ in $\mathbb{P}(\mathbb{R}^{4,2})$. Therefore, the four elements $\hat{p}, \hat{p}_1, \hat{p}_2, \hat{p}_{12}$ also lie in a plane.

In view of Theorems 7.1, 7.2, it is natural to ask whether, given a circular net $x : \mathbb{Z}^2 \to \mathbb{R}^3$, or a conical net $P : \mathbb{Z}^2 \to \{\text{oriented planes in } \mathbb{R}^3\}$, there exists a principal contact element net

$$(x, P) : \mathbb{Z}^2 \to \{\text{contact elements of surfaces in } \mathbb{R}^3\},$$

with half of the data (x or P) prescribed. A positive answer to this question is a corollary of the following general theorem.

Theorem 7.3. (Extending R-congruences of spheres to principal contact element nets) *Given a discrete R-congruence of spheres*

$$S : \mathbb{Z}^2 \to \{\text{oriented spheres in } \mathbb{R}^3\},$$

there exists a two-parameter family of principal contact element nets

$$(x, P) : \mathbb{Z}^2 \to \{\text{contact elements of surfaces in } \mathbb{R}^3\}$$

such that S *belongs to the contact element* (x, P)*, i.e.,* P *is the tangent plane to* S *at the point* $x \in S$*, for all* $u \in \mathbb{Z}^2$*. Such a principal contact element net is uniquely determined by prescribing a contact element* $(x, P)(0, 0)$ *containing the sphere* $S(0, 0)$*.*

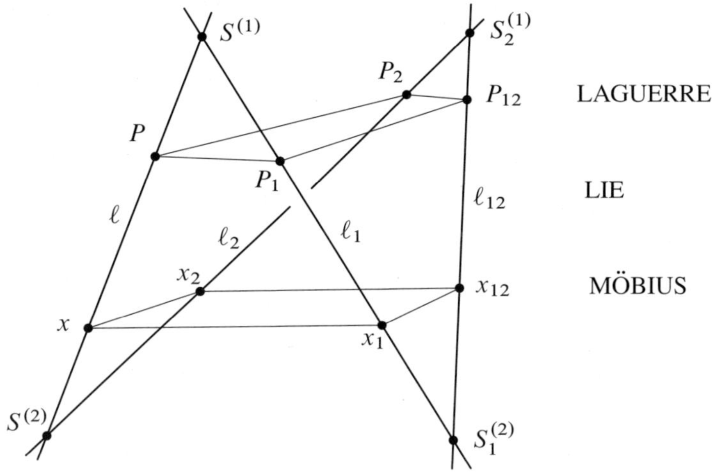

Figure 10.14. Elementary quadrilateral of a curvature line parametrized surface with vertices x and tangent planes P in the projective model. The vertices x build a circular net (Möbius geometry), and lie in the planes P building a conical net (Laguerre geometry). Contact elements (x, P) are represented by isotropic lines ℓ (Lie geometry). Principal curvature spheres $S^{(i)}$ pass through pairs of neighboring points x, x_i and are tangent to the corresponding pairs of planes P, P_i.

Since the representatives $\hat{x}$ in $\mathbb{P}(\mathbb{L}^{4,2})$ of a circular net $x : \mathbb{Z}^2 \to \mathbb{R}^3$ form a Q-net in $\mathbb{P}(\mathbb{R}^{4,2})$, and the same holds for the representatives $\hat{p}$ in $\mathbb{P}(\mathbb{L}^{4,2})$ of a conical net $P : \mathbb{Z}^2 \to \{\text{oriented planes in } \mathbb{R}^3\}$, we come to the following conclusion.

Corollary 7.4. (Extending circular and conical nets to principal contact element nets)

1. *Given a circular net $x : \mathbb{Z}^2 \to \mathbb{R}^3$, there exists a two-parameter family of conical nets $P : \mathbb{Z}^2 \to \{\text{planes in } \mathbb{R}^3\}$ such that $x \in P$ for all $u \in \mathbb{Z}^2$, and the contact element net*

$$(x, P) : \mathbb{Z}^2 \to \{\text{contact elements of surfaces in } \mathbb{R}^3\}$$

is principal. Such a conical net is uniquely determined by prescribing a plane $P(0,0)$ through the point $x(0,0)$.

2. *Given a conical net $P : \mathbb{Z}^2 \to \{\text{oriented planes in } \mathbb{R}^3\}$, there exists a two-parameter family of circular nets $x : \mathbb{Z}^2 \to \mathbb{R}^3$ such that $x \in P$ for all $u \in \mathbb{Z}^2$, and the contact element net*

$$(x, P) : \mathbb{Z}^2 \to \{\text{contact elements of surfaces in } \mathbb{R}^3\}$$

is principal. Such a circular net is uniquely determined by prescribing a point $x(0,0)$ in the plane $P(0,0)$.

These relations are summarized in Figure 10.14.

8 Conclusion

In this chapter, we have demonstrated several examples illustrating a spectacular application of the classical but ever young ideas of Felix Klein's Erlangen program to an active research area of discrete differential geometry. The combination of the transformation group principle with the multidimensional consistency principle can serve as the guiding and organizing principles of this new area, allowing one to systematically discover correct discretizations of the complicated constructions of classical differential geometry. On this way, one achieves a tremendous simplification and a deeper understanding of both the geometry and the accompanying integrable systems. Needless to say that the examples here by no means exhaust the area; on the contrary, many more are already known (and can be looked up, for instance, in [12]), and still more await to be discovered. The ideas of the Erlangen program will guide us on this way for years to come.

Bibliography

[1] N. A'Campo, A. Papadopoulos, On Klein's *So-called Non-Euclidean geometry*. In *Sophus Lie and Felix Klein: The Erlangen program and its impact in mathematics and physics*. Ed. L. Ji and A. Papadopoulos, Zürich, European Mathematical Society (EMS), 2015, 91–136.

[2] V. E. Adler, A. I. Bobenko and Yu. B. Suris, Classification of integrable equations on quad-graphs. The consistency approach. *Commun. Math. Phys.*, 2003, 233, 513–543.

[3] M. Berger, *Geometry.* Berlin etc.: Springer-Verlag, 1987.

[4] L. Bianchi, *Lezioni di geometria differenziale.* 3rd edition. Pisa: Enrico Spoerri, 1923 (Italian). iv+806, xi+832 pp.

[5] W. Blaschke, *Vorlesungen über Differentialgeometrie III: Differentialgeometrie der Kreise und Kugeln*. Bearbeitet von G. Thomsen. Berlin: Springer, 1929 (German). x+474 pp.

[6] W. Blaschke, *Projektive Geometrie*. 3rd edition. Basel etc.: Birkhäuser, 1954. 197 pp.

[7] A. I. Bobenko, Discrete conformal maps and surfaces. In *Symmetries and integrability of difference equations* (Canterbury, 1996), CRM Proc. Lecture Notes, 255 (1999), 97–108.

[8] A. I. Bobenko and U. Pinkall, Discrete surfaces with constant negative Gaussian curvature and the Hirota equation. *J. Differential Geom.* 43 (1996), no. 3, 527–611.

[9] A. I. Bobenko and U. Pinkall, Discrete isothermic surfaces. *J. Reine Angew. Math.* 475 (1996), 187–208.

[10] A. I. Bobenko and U. Pinkall, Discretization of surfaces and integrable systems. In *Discrete integrable geometry and physics*. Eds. A. I. Bobenko and R. Seiler, Oxford: Clarendon Press, 1999, 3–58.

[11] A. I. Bobenko and Yu. B. Suris, Integrable systems on quad-graphs. *Internat. Math. Res. Notices* 2002, Nr. 11, 573–611.

[12] A. I. Bobenko and Yu. B. Suris, *Discrete Differential Geometry. Integrable Structure.* Graduate Studies in Mathematics , Vol. 98. AMS, 2008. xxiv+404 pp.

[13] J. Cieslinski, A. Doliwa and P. M. Santini, The integrable discrete analogues of orthogonal coordinate systems are multi-dimensional circular lattices. *Phys. Lett. A* 235 (1997), 480–488.

[14] G. Darboux, *Leçons sur la théorie générale des surfaces et les applications géométriques du calcul infinitésimal.* T.I–IV. 3rd edition. Paris: Gauthier-Villars, 1914–1927.

[15] G. Darboux, *Leçons sur les systèmes orthogonaux et les coordonnées curviligne*, Paris: Gauthier-Villars, 1910.

[16] A. Doliwa, Quadratic reductions of quadrilateral lattices. *J. Geom. Phys.* 30 (1999), 169–186.

[17] A. Doliwa, Discrete asymptotic nets and W-congruences in Plücker line geometry. *J. Geom. Phys.* 39 (2001), no. 1, 9–29.

[18] A. Doliwa and P. M. Santini, Multidimensional quadrilateral lattices are integrable. *Phys. Lett. A*, 233 (1997), 265–372.

[19] A. Doliwa, P. M. Santini and M. Mañas, Transformations of quadrilateral lattices. *J. Math. Phys.* 41 (2000), 944–990.

[20] L. P. Eisenhart, *Transformations of surfaces.* Princeton University Press, 1923. ix+379 pp.

[21] B. G. Konopelchenko and W. K. Schief, Three-dimensional integrable lattices in Euclidean spaces: conjugacy and orthogonality. *R. Soc. Lond. Proc. Ser. A*, 454 (1998), 3075–3104.

[22] Y. Liu, H. Pottmann, J. Wallner, Y. Yang, and W. Wang, Geometric modeling with conical meshes and developable surfaces. *ACM Trans. Graphics*, 25:3 (2006), 681–689.

[23] R. Sauer, *Projektive Liniengeometrie*. Berlin: W. de Gruyter & Co., 1937, 194 pp.

[24] R. Sauer, *Differenzengeometrie*. Berlin etc.: Springer, 1970.

The University of Erlangen-Nürnberg (photo C. Meusburger).

The University of Erlangen-Nürnberg (photo C. Meusburger).

Chapter 11

Three-dimensional gravity – an application of Felix Klein's ideas in physics

Catherine Meusburger

Contents

1 Introduction

Felix Klein's Erlangen programme, which was introduced in 1872 in his work *Vergleichende Betrachtungen über neuere geometrische Forschungen* [32] has proven influential in the development of geometry and also found numerous applications in mathematical physics. In particular, his idea to characterise different geometries in terms of group invariants led to a common description of the geometrical spaces known in his time in the common framework of homogeneous spaces, and made the notion of symmetries gain importance in physics.

Applications of this idea in physics were realised and investigated by himself later, in particular in the context of Einstein's theory of special and general relativity. In his article *Über die Differentialgesetze für Erhaltung von Impuls und Energie in der Einsteinschen Gravitationstheorie* [34], he related the notion of invariants to the conservation of momentum and energy in general relativity. In his work *Über die geometrischen Grundlagen der Lorentzgruppe* [33] he linked the physical concept of relativity in special relativity to the theory of invariants of the Lorentz group:

> Was die modernen Physiker Relativitätstheorie nennen, ist die Invariantentheorie des vierdimensionalen Raum-Zeit-Gebietes, x, y, z, t (der Minkowskischen Welt)

> gegenüber einer bestimmten Gruppe von Kollineationen, eben der "Lorentzgruppe"; oder allgemeiner, und nach der anderen Seite gewandt: "Man könnte, wenn man Wert darauf legen will, den Namen 'Invariantentheorie relativ zu einer Gruppe von Transformationen' sehr wohl durch das Wort ,Relativitätstheorie bezüglich einer Gruppe' ersetzen.

A systematic understanding of the relation between symmetries, conserved quantities and invariants in physics was established in 1918 by Emmy Noether in her article *Invariante Variationsprobleme* [43]. In this article, she derived the relation between symmetries of physical systems and conserved quantities in a general setting, namely for physical systems governed by an action functional. Her work [43] uses the mathematical framework of Lie's transformation groups as well as the concept of invariants and also cites Klein's articles on relativity [33, 34].

The notion of symmetry remains central in modern mathematical physics, although in a more general and larger sense than the physical symmetries considered by Noether and Klein. Examples of this are the gauge theories which describe the electromagnetic, weak and strong interaction. In this setting, the relevant symmetries are Lie groups which act as structure groups of bundles, namely the abelian group $U(1)$ in electromagnetism and the non-abelian groups $SU(2)$ and $SU(3)$ for, respectively, the weak and strong interaction. The systematic formulation of gauge theories for non-abelian gauge groups is due to Yang and Mills [48].

In contrast to the physical symmetries considered by Noether and Klein, the gauge symmetries in these theories do not relate different physical states but equivalent descriptions of the *same* physical state. The physical or gauge invariant phase space of these theories, e.g. the space of physical states, is not the set of solutions of the equations of motion but the quotient of this space by the group of gauge symmetries.

A more general notion of such non-physical or "gauge" symmetries relating different descriptions of a physical state was introduced by Dirac in his work on constrained mechanical systems [24, 25], which showed that the notion of a gauge symmetry is tied to the presence of constraints on the variables in the action functional. The framework of constrained systems encompasses finite-dimensional mechanical systems as well as gauge and field theories and Einstein's theory general relativity. (For an overview see [31] and [30]. An accessible introduction to symmetries in gauge theories and in general relativity from the perspective of differential geometry is given in [28].)

Many theories in modern mathematical physics are examples of such constrained systems. In particular this includes topological field theories such as Chern–Simons gauge theory, certain conformal field theories and three-dimensional gravity. In all of these cases, the underlying theory is a gauge field theory, whose group of gauge symmetries is so large that the physically distinct solutions of the field equations can be classified completely. Despite the fact that they are field theories, these theories exhibit finite-dimensional gauge invariant phase spaces. This has important implications in their quantisation and gives rise to interesting mathematical structures. Examples are the appearance of knot polynomials and manifold invariants in the quantisation of Chern–Simons theories [46, 44], the representation theoretical structures in conformal field theories and recent results triggered by the quantisation of moduli spaces such as Lie group-valued moment maps [5]. A common feature of these developments is that they establish interesting relations between purely algebraic objects such as quantum groups or representations of fundamental groups of surfaces and the geometrical properties of the associated field theories.

In this chapter, we discuss applications of Klein's ideas in the context of three-dimensional (3d) gravity. Three-dimensional gravity is Einstein's theory of general relativity with one time and two space dimensions and is closely related to Chern–Simons gauge theory [1, 47]. As its phase space is finite-dimensional, this theory serves as a toy model for quantum gravity in higher dimensions and allows one to investigate conceptual questions of quantum gravity in a theory that can be quantised rigorously. Besides this physical applications, the theory is also interesting in its own right due to its rich mathematical structure and its relation with Chern–Simons gauge theory [1, 47], moduli spaces of flat connections and Teichmüller theory [35].

Three-dimensional gravity is related to Klein's ideas insofar as any vacuum spacetime is obtained as a quotient of (certain subsets of) homogeneous spaces by a discrete subgroup of their isometry groups [35]. This description in terms of homogeneous spaces establishes a link with Klein's Erlangen programme [32]. Moreover, in the case of vanishing cosmological constant, the relevant homogeneous space is three-dimensional Minkowski space. This implies that in that case, diffeomorphism equivalence classes of vacuum solutions of Einstein equations can be described entirely within the framework of *special relativity*, which is the context of Klein's work [33]. A further link with Klein's ideas is the fact that many physics problems in this theory involve the relation of group-theoretical or algebraic data, which classifies the spacetimes and serves as the fundamental building block in quantisation, to the geometry of the spacetimes, which is linked to physical measurements by observers.

The chapter is structured as follows. Section 2 introduces the relevant background on the geometry of spacetimes in 3d gravity and their classification. It also discusses the relation of the gauge invariant phase space of 3d gravity to certain moduli spaces of flat connections and summarises known results on its symplectic structure.

In section 3, we discuss examples of physics questions arising in three- and higher-dimensional quantum gravity, namely the question of reconstructing spacetime geometry and the outcome of measurements made by observers from algebraic data and the question about the role of quantum group symmetries in a quantum theory of gravity. These questions are addressed in the framework of 3d gravity in Sections 4 and 5, respectively. Section 4 explains how measurements made by observers such as frequency shifts or return times of lightrays are related to the classifying data of the theory and allows one to reconstruct it. Section 5 discusses how a gauge fixing procedure for the phase space of 3d gravity gives rise to dynamical quantum group symmetries.

2 Gravity in three dimensions

2.1 Geometry of spacetimes As in higher dimensions, the dynamical variable of 3d gravity is a Lorentz metric on a three-dimensional manifold M, and the equations of motion are the Einstein equations. These equations take the same form as in higher dimensions

$$\mathrm{Ric}_{\mu\nu} - \tfrac{1}{2} g_{\mu\nu} R + \Lambda g_{\mu\nu} = 8\pi G T_{\mu\nu}, \tag{11.1}$$

where $g_{\mu\nu}$ denotes a metric of signature $(-1,1,1)$ on M, $\mathrm{Ric}_{\mu\nu}$ its Ricci curvature, R the scalar curvature, Λ and G, respectively, the cosmological and the gravitational constant and $T_{\mu\nu}$ the stress-energy tensor.

The distinguishing feature in three dimensions is that the Ricci curvature of a three-dimensional (semi-)Riemannian manifold determines its sectional and Riemann curvature. This implies that the theory has no local gravitational degrees of freedom and only a finite number of global degrees of freedom due to the matter content and topology of the spacetime. In particular, any vacuum solution of the Einstein equations, i.e. a solution for vanishing of the stress-energy tensor $T_{\mu\nu}$, is locally isometric to one of three model spacetimes. Depending on the cosmological constant Λ, these model spacetimes are three-dimensional de Sitter space dS_3 ($\Lambda > 0$), Minkowski space $\mathbb{M}_3$ ($\Lambda = 0$) and three-dimensional anti-de Sitter space AdS_3 ($\Lambda < 0$).

Note that all of these model spacetimes are homogeneous spaces (see Table 11.1), which implies that in the spirit of Klein, their geometric properties can be formulated entirely in terms of Lie groups. In all cases the corresponding quotients are taken with respect to the subgroup $\mathrm{SO}_0(2,1) \cong \mathrm{PSL}(2,\mathbb{R})$, which is the proper orthochronous Lorentz group in three dimensions. Consequently, all isometry groups of the model spacetimes contain the group $\mathrm{SO}_0(2,1)$ as a subgroup, where the inclusion $\mathrm{SO}_0(2,1) \to G$ is the obvious one for $\Lambda \geq 0$ and the diagonal one for $\Lambda < 0$.

Under certain additional assumptions on their causality properties, vacuum spacetimes can be classified completely. The following classification theorem is due to Mess [35].

Theorem 2.1. [35, 7] *For $\Lambda \leq 0$, any maximally globally hyperbolic vacuum spacetime with a compact Cauchy surface S of genus $g \geq 2$ is determined uniquely by a group homomorphism $h : \pi_1(S) \to G$ whose Lorentzian component defines a regular representation of $\pi_1(S)$. Conversely, two vacuum solutions are isometric if and only if the associated group homomorphisms are related by conjugation. The set of isometry classes of vacuum solutions of the Einstein equations is therefore given by*

$$\mathcal{M}_0^{G,S} = Hom_0(\pi_1(S), G)/G$$

where the index 0 stands for the restriction to regular representations and G denotes the isometry group from Table 11.1.

Note that a similar classification result holds for $\Lambda > 0$, but in that case, a discrete parameter is needed in addition to the group homomorphism to classify the vacuum

	model spacetime X	isometry group G	homogeneous space
$\Lambda > 0$	dS_3	$SO_0(3,1)$	$SO_0(3,1)/SO_0(2,1)$
$\Lambda = 0$	$\mathbb{M}_3$	$ISO(2,1)$	$ISO(2,1)/SO_0(2,1)$
$\Lambda < 0$	AdS_3	$SO_0(2,1)\times SO_0(2,1)$	$SO_0(2,2)/SO_0(2,1)$

Table 11.1. Model spacetimes for 3d gravity

solutions completely. There also exist weaker classification results for spacetimes with point particles, which correspond to Lorentzian manifolds $\mathbb{R}\times S$, where S is a surface with punctures or marked points [10, 11, 15]. However, a complete classification is still missing in that situation.

Due to the classification result in Theorem 2.1, the moduli space $\mathcal{M}_0^{G,S}$ of vacuum solutions can be viewed as the physical or gauge invariant phase space of gravity in three dimensions for a fixed topology of the spacetime and a fixed cosmological constant. Each phase space point represents an isometry class of Lorentzian constant curvature spacetimes and hence a universe with two space and one time dimension. A feature that is important for the quantisation of the theory is that the moduli space $\mathcal{M}_0^{G,S}$ of vacuum solutions is a connected component of the moduli space $\mathcal{M}^{G,S} = \mathrm{Hom}(\pi_1(S),G)/G$ of flat G-connections on S, for which the restriction to *regular* representations is not imposed. This places the quantisation of 3d gravity in the context of the quantisation of moduli spaces of flat connections.

The moduli space of flat G connections on S arises as the gauge invariant phase space of Chern–Simons theory with gauge group G on the three-manifold $\mathbb{R}\times S$. This description can be easily extended to the case where S is a surface with marked points and, in this case, requires the assignment of a conjugacy class $\mathcal{C}_i\subset G$ to each marked point p_i. In the application to 3d gravity, these conjugacy classes correspond to two-dimensional Cartan subalgebras of $\mathfrak{g}$ and are chosen in such a way that they involve only elliptic elements of the Lorentz group $SO_0(2,1)\subset G$. The associated moduli space then describes spacetimes with point particles for which curvature and torsion become singular along one-dimensional submanifolds corresponding to their worldlines. The associated moduli space of flat G-connections is given by

$$\mathcal{M}^{G,S} = \mathrm{Hom}_{\mathcal{C}_1,\dots,\mathcal{C}_n}(\pi_1(S),G)/G \tag{11.2}$$

where the group homomorphisms $h:\pi_1(S_{g,n})\to G$ are required to map the loops around the ith puncture to the associated conjugacy class $\mathcal{C}_i$ [8].

2.2 Symplectic structure Moduli spaces of flat connections have rich mathematical properties. In particular, they carry a canonical symplectic structure [8, 29], which for the relevant groups G from Table 11.1 induces a symplectic structure on $\mathcal{M}_0^{G,S}\subset\mathcal{M}^{G,S}$. The symplectic structure depends on the choice of an Ad-invariant symmetric non-degenerate bilinear form $\langle\ ,\ \rangle$ on $\mathfrak{g}=\mathrm{Lie}(G)$, which enters the definition of the Chern–Simons action.

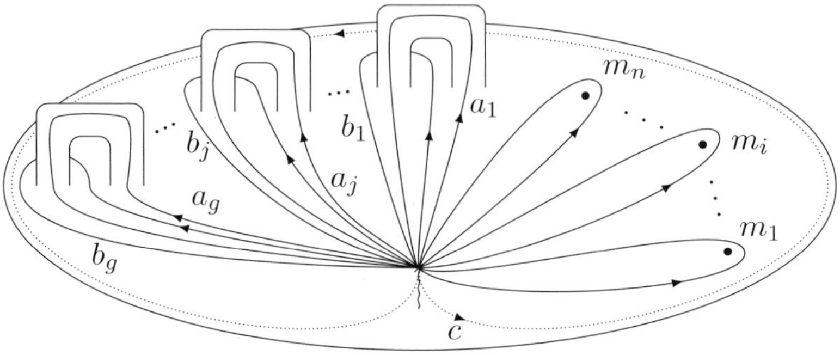

Figure 11.1. Generators of the fundamental group $\pi_1(S)$ for a surface S of genus g with n marked points.

Functions on the moduli space $\mathcal{M}^{G,S}$ can then be identified with conjugation invariant functions on $\mathrm{Hom}_{\mathcal{C}_1,\ldots,\mathcal{C}_n}(\pi_1(S), G)$. In particular, every class function f on G and every element $\lambda \in \pi_1(S)$ gives rise to a function on $\mathcal{M}^{G,S}$ that assigns to a group homomorphism $h : \pi_1(S) \to G$ the value $f \circ h(\lambda)$. These functions are known as Wilson loop observables and play an important role in the description of the symplectic structure on the moduli space. In particular, it was shown by Goldman [29] that the Poisson bracket of two Wilson loop observables yields a function of certain Wilson loop observables, which depends only on the chosen class functions f on G and the intersection behaviour of the curves representing $\lambda \in \pi_1(S)$.

A convenient parametrisation of this symplectic structure, which is adapted to the description in terms of group homomorphisms and will be used in the following, is due to Fock and Rosly [27] and Alekseev and Malkin [4]. It describes the symplectic structure on the moduli space in terms of a (non-canonical) Poisson structure on a larger ambient space. The key idea is to parametrise group homomorphisms $h : \pi_1(S) \to G$ in terms of the images of a set of generators of $\pi_1(S)$.

The fundamental group $\pi_1(S)$ of a surface S of genus g and with n marked points can be presented in terms of the a- and b-cycles a_j, b_j of each handle and loops m_i around each puncture as

$$\pi_1(S) = \langle m_1, \ldots, m_n, a_1, b_1, \ldots, a_g, b_g : [b_g, a_g^{-1}] \cdots [b_1, a_1^{-1}] m_n \cdots m_1 = 1 \rangle. \tag{11.3}$$

A set of representing curves is shown in Figure 11.1. If one characterises the group homomorphisms $h : \pi_1(S_{g,n}) \to G$ by the images of the generators, one can therefore identify the moduli space $\mathcal{M}^{G,S}$ with the set

$$\begin{aligned}\mathcal{M}^{G,S} = \{&(M_1, \ldots, B_g) \in G^{n+2g} :\\ &M_i \in \mathcal{C}_i,\ [B_g, A_g^{-1}] \cdots [B_1, A_1^{-1}] \cdot M_n \cdots M_1 = 1\}/G.\end{aligned} \tag{11.4}$$

This allows one to describe the canonical symplectic structure on $\mathcal{M}^{G,S}$ in terms of a Poisson structure on G^{n+2g}, where the different copies of G correspond to the group elements assigned to the generators of $\pi_1(S)$ by a group homomorphism $h : \pi_1(S) \to G$. The central ingredient in this description is a *classical r-matrix* for the group G. This is an element $r \in \mathfrak{g} \otimes \mathfrak{g}$ that satisfies the classical Yang–Baxter equation (CYBE)

$$[[r, r]] = [r_{12}, r_{13}] + [r_{12}, r_{23}] + [r_{13}, r_{23}] = 0.$$

If one denotes by L^i_α and R^i_α the right- and left-invariant vector fields on the different copies of G associated with a basis $\{T_\alpha\}$ of $\mathfrak{g}$

$$\begin{aligned} L^i_\alpha f(u_1, \dots, u_{n+2g}) &= \frac{d}{dt}\Big|_{t=0} f(u_1, \dots, \exp(-tT_\alpha) \cdot u_i, \dots, u_{n+2g}), \\ R^i_\alpha f(u_1, \dots, u_{n+2g}) &= \frac{d}{dt}\Big|_{t=0} f(u_1, \dots, u_i \cdot \exp(tT_\alpha), \dots, u_{n+2g}), \end{aligned}$$

then one obtains a Poisson structure on G^{n+2g} for which the classical r-matrix plays the role of a structure constant.

Theorem 2.2. [27, 4] *Let G be a Lie group, $\{T_\alpha\}_{\alpha=1,\dots,m}$ a basis of $\mathfrak{g}$ and $r = \sum_{\alpha,\beta} r^{\alpha\beta} T_\alpha \otimes T_\beta \in \mathfrak{g} \otimes \mathfrak{g}$ a solution of the CYBE with symmetric component $r_{(s)}$ and antisymmetric component $r_{(a)}$. Then the bivector*

$$\begin{aligned} B^{n,g}_r &= \tfrac{1}{2} r^{\alpha\beta}_{(a)} \left(\sum_{i=1}^{n+2g} (R^i_\alpha + L^i_\alpha) \right) \otimes \left(\sum_{j=1}^{n+2g} \left(R^j_\beta + L^j_\beta \right) \right) \\ &+ \tfrac{1}{2} r^{\alpha\beta}_{(s)} \sum_{j=1}^{g} \Big(L^{n+2j}_\alpha \wedge L^{n+2j}_\beta + L^{n+2j}_\alpha \wedge L^{n+2j-1}_\beta + R^{n+2j}_\alpha \wedge L^{n+2j-1}_\beta \\ &\qquad + R^{n+2j}_\alpha \wedge R^{n+2j-1}_\beta \Big) \\ &+ \tfrac{1}{2} r^{\alpha\beta}_{(s)} \sum_{i=1}^{n} R^i_\alpha \wedge R^i_\beta + \tfrac{1}{2} r^{\alpha\beta}_{(s)} \sum_{1 \le i < j \le n+2g} (L^i_\alpha + R^i_\alpha) \wedge (L^j_\beta + R^j_\beta) \end{aligned} \tag{11.5}$$

defines a Poisson-structure $\{\ ,\ \}_r$ on G^{n+2g}. If $r_{(s)}$ is dual to the chosen Ad-invariant symmetric bilinear form $\langle\ ,\ \rangle$, this Poisson structure induces the canonical Poisson structure on $\mathcal{M}^{G,S}$.

The symplectic structure on the moduli space $\mathcal{M}^{G,S}$ is obtained from the one on G^{n+2g} by Poisson reduction [27, 4]. This induces in particular a symplectic structure on $\mathcal{M}^{G,S}_0$ which for the appropriate groups G from Table 11.1 can be viewed as the symplectic structure on the phase space of three-dimensional gravity.

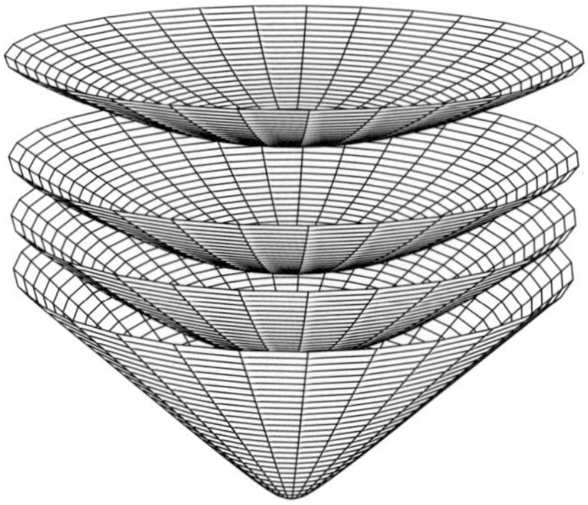

Figure 11.2. Foliation of a lightcone by hyperboloids.

2.3 Spacetime geometry and universal covers The relation between the group homomorphisms in Theorems 2.1 and 2.2 and the geometry of the spacetimes is given by the universal cover of the spacetimes. It is shown in [9, 12, 13] that under the assumptions of Theorem 2.1 the universal cover of the spacetimes is a regular domain $D \subset X$ in the model spacetime from Table 11.1 which has an initial singularity and which is bounded by lightlike hyperbolic planes. For $\Lambda < 0$, the domain also has a final singularity, in the other two cases, it is future-complete.

The domains are equipped with cosmological time functions $T : D \mapsto (0, \infty)$ [14], which give the geodesic distance of points in the spacetime from the initial singularity and induce a foliation of D by surfaces D_T of constant cosmological time T. The fundamental group $\pi_1(S)$ acts on the domains via group homomorphisms $h : \pi_1(S) \to G$ in such a way that each surface of constant cosmological time is preserved. The spacetime is obtained by taking the quotient of D with respect to this group action. It therefore inherits a metric of signature $(-1, 1, 1)$ as well as a cosmological time function and a foliation by surfaces of constant cosmological time.

The simplest cases are the so-called conformally static spacetimes. In this case, the universal cover is a lightcone in the model spacetime, i.e. the set of points in X that can be reached from a given point by future directed timelike geodesics. The cosmological time function gives the distance of a point $p \in D$ from the tip of the lightcone, and all surfaces of constant cosmological time D_T are (up to a conformal scaling factor depending on T and the cosmological constant) copies of two-dimensional hyperbolic space. The case $\Lambda = 0$ is depicted in Figure 11.2. In this case, one obtains a lightcone in three-dimensional Minkowski space and the standard foliation of a lightcone by hyperboloids. This implies that – up to a global conjugation – each group homomorphism $\pi_1(S) \to G$ which preserves the surfaces D_T must take values in the subgroup $SO_0(2, 1) \cong PSL(2, \mathbb{R}) \subset G$, and the resulting action of $\pi_1(S)$ on D_T coincides with the action of a Fuchsian group in $PSL(2, \mathbb{R})$ on the upper half-plane.

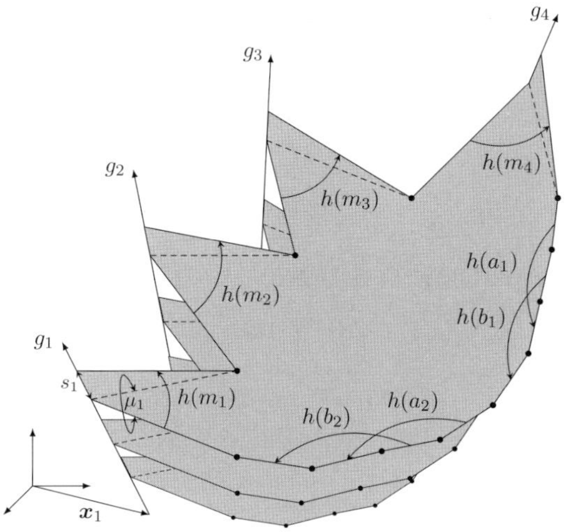

Figure 11.3. The construction of a spacetime with particles by gluing the boundary of a domain $D \subset \mathbb{M}^3$.

Due to the restriction to regular representations, the quotient of each surface D_T by this action of $\pi_1(S)$ defines a hyperbolic structure g_S on S, and the metric of the associated spacetime $M = D/\pi_1(S)_h$ takes the form

$$g = -dT^2 + s_\Lambda(T)^2 g_S \qquad s_\Lambda(T) = \begin{cases} \sin T & \Lambda < 0 \\ T & \Lambda = 0 \\ \sinh(T) & \Lambda > 0 \end{cases},$$

where T denotes the cosmological time. As the geometry of the cosmological time surfaces in M changes with the cosmological time only through a rescaling, such spacetimes are called conformally static. For spacetimes that are not conformally static, the domains $D \subset X$ and the cosmological time surfaces take a more complicated form, and the group homomorphisms $h : \pi_1(S) \to G$ which determine the action of $\pi_1(S)$ on D no longer take values in the subgroup $\mathrm{SO}_0(2,1) \subset G$.

The case of spacetimes with point particles is more involved. On the one hand, this is due to the lack of classification results and to the absence of the global hyperbolicity and of a cosmological time. On the other hand, this is related to the fact that the group action of $\pi_1(S)$ on the universal cover $D \subset X$ involves elliptic elements of $\mathrm{SO}_0(2,1)$. However, spacetimes with point particles can still be obtained by gluing domains in the model spacetimes. The relevant gluing pattern is illustrated in Figure 11.3. For a surface S of genus g and with n marked points corresponding to point particle singularities, the relevant domain can be chosen in such a way that it is foliated by $4g + 2n$-gons as shown in Figure 11.3. The group homomorphism $h : \pi_1(S) \to G$ then describes the identification of the boundaries of the domain as shown in Figure 11.3 and stabilises the geodesics associated with the particles. As the lines must correspond to worldlines of particles, they are required to be timelike geodesics in the model spacetime, and the associated group elements $M_i \in G$ that stabilise these timelike geodesics have elliptic Lorentzian components.

3 Three-dimensional gravity as a model for quantum gravity

As illustrated in the previous sections, 3d gravity is a theory that follows the spirit of Klein insofar as all of its physical and geometrical content is determined by certain Lie groups and discrete subgroups thereof. On the one hand, these are the Lie groups which define the model spacetimes, and the associated homogeneous spaces. One the other hand, Klein's philosophy of characterising geometry in terms of groups is reflected in the group homomorphisms $h : \pi_1(S) \to G$ that classify the spacetimes. By Theorem 2.1, these group homomorphisms determine the geometry of vacuum spacetimes completely and hence contain all geometrical and physical data of the theory.

The fact that such a purely algebraic description of the solutions of Einstein's equations is possible in three dimensions is tied closely to the role of 3d gravity as a toy model for quantum gravity in higher dimensions. As its phase space is given by the finite-dimensional manifold $\mathcal{M}_0^{G,S} = \mathrm{Hom}_0(\pi_1(S), G)/G$ and there is an explicit description of its symplectic structure, the quantisation of the theory is much simpler than in four dimensions. 3d gravity thus offers the prospect of investigating conceptual questions of quantum gravity in a fully quantised theory.

On the one hand, these are questions about the role of time and space in a quantum theory of gravity and an appropriate notion of observers, which are a fundamental concept in general relativity and must be included appropriately in the quantum theory. Another set of questions concerns the relation between measurements and observables in 3d gravity. While the physical observables of the theory are by definition diffeomorphism invariant quantities, most realistic measurements made by an observer would depend on a notion of time such as the eigentime of the observer or a cosmological time. This indicates that the relation between observables and measurements is less direct than in other quantum theories, where the former describe the latter. Rather, it is argued in [45] that observables in quantum gravity relate different quantities measured by observers. Further issues of this type are questions about the different notions of time in the theory, questions about non-commutative aspects of spacetime geometry and questions about the spectra of of geometrical operators; for an overview, see [20].

These issues are also directly apparent in three-dimensional gravity. As shown in the previous section, the diffeomorphism invariant observables of 3d gravity can be identified with conjugation invariant functions on the space $\mathrm{Hom}_0(\pi_1(S), G)$ of regular representations of the fundamental group. By Theorem 2.1, these regular representations determine the geometry of vacuum spacetimes completely and hence encode the entire geometrical and physical content of the theory. This implies in particular that the outcome of all physical measurements made by observers in such a spacetime should be given in terms of these group homomorphisms.

Another important question related to the quantisation of (3d) gravity is the question about the symmetries of the quantum theory. While it is conjectured from phenomenological considerations that quantum groups would play an important role as symmetries of quantum gravity, it is currently not possible to confirm or refute this argument for the four-dimensional theory. In contrast, the relation between quantum group symmetries and the quantisation of gravity is more direct in three dimensions.

Many established quantisation formalisms for moduli spaces of flat G-connections such as combinatorial quantisation [2, 3, 6, 18, 17, 42, 39, 38] or the Reshetikhin–Turaev invariants [44] which arise in the quantisation of Chern–Simons theory are constructed via the representation theory of certain quantum groups associated with G. Moreover, the appearance of such quantum group symmetries in the quantisation of the theory is natural, since the description of the symplectic structure on the moduli space of flat connections in Theorem 2.2 involves a classical r-matrix.

Classical r-matrices play an important role in the description of Poisson–Lie groups and can be viewed as the classical counterpart of the quantum R-matrix in a quantum group. The appearance of classical r-matrices in the description of the symplectic structure on moduli spaces of flat connections thus suggests that quantum groups should arise in its quantisation. However, since these Poisson–Lie symmetries act on a larger ambient space from which the moduli space of flat connections is obtained via Poisson reduction, it remains unclear to what degree they survive the reduction process. Consequently, it is not readily apparent whether the associated quantum group symmetries have a physical interpretation or whether they merely serve as technical tools for the construction of the quantum theory. Although there is much discussion of this question on a phenomenological level, this question cannot be given a final answer unless one explicitly performs this Poisson reduction to the phase space of 3d gravity and investigates the mathematical structures arising from it.

These two sets of questions are examples of fundamental questions of quantum gravity that can be addressed in 3d gravity and that are intimately tied to the algebraic nature of the theory, namely its characterisation in terms of group homomorphisms. While they are of high relevance for the quantisation of 3d gravity, they are already present in the classical theory and useful conclusions can be drawn by investigating them in the classical framework. In the following sections, we illustrate this with two examples. The first is the relation between the group theoretical data that parametrises the phase space of 3d gravity and the outcome of concrete measurements made by observers in Section 4. The second is the physical interpretation of quantum group symmetries and their classical counterparts, Poisson–Lie symmetries, and the associated mathematical structures in Section 5.

4 Measurements from algebraic data

In this section, we show how the group-theoretical data that characterises the spacetimes by Theorem 2.1 is related to concrete and realistic measurements made by observers and how it encodes the general relativistic content of the theory. For this, we focus on maximally globally hyperbolic vacuum spacetimes of genus $g \geq 2$ for vanishing cosmological constant, although similar considerations are possible for all values of Λ.

We consider an observer in free fall in a spacetime who attempts to determine its geometry by sending and receiving lightray. In the absence of matter, the only quantities that could be measured by such an observer is the geometry of the universe itself. The key idea to obtain physically meaningful measurements is to note that the geometry of the surfaces of constant cosmological time manifests itself in the presence of returning lightrays – light signals that are sent by the observer and return to him at a later time. The observer can then measure several quantities associated with such returning lightrays. For instance, he can determine their return time, i.e. the time elapsed between the emission and return of a light signal, the direction from which the light returns or in which it needs to be sent to return and the frequency shift of the returning lightray.

To derive explicit expressions for the measurements associated with returning lightrays it is advantageous to work in the universal cover. We consider a flat maximally globally hyperbolic 3d Lorentzian manifold M with a compact Cauchy surface S of genus $g \geq 2$ together with the regular domain $D \subset \mathbb{M}^3$ and group homomorphism $h : \pi_1(S) \to \mathrm{ISO}(2,1)$ as in Theorem 2.1. We start with a precise formulation of the relevant physics concepts.

Definition 4.1 ([36]).

1. An *observer* in M is determined by a timelike, future-directed geodesic $g : [a,\infty) \to M$, his *worldline* or, equivalently, by a $\pi_1(S)$-equivalence class of timelike, future-directed geodesics $\tilde{g} : [a,\infty) \to D$ in the universal cover. The worldline $g : [a,\infty) \to M$ is called *parametrised according to eigentime* if $\dot{g}(t)^2 = -1\ \forall t \in [a,\infty)$.
2. A *lightray* in M is a lightlike future-directed geodesic $c : [p,\infty) \to M$ or, equivalently, a $\pi_1(S)$-equivalence class of lightlike future-directed geodesics in D.
3. A lightray emitted (received) by an observer with worldline $g : [a,\infty) \to M$ at eigentime t is a lightlike future-directed geodesic $c : [p,q] \to M$ with $c(p) = g(t)$ $(c(q) = g(t))$ or, equivalently, the $\pi_1(S)$-equivalence class of lightlike future-directed geodesics $\tilde{c} : [p,q] \to D$ for which there exists a lift $\tilde{g} : [a,\infty) \to \tilde{M}$ of g such that $\tilde{c}(p) = \tilde{g}(t)$ $(\tilde{c}(q) = \tilde{g}(t))$.

Note that any future directed timelike geodesic in the universal cover $D \subset \mathbb{M}^3$ can be parametrised in terms of an element $\boldsymbol{x} \in \mathbb{H}^2$, the *velocity vector* of the observer,

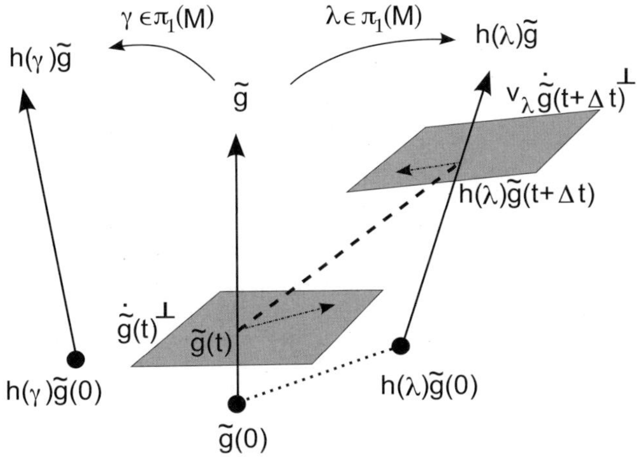

Figure 11.4. Lifts of the observer's worldline to D with returning lightray (dashed line), orthogonal complement $\boldsymbol{x}^{\perp} = \dot{\tilde{g}}(t)^{\perp}$ and its image (grey planes) and projection of the returning lightray to $\boldsymbol{x}^{\perp} = \dot{\tilde{g}}(t)^{\perp}$ and its image (dashed arrows).

and a vector $\boldsymbol{x}_0 \in D$, the observer's *initial position* as

$$\tilde{g}(t) = t\boldsymbol{x} + \boldsymbol{x}_0. \tag{11.6}$$

The parameter t gives the time as perceived by the observer, i.e. the time that would be shown by a clock carried by the observer. Similarly, each lightlike future-directed geodesic $\tilde{c} : [0, \infty) \to D$ is given by a vector $\boldsymbol{y}$ in the future lightcone and an initial position vector $\boldsymbol{y}_0 \in D$

$$\tilde{c}(s) = s\boldsymbol{y} + \boldsymbol{y}_0. \tag{11.7}$$

It is directly apparent that Definition 4.1 also captures the phenomenon of *returning lightrays*, lightrays that are emitted by an observer and return to him at a later time. Such a returning lightray corresponds to a lightlike future-directed geodesic $c : [p, q] \to M$ that intersects g in $c(p)$ and $c(q)$. Equivalently, a returning lightray can be described as a $\pi_1(S)$-equivalence class of lightlike future-directed geodesics $\tilde{c} : [p, q] \to D$ for which there is an element $\lambda \in \pi_1(S)$ and a lift $\tilde{g} : [a, \infty) \to D$ of g with $\tilde{c}(p) \in \tilde{g}$, $\tilde{c}(q) \in h(\lambda)\tilde{g}$.

This description of returning lightrays in the universal cover D is shown in Figure 11.4. As a returning lightray relates a lift of the observer's worldline to one of its images under the action of $\pi_1(S)$, it defines a unique element $\lambda \in \pi_1(S)$. However, it is a priori not guaranteed that for each observer and each element of $\pi_1(S)$ there exists a returning lightray. This is a consequence of the geometrical properties of Minkowski space and the future-completeness of the domains D and follows by a direct computation.

Lemma 4.1 ([36]). *Let $g : [a, \infty) \to M$ be the worldline of an observer. Then for all $t \in [a, \infty)$ the returning lightrays $c : [p, q] \to M$ with $c(p) = g(t)$ are in one-to-one correspondence with elements of the fundamental group $\pi_1(S)$.*

The formulation of the relevant physics concepts in terms of the universal cover $D \subset \mathbb{M}^3$ allows one to explicitly compute the measurements associated with returning lightrays in terms of the group homomorphisms $h : \pi_1(S) \to \mathrm{ISO}(2,1)$. We start with the return time, the interval of eigentime elapsed between the emission of a returning lightrays and its return as measured by the observer.

Definition 4.2 ([36]). Let $\tilde{g} : [a, \infty) \to D$ a lift of an observer's worldline parametrised as in (11.6). Then for each $t_e \in [a, \infty)$, $\lambda \in \pi_1(S)$, there exists a unique $t_r \in (t_e, \infty)$ and a unique lightlike geodesic $\tilde{c}_\lambda : [0, 1] \to D$ with $\tilde{c}_\lambda(0) = \tilde{g}(t_e)$ and $\tilde{c}_\lambda(1) = h(\lambda)\tilde{g}(t_r)$. The return time $\Delta t = t_r - t_e$ is the unique positive solution of the quadratic equation

$$(h(\lambda)\tilde{g}(t_e + \Delta t) - \tilde{g}(t_e))^2 = 0. \tag{11.8}$$

To obtain the directions in which the light needs to be emitted in order to return to the observer, we recall that the directions an observer perceives as "spatial" are given by the orthogonal complement $\boldsymbol{x}^\perp = \dot{\tilde{g}}(t)^\perp$, where $\tilde{g} : [a, \infty) \to D$ is the lift observer's worldline parametrised as in (11.6). To determine the relative frequencies of a returning lightray at its emission and return, one works in the universal cover and performs a computation similar to the relativistic Doppler effect. This yields the following definition.

Definition 4.3 ([36]). Let $\tilde{g} : [a, \infty) \to D$ be a lift of an observer's worldline as in (11.6) and $\tilde{c} : [p, q] \to D$ a lightlike geodesic associated with a returning lightray that satisfies $\tilde{c}(p) = \tilde{g}(t_e)$, $\tilde{c}(q) = h(\lambda)\tilde{g}(t_r)$ for an element $\lambda \in \pi_1(S)$.

1. The direction into which the lightray associated with $\tilde{c}$ is emitted as perceived by the observer is given by the spacelike unit vector

$$\hat{\boldsymbol{p}}_e = \Pi_{\boldsymbol{x}^\perp}(\dot{\tilde{c}}(p))/|\Pi_{\boldsymbol{x}^\perp}(\dot{\tilde{c}}(p))| \tag{11.9}$$

where $\Pi_{\boldsymbol{x}\perp} : \mathbb{M}^3 \to \boldsymbol{x}^\perp$ denotes the projection on the orthogonal complement $\boldsymbol{x}^\perp$.

2. The quotient of frequencies of the lightray at its emission and return as measured by the observer is given by

$$\frac{f_r}{f_e} = \frac{h(\lambda)\boldsymbol{x} \cdot (h(\lambda)\tilde{g}(t_r) - \tilde{g}(t_e))}{\boldsymbol{x} \cdot (h(\lambda)\tilde{g}(t_r) - \tilde{g}(t_e))}. \tag{11.10}$$

To obtain explicit results for the return time, the directions of emission and the frequency shift, it is advantageous to introduce additional parameters, which are given as functions of the velocity vector $\boldsymbol{x}$, the initial position $\boldsymbol{x}_0$ and the group elements

$h(\lambda) = (v_\lambda, \boldsymbol{a}_\lambda) \in \mathrm{ISO}(2,1)$ with $v_\lambda \in \mathrm{SO}(2,1)_0$, $\boldsymbol{a}_\lambda \in \mathbb{R}^3$. For any element $\lambda \in \pi_1(S)$ and any geodesic $\tilde{g}$ parametrised as in (11.6), we define

$$\begin{aligned} \cosh \rho_\lambda &= -\boldsymbol{x} \cdot v_\lambda \boldsymbol{x} \\ h(\lambda)\tilde{g}(0) - \tilde{g}(0) &= \sigma_\lambda (v_\lambda \boldsymbol{x} - \boldsymbol{x}) + \tau_\lambda v_\lambda \boldsymbol{x} + \nu_\lambda \boldsymbol{x} \wedge v_\lambda \boldsymbol{x}. \end{aligned} \tag{11.11}$$

The parameter ρ_λ, which depends only on the velocity vector $\boldsymbol{x}$ and the Lorentzian component of $h(\lambda)$ has a direct interpretation as the geodesic distance of $\boldsymbol{x}$ and its image in hyperbolic space $\mathbb{H}^2$. The parameters $\sigma_\lambda, \tau_\lambda, \nu_\lambda$ characterise the relative initial position of the geodesic $\tilde{g}$ and its image $h(\lambda)\tilde{g}$. They depend on the velocity vector $\boldsymbol{x}$, the initial position $\boldsymbol{x}_0$ as well as the group homomorphism $h : \pi_1(S) \to \mathrm{ISO}(2,1)$. Using Definitions 4.2, 4.3, one can derive explicit expressions for the measurements associated with returning lightrays in terms of the parameters $\rho_\lambda, \sigma_\lambda, \tau_\lambda, \nu_\lambda$, which are summarised in the following theorem.

Theorem 4.4 ([36]). *Let $\tilde{g} : [a, \infty) \to \tilde{M}$ be a lift of the worldline parametrised as in* (11.6)*. Consider a returning lightray associated with an element $\lambda \in \pi_1(S)$ that is emitted by the observer at eigentime t and returns at $t + \Delta t$. Then the eigentime Δt elapsed between the emission and return of the lightray is given by*

$$\Delta t(t, \boldsymbol{x}, \boldsymbol{x}_0, h(\lambda)) = (t + \sigma_\lambda)(\cosh \rho_\lambda - 1) - \tau_\lambda + \sinh \rho_\lambda \sqrt{(t + \sigma_\lambda)^2 + \nu_\lambda^2}, \tag{11.12}$$

where $\rho_\lambda, \sigma_\lambda, \tau_\lambda, \nu_\lambda$ are functions of $\boldsymbol{x}, \boldsymbol{x}_0$ and $h(\lambda)$ defined by (11.11)*. The direction into which the lightray is emitted is given by the spacelike unit vector*

$$\begin{aligned} \hat{\boldsymbol{p}}^e_\lambda &= \cos \phi_e \frac{v_\lambda \boldsymbol{x} + (\boldsymbol{x} \cdot v_\lambda \boldsymbol{x})\boldsymbol{x}}{|v_\lambda \boldsymbol{x} + (\boldsymbol{x} \cdot v_\lambda \boldsymbol{x})\boldsymbol{x}|} + \sin \phi_e \frac{\boldsymbol{x} \wedge v_\lambda \boldsymbol{x}}{|\boldsymbol{x} \wedge v_\lambda \boldsymbol{x}|} \\ \textit{with} \quad \tan \phi_e(t, \boldsymbol{x}, \boldsymbol{x}_0, h(\lambda)) &= \frac{\nu_\lambda}{\sinh \rho_\lambda \sqrt{(t + \sigma_\lambda)^2 + \nu_\lambda^2}}. \end{aligned} \tag{11.13}$$

The relative frequencies of the lightray at its emission and return are given by

$$f_r / f_e(t, \boldsymbol{x}, \boldsymbol{x}_0, h(\lambda)) = \frac{\sqrt{(t + \sigma_\lambda)^2 + \nu_\lambda^2}}{\cosh \rho_\lambda \sqrt{(t + \sigma_\lambda)^2 + \nu_\lambda^2} + \sinh \rho_\lambda (t + \sigma_\lambda)} < 1. \tag{11.14}$$

The description of a flat maximally globally hyperbolic three-dimensional spacetime with a compact Cauchy surface S of genus $g \geq 2$ by a group homomorphism $h : \pi_1(S) \to \mathrm{ISO}(2,1)$ thus allows one to directly determine the outcome of measurements by an observer as a function of the group homomorphism h. In particular, it is shown in [37] that such measurements are sufficient to determine the group homomorphism $h : \pi_1(S) \to \mathrm{ISO}(2,1)$ associated with a vacuum spacetime up to conjugation in *finite eigentime*.

5 Gauge fixing and observers in quantum gravity

5.1 Three-dimensional gravity as a constrained system In this section, we explain how three-dimensional gravity can be viewed as a constrained system in the sense of Dirac [24, 25] and investigate the mathematical structures with gauge fixing. In the following, we restrict our attention to the case of vanishing cosmological constant and spacetimes $\mathbb{R} \times S$, where S is a surface of genus g with $n \geq 2$ marked points. We consider the description of the associated moduli space $\mathcal{M}^{G,S}$ with $G = \mathrm{ISO}(2,1)$ in terms of group homomorphisms $h : \pi_1(S) \to \mathrm{ISO}(2,1)$ and the associated description of the symplectic structure from Theorem 2.2 in terms of a Poisson structure on $\mathrm{ISO}(2,1)^{n+2g}$. According to (11.4), the moduli space of flat $\mathrm{ISO}(2,1)$-connections on S is then given by

$$\begin{aligned}\mathcal{M}^{\mathrm{ISO}(2,1),S} &= \mathrm{Hom}_{\mathcal{C}_1,\dots,\mathcal{C}_n}(\pi_1(S), \mathrm{ISO}(2,1))/\mathrm{ISO}(2,1)\\ &= \{(M_1,\dots,B_g) \in \mathrm{ISO}(2,1)^{n+2g} :\\ &\qquad M_i \in \mathcal{C}_i, [B_g, A_g^{-1}]\cdots[B_1, A_1^{-1}]\cdot M_n\cdots M_1 = 1\}/\mathrm{ISO}(2,1).\end{aligned}$$

The moduli space is thus obtained from $\mathrm{ISO}(2,1)^{n+2g}$ by restricting the group elements M_i for the loops around punctures to fixed conjugacy classes, by imposing the relation

$$C = [B_g, A_g^{-1}]\cdots[B_1, A_1^{-1}]\cdot M_n\cdots M_1 = 1, \tag{11.15}$$

which arises from the defining relation of the fundamental group $\pi_1(S)$, and by identifying points which are related by global conjugation.

In the application to 3d gravity, this has a direct interpretation in terms of spacetime geometry. Equation (11.15) ensures that the construction of the spacetime M by gluing the boundary of a domain $D \subset \mathbb{M}^3$ yields a spacetime of topology $\mathbb{R} \times S_{g,n}$, i.e. that the black dots in Figure 11.3 are identified to a single point. The restriction of the group elements M_i to fixed conjugacy classes $\mathcal{C}_i \subset \mathrm{ISO}(2,1)$ is related to the interpretation of the punctures as point particles in 3d gravity. If one considers conjugacy classes which contain only elliptic elements of $\mathrm{SO}_0(2,1) \cong \mathrm{PSL}(2,\mathbb{R})$, the geodesics in $\mathbb{M}^3$ that are stabilised by the group elements M_i are timelike and hence can be viewed as worldlines of massive particles. The two parameters which characterise the conjugacy classes $\mathcal{C}_i$ are then an angle $\mu_i \in [0, 2\pi)$, which describes the mass of the point particle and a translation parameter $s_i \in \mathbb{R}$, which arises from the particle's internal angular momentum or spin [21, 22, 23].

Group homomorphisms $h : \pi_1(S_{g,n}) \to \mathrm{ISO}(2,1)$ that are related by conjugation with a fixed element of $\mathrm{ISO}(2,1)$ correspond to a Poincaré transformation applied to the domain $D \subset \mathbb{M}^3$. As this is an isometry of $\mathbb{M}^3$, the resulting quotient spacetimes with the induced metric are isometric. Such Poincaré transformations thus play the role of gauge transformations that relate equivalent descriptions of the same spacetime.

This ambiguity in the description of the spacetimes is linked to the fact that the Poisson structure in Theorem 2.2 is a constrained mechanical system in the sense of Dirac [24, 25], and the relation (11.15) can be considered as a set of first-class constraints. If one considers for $f \in C^\infty(\mathrm{ISO}(2,1))$ the associated functions

$$f_C : \mathrm{ISO}(2,1)^{n+2g} \to \mathbb{R}, \quad f_C(M_1, \ldots, B_g) = f(C),$$

where C is given by (11.15), then it follows from (11.5) that for any $f, g \in C^\infty(\mathrm{ISO}(2,1))$, the Poisson bracket $\{f_C, g_C\}$ vanishes on the constraint surface

$$\Sigma = \{(M_1, \ldots, B_g) \in \mathrm{ISO}(2,1)^{n+2g} : [B_g, A_g^{-1}] \cdots [B_1, A_1^{-1}] \cdot M_n \cdots M_1 = 1\}, \tag{11.16}$$

and the diagonal action of $\mathrm{ISO}(2,1)$ on $\mathrm{ISO}(2,1)^{n+2g}$ is generated by the functions f_C

$$\{f_C, g\} = \frac{d}{dt}\Big|_{t=0} g \circ \phi^t_f \quad \forall g \in C^\infty(\mathrm{ISO}(2,1)^{n+2g}), \text{ where}$$
$$\phi^t_f(M_1, \ldots, B_g) = (\exp(t\mathbf{x}_f) \cdot M_1 \cdot \exp(-t\mathbf{x}_f), \ldots, \exp(t\mathbf{x}_f) \cdot B_g \cdot \exp(-t\mathbf{x}_f))$$

with $\mathbf{x}_f \in \mathfrak{iso}(2,1)$ determined by f. By choosing an appropriate set of functions $f_1, \ldots, f_6 \in C^\infty(\mathrm{ISO}(2,1))$ such that $\Sigma = f_{1,C}^{-1}(0) \cap \ldots \cap f_{6,C}^{-1}(0)$, one can thus interpret the relation (11.15) as a set of six first-class constraints which generate gauge transformations by simultaneous conjugation.

In contrast, the restriction of the group elements M_i to fixed conjugacy classes $\mathcal{C}_i$ does not correspond to any gauge freedom. One can show that for any class function $g \in C^\infty(\mathrm{ISO}(2,1))$, the functions $g^i : \mathrm{ISO}(2,1)^{n+2g} \to \mathbb{R}$, $g^i(M_1, \ldots, B_g) = g(M_i)$ are Casimir functions of the Poisson structure (11.5).

5.2 Gauge fixing and dynamical r-matrices In the context of three-dimensional gravity, the gauge freedom in the description of the moduli space of flat connections is directly related to the implementation of an observer into the theory [40]. As explained in the previous section, an observer in free fall corresponds to a $\pi_1(S)$-equivalence class of timelike geodesics in the domain $D \subset \mathbb{M}^3$, and any two observers are related by a Poincaré transformation. One can thus interpret the conjugation of a group homomorphism $h : \pi_1(S) \to \mathrm{ISO}(2,1)$ by an element of $\mathrm{ISO}(2,1)$ as the transition between two observers.

To eliminate this gauge freedom and to select an observer, one must characterise an observer with respect to the geometry of the spacetime. The most direct way of doing so is to relate his reference frame to the motion of the point particles in M. For instance, one can impose that one of the point particles in the spacetime is at rest at the origin and another one moves in a fixed direction in such a way that its distance from the first particle is minimal at a fixed time. If the lifts of the worldlines of the two chosen particles to D are not parallel, this eliminates the gauge freedom by Poincaré transformations.

In the following, we impose such gauge fixing conditions based on two point particles and analyse the resulting Poisson structures. For this we employ Dirac's gauge fixing procedure [24, 25]. In mathematical terms, this procedure amounts to selecting a representative in each ISO(2, 1)-orbit on Σ and describing the symplectic structure on $\mathcal{M}^{\mathrm{ISO}(2,1),S}$ in terms of a Poisson structure on the space of representatives. This approach can be viewed as the Poisson counterpart of symplectic reduction and is summarised as follows.

Definition 5.1 ([24, 25]). Let $(M, \{\,,\,\})$ be a Poisson manifold.

1. A set of first class constraints for $(M, \{\,,\,\})$ is a function $\Phi = (\phi_1, \ldots, \phi_k) : M \to \mathbb{R}^k$ such that 0 is a regular value of Φ and $\{\phi_i, \phi_j\}|_{\Phi^{-1}(0)} = 0$.
2. A gauge fixing condition for a set of first class constraints Φ is a function $\Psi = (\psi_1, \ldots, \psi_k) : M \to \mathbb{R}^k$ such that
   - 0 is a regular value of $C = (\phi_1, \ldots, \phi_k, \psi_1, \ldots, \psi_k) : M \to \mathbb{R}^{2k}$.
   - The matrix $(\{\phi_i, \psi_j\})_{i,j=1,\ldots,k} : M \to \mathrm{Mat}(k, \mathbb{R})$ is invertible everywhere on $C^{-1}(0)$.
   - Every point on $\Phi^{-1}(0)$ can be mapped to one point on $C^{-1}(0)$ via the flows generated by the constraint functions ϕ_i.

Theorem 5.2 ([24, 25]). *Let $(M, \{\,,\,\})$ a Poisson manifold, $\Phi : M \to \mathbb{R}^k$ a set of first class constraints and $\Psi : M \to \mathbb{R}^k$ a gauge fixing condition. Then the matrix valued function $D = (\{C_i, C_j\})_{i,j=1,\ldots,2k} : M \to \mathrm{Mat}(2k, \mathbb{R})$ is invertible on $C^{-1}(0)$ and*

$$\{f, g\}_D = \{\tilde{f}, \tilde{g}\}|_{C^{-1}(0)} + \sum_{i,j=1}^{2k} \{\tilde{f}, C_i\}|_{C^{-1}(0)} (D|^{-1}_{C^{-1}(0)})_{ij} \{\tilde{g}, C_j\}|_{C^{-1}(0)} \tag{11.17}$$

for $f, g \in C^\infty(C^{-1}(0))$ and arbitrary extensions $\tilde{f}, \tilde{g} \in C^\infty(M)$ defines a Poisson structure on $C^{-1}(0)$.

We now apply Dirac's gauge fixing formalism to the description of the moduli space of flat ISO(2, 1)-connections. In this case, we have $M = \mathrm{ISO}(2,1)^{n+2g}$ with the Poisson structure from Theorem 2.2. The choice of classical r-matrix for ISO(2, 1) is non-unique. For computational reasons, it is advantageous to work with the classical r-matrix that realises ISO(2, 1) as the classical double of SO(2, 1) with the trivial cocommutator. If one chooses a basis $\{J_a, P_a\}_{a=0,1,2}$ of $\mathfrak{iso}(2,1)$, in which the Lie bracket takes the form

$$\begin{array}{lll}
[J_0, J_1] = J_2 & [J_0, J_2] = -J_1 & [J_1, J_2] = -J_0 \\
[J_0, P_0] = 0 & [J_0, P_1] = P_2 & [J_0, P_2] = -P_1 \\
[J_1, P_0] = -P_2 & [J_1, P_1] = 0 & [J_1, P_2] = -P_0 \\
[J_2, P_0] = -P_1 & [J_2, P_1] = P_0 & [J_2, P_2] = 0 \\
[P_0, P_1] = 0 & [P_0, P_2] = 0 & [P_1, P_2] = 0,
\end{array}$$

this classical r-matrix is given by $r = -P_0 \otimes J_0 + P_1 \otimes J_1 + P_2 \otimes J_2$. The constraints correspond to the ISO(2, 1)-valued constraint (11.15) and the associated gauge transformations are given by the ISO(2, 1)-action on $\mathrm{ISO}(2,1)^{n+2g}$.

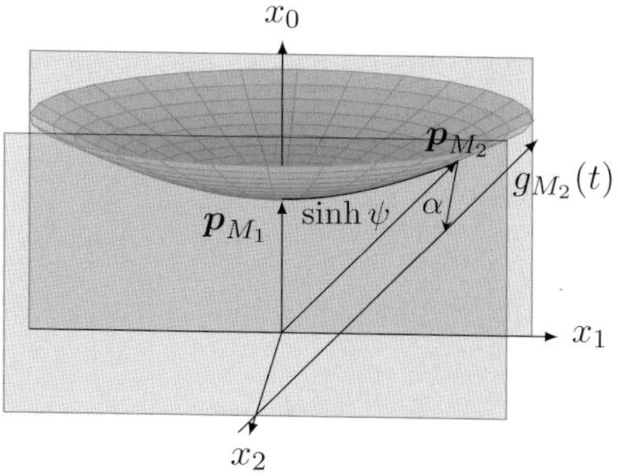

Figure 11.5. Gauge fixing condition for point particles

We now impose a set of six gauge fixing conditions that eliminates this gauge freedom by restricting the group elements for two point particles. As different orderings of the particles correspond to a braid group action on $\mathrm{ISO}(2,1)^{n+2g}$ which is a Poisson isomorphism with respect to the Poisson structure from Theorem 2.2, one can restrict attention to gauge fixing conditions which restrict only the group elements M_1, M_2. Moreover, the constraints and gauge fixing conditions should respect the natural $\mathbb{N}$-grading of the Poisson-structure that corresponds to a physical dimension of $\hbar$. This leads one to consider that a set of constraints and gauge fixing conditions consists of six functions that depend only on the Lorentzian components of the group elements $M_1, M_2 \in \mathrm{ISO}(2,1)$ and six functions that depend linearly on their translational components.

Subject to these restrictions, the group elements $M_1, M_2 \in \mathrm{ISO}(2,1)$ are determined uniquely by two Poincaré invariant quantities, the angle ψ between the associated geodesics in Minkowski space, which is related to their relative velocity, and their minimal distance α as shown in Figure 11.5. With the formula in Theorem 5.2, one can then compute the Dirac bracket for the associated constraints and gauge fixing conditions. If one parametrises the surface $C^{-1}(0)$ in term of these parameters ψ and α and the residual, non-gauge fixed group elements $M_3, \ldots, B_g \in \mathrm{ISO}(2,1)$, this determines a Poisson structure on $\mathbb{R}^2 \times \mathrm{ISO}(2,1)^{n+2g-2}$, for which the condition (11.15) acts as a Casimir function.

Theorem 5.3 ([41]). *For every set of constraints and gauge fixing conditions as above, the Dirac bracket* (11.17) *determines a Poisson structure* $\{\ ,\ \}_{r,x}$ *on* $\mathbb{R}^2 \times \mathrm{ISO}(2,1)^{n+2g-2}$ *with*

$$\{\psi,\alpha\}_{r,x} = 0,\ \{\psi, f\}_{r,x} = \frac{d}{dt}\bigg|_{t=0} f \circ \phi^t_\psi,$$
$$\{\alpha, f\}_{r,x} = \frac{d}{dt}\bigg|_{t=0} f \circ \phi^t_\alpha,\ \{f,g\}_{r,x} = \{f,g\}_r$$

for $f, g \in C^\infty(\mathrm{ISO}(2,1)^{n+2g-2})$ *and*

$$\phi^t_\psi(M_3, \ldots, B_g) = (\exp(t\,\mathbf{x}_\psi) \cdot M_3 \cdot \exp(-t\mathbf{x}_\psi), \ldots, \exp(t\,\mathbf{x}_\psi) \cdot B_g \cdot \exp(-t\,\mathbf{x}_\psi))$$
$$\phi^t_\alpha(M_3, \ldots, B_g) = (\exp(t\,\mathbf{x}_\alpha) \cdot M_3 \cdot \exp(-t\mathbf{x}_\alpha), \ldots, \exp(t\,\mathbf{x}_\alpha) \cdot B_g \cdot \exp(-t\,\mathbf{x}_\alpha))$$

with $\mathbf{x}_\psi : \mathbb{R}^2 \to \mathbb{R}^3 \subset \mathfrak{iso}(2,1)$*,* $\mathbf{x}_\alpha : \mathbb{R}^2 \to \mathfrak{iso}(2,1)$*, and where* $\{\ ,\ \}_r$ *is the Poisson bracket* (11.5) *on* $\mathrm{ISO}(2,1)^{n+2g-2}$ *for* $r : \mathbb{R}^2 \to \mathfrak{iso}(2,1) \otimes \mathfrak{iso}(2,1)$*. For all* $(\psi, \alpha) \in \mathbb{R}^2$ $\mathbf{x}_\psi(\psi,\alpha)$*,* $\mathbf{x}_\alpha(\psi,\alpha)$ *span an abelian Lie subalgebra* $\mathfrak{h}(\psi,\alpha) \subset \mathfrak{iso}(2,1)$*, and r is a solution of the classical dynamical Yang–Baxter equation (DCYBE)*

$$[[r,r]] = \mathbf{x}^{(1)}_\psi \frac{\partial r_{23}}{\partial \psi} - \mathbf{x}^{(2)}_\psi \frac{\partial r_{13}}{\partial \psi} + \mathbf{x}^{(3)}_\psi \frac{\partial r_{12}}{\partial \psi} + \mathbf{x}^{(1)}_\alpha \frac{\partial r_{23}}{\partial \alpha} - \mathbf{x}^{(2)}_\alpha \frac{\partial r_{13}}{\partial \alpha} + \mathbf{x}^{(3)}_\alpha \frac{\partial r_{12}}{\partial \alpha}.$$

The gauge fixed Poisson structure from Theorem 5.3 is thus obtained from the Poisson structure (11.5) on $\mathrm{ISO}(2,1)^{n+2g}$ by removing the first two arguments and replacing the classical r-matrix by a solution $r : \mathbb{R}^2 \to \mathfrak{iso}(2,1) \otimes \mathfrak{iso}(2,1)$ of the *dynamical* classical Yang–Baxter equation (DCYBE). The two dynamical parameters are related to the relative velocity and the minimal distance of the two gauge fixed point particles.

It remains to clarify how the concrete choice of gauge fixing conditions manifests itself in this description. For this, note that for different gauge fixing conditions based on two point particles, the associated group elements M_1, M_2 are always related by a Poincaré transformation which depends on the variables ψ and α. The transition between different gauge fixing conditions is therefore given by transformations of the form

$$P : (\psi, \alpha, M_3, \ldots, B_g) \mapsto (\psi, \alpha, p(\psi,\alpha) M_3 p(\psi,\alpha)^{-1}, \ldots, p(\psi,\alpha) B_g\, p(\psi,\alpha)^{-1}) \tag{11.18}$$

with $p \in C^\infty(\mathbb{R}^2, \mathrm{ISO}(2,1))$. One can show that such dynamical Poincaré transformations induce transformations of the associated dynamical r-matrices.

Theorem 5.4 ([41]). *Under a dynamical Poincaré transformation* (11.18) *the Poisson bracket from Theorem 5.3 transforms according to*

$$\begin{aligned} &\{f \circ P, g \circ P\}_{r,x} = \{f, g\}_{r^P, x^P} \circ P, \\ &\textit{where } \mathbf{x}^P_\alpha = Ad(p)\mathbf{x}_\alpha, \quad \mathbf{x}^P_\psi = Ad(p)\mathbf{x}_\psi, \end{aligned} \tag{11.19}$$

$$r^P = (\mathrm{Ad}(p) \otimes \mathrm{Ad}(p))(r + \bar{\eta} - \bar{\eta}_{21})$$

with

$$\bar{\eta} = \mathbf{x}_\psi \otimes p^{-1}\partial_\psi p + \mathbf{x}_\alpha \otimes p^{-1}\partial_\alpha p.$$

The transformations (11.19) generalise the gauge transformations of dynamical r-matrices in [26] for a fixed abelian subalgebra $\mathfrak{h}$. As the latter are used in the classification of dynamical r-matrices, it seems plausible that by applying such transformations together with a rescaling $\psi \to f(\psi)$, $\alpha \to g(\psi)\alpha + h(\psi)$, it should be possible to classify all gauge fixed Poisson structures in Theorem 5.3. This is indeed possible locally, for those values of the parameters ψ, α for which $\mathfrak{h}(\psi, \alpha)$ does not contain parabolic elements of $\mathfrak{so}(2,1) \subset \mathfrak{iso}(2,1)$.

Theorem 5.5 ([41]). *Let (ψ_0, α_0) be a point for which $\mathfrak{h}(\psi_0, \alpha_0)$ does not contain parabolic elements of $\mathfrak{so}(2,1) \subset \mathfrak{iso}(2,1)$. Then there exists a neighbourhood $U \subset \mathbb{R}^2$ of (ψ_0, α_0) in which by a transformation* (11.18) *and a rescaling of the parameters ψ, α, the dynamical r-matrix and the functions $\mathbf{x}_\psi, \mathbf{x}_\alpha : \mathbb{R}^2 \to \mathfrak{iso}(2,1)$ from Theorem* 5.3 *can be brought into the form: $\mathbf{x}_\psi, \mathbf{x}_\alpha : U \to \mathrm{Span}\{P_0, J_0\}$ and*

$$r(\psi, \alpha) = r_{(s)} + \tfrac{1}{2}\tan(\tfrac{\psi}{2})(P_1 \wedge J_2 - P_2 \wedge J_1) + \frac{\alpha}{4\cos^2(\frac{\psi}{2})} P_1 \wedge P_2$$

or $\mathbf{x}_\psi, \mathbf{x}_\alpha : U \to \mathrm{Span}\{P_1, J_1\}$ and

$$r(\psi, \alpha) = r_{(s)} + \tfrac{1}{2}\tanh(\tfrac{\psi}{2})(P_2 \wedge J_0 - P_0 \wedge J_2) + \frac{\alpha}{4\cosh^2(\frac{\psi}{2})} P_2 \wedge P_0.$$

with

$$r_{(s)} = \tfrac{1}{2}(-P_0 \otimes J^0 - J^0 \otimes P_0 + P_1 \otimes J^1 + J^1 \otimes P_1 + P_2 \otimes J^2 + J^2 \otimes P_2).$$

It is shown in [41] that these two solutions have a direct physical interpretation. The first corresponds to the centre of mass frame of an asymptotically conical universe, in which the centre of mass of the spacetime behaves like a particle which is at rest at the origin with respect to the observer. The second solution corresponds to a spacetime that asymptotically has the geometry of a torus universe. The two dynamical parameters ψ and α correspond, respectively, to the total energy and total angular momentum of the spacetime as measured by the observer.

This investigation of the classical Poisson–Lie symmetries on the gauge invariant phase space of 3d gravity allows one to draw conclusions about the quantum group symmetries arising in its quantisation. The preceding discussion shows that the question about the quantum group symmetries in 3d gravity is subtle. Specifically, the relevant quantum groups and Poisson–Lie groups change depending on whether one considers the quantisation of a larger ambient space or the gauge invariant phase space obtained via a gauge fixing procedure. On the former, the relevant quantum groups are the ones that arise from solutions of the CYBE whose symmetric component is dual to the Ad-invariant bilinear form in the Chern–Simons action. On the latter, the relevant Poisson–Lie and quantum group symmetries are *dynamical* ones determined by the classical r-matrix in Theorem 5.3.

As gauge fixing amounts to selecting an observer whose motion is specified with respect to two point particles in the spacetime, the dynamical quantum group symmetries can be viewed as structures associated with the inclusion of an observer in the quantum theory. This interpretation is in agreement with the fact that the two dynamical parameters have the interpretation of energy and angular momentum, whose definition requires the specification of a reference frame. Such a link between dynamical r-matrices and observers in 3d gravity was obtained by very different methods in [16, 19], which is strong evidence that dynamical quantum group symmetries should be considered as the correct *physical* symmetries for 3d gravity.

Acknowledgements. The author thanks the organisers of the *90e rencontre entre physiciens théoriciens et mathématiciens: Lie et Klein; le programme d'Erlangen et son impact en physique et en mathématique*, September 20–22 2012, IRMA, Strasbourg, where this work was presented. The research was funded by the German Research Foundation (DFG) through the Emmy Noether fellowship ME 3425/1-1.

Bibliography

[1] A. Achucarro and P. Townsend, A Chern–Simons action for three-dimensional anti-de Sitter supergravity theories. *Phys. Lett.* B 180 (1986), 85–100.

[2] A. Y. Alekseev, H. Grosse and V. Schomerus, Combinatorial quantization of the Hamiltonian Chern–Simons theory I. *Commun. Math. Phys.* 172 (1995), 317–358.

[3] A. Y. Alekseev, H. Grosse and V. Schomerus, Combinatorial quantization of the Hamiltonian Chern–Simons theory II. *Commun. Math. Phys.* 174 (1996), 561–604.

[4] A. Y. Alekseev and A. Z. Malkin. Symplectic structure of the moduli space of flat connection on a Riemann surface. *Commun. Math. Phys.* 169 (1995), 99–119.

[5] A. Alekseev, A. Malkin and E. Meinrenken, Lie group valued moment maps. *J. Differ. Geom.* 48, Vol. 3 (1998), 445–495.

[6] A. Y. Alekseev and V. Schomerus, Representation theory of Chern–Simons observables. *Duke Math. J.* 85 (1996), 447–510.

[7] L. Andersson, T. Barbot, R. Benedetti, F. Bonsante, W. M. Goldman, F. Labourie, K. P. Scannell and J.-M. Schlenker, Notes on a paper of Mess. *Geometriae Dedicata* 126:1 (2007), 47–70.

[8] M. Atiyah and R. Bott, The Yang–Mills equations over Riemann surfaces. *Philos. T. R. Soc. S.-A*54 (1983), 523–615.

[9] T. Barbot, Globally hyperbolic flat spacetimes. *Journ. Geom. Phys.* 53 (2005), 123–165.

[10] T. Barbot, F. Bonsante and J.-M. Schlenker, Collisions of particles in locally AdS spacetimes I. Local description and global examples. *Commun. Math. Phys.* 308 (2011), 147–200.

[11] T. Barbot, F. Bonsante and J.-M. Schlenker, Collisions of particles in locally AdS spacetimes II Moduli of globally hyperbolic spaces. Preprint, 2012; arXiv:1202.5753[math.DG].

[12] R. Benedetti and F. Bonsante, Canonical Wick rotations in 3-dimensional gravity. *Mem. Am. Math. Soc.* 926, Vol 198 (2009).

[13] R. Benedetti and F. Bonsante, (2 + 1) Einstein spacetimes of finite type. In *Handbook of Teichmüller theory*, Volume II, A. Papadopoulos, (ed.), Zürich, European Mathematical Society (EMS). IRMA Lectures in Mathematics and Theoretical Physics 13, 533–609 (2009).

[14] R. Benedetti and E. Guadagnini, Cosmological time in (2+1)-gravity. *Nucl. Phys. B* 613 (2001), 330–352.

[15] F. Bonsante and J.-M. Schlenker, AdS manifolds with particles and earthquakes on singular surfaces. *Geom. Funct. Anal.* 19.1 (2009), 41–82.

[16] E. Buffenoir and K. Noui, Unfashionable observations about three-dimensional gravity. Preprint, 2003; arXiv: gr-qc/0305079.

[17] E. Buffenoir, K. Noui and Ph. Roche, Hamiltonian Quantization of Chern–Simons theory with $SL(2,\mathbb{C})$ Group. *Class. Quant. Grav.* 19 (2002), 4953–5016.

[18] E. Buffenoir and P. Roche, Two dimensional lattice gauge theory based on a quantum group. *Commun. Math. Phys.* 170 (1995), 669–698.

[19] E. Buffenoir and P. Roche, Chern–Simons Theory with Sources and Dynamical Quantum Groups I: Canonical Analysis and Algebraic Structures. Preprint, 2005; arXiv: hep-th/0505239.

[20] S. Carlip, *Quantum gravity in 2+1 dimensions*. Cambridge University Press, Cambridge, 1998.

[21] S. Deser, R. Jackiw and G. 't Hooft, Three-dimensional Einstein gravity: Dynamics of flat space. *Ann. Phys.-New York* 152 (1984), 220–235.

[22] S. Deser and R. Jackiw, Classical and quantum scattering on a cone. *Commun. Math. Phys.* 118 (1988), 495–509.

[23] P. de Sousa Gerbert, On spin and (quantum) gravity in 2 + 1 dimensions. *Nucl. Phys. B* 346 (1990), 440–472.

[24] P. A. M. Dirac, Forms of Relativistic Dynamics. *Rev. Mod. Phys.* 21 (1949), 392.

[25] P. A. M. Dirac, Generalized Hamiltonian dynamics. *Canadian J. Math.* 2 (1950), 129–148.

[26] P. Etingof and A. Varchenko, Geometry and Classification of Solutions of the Classical Dynamical Yang–Baxter Equation. *Commun. Math. Phys.* 192 (1998), 77–120.

[27] V. V. Fock and A. A. Rosly, Poisson structure on moduli of flat connections on Riemann surfaces and r-matrix. *Am. Math. Soc. Transl.* 191 (1999), 67–86.

[28] M. Göckeler and T. Schücker, *Differential geometry, gauge theories, and gravity*. Cambridge University Press, Cambridge, 1989.

[29] W. Goldman, The symplectic nature of fundamental groups of surfaces, *Adv. Math.* 54 (1984), 200–225.

[30] A. Hanson, T. Regge and C. Teitelboim, *Constrained hamiltonian systems*. Accademia Nazionale dei Lincei, 1976.

[31] M. Henneaux and C. Teitelboim, *Quantization of gauge systems*. Princeton University press, Princeton, 1992.

[32] F. Klein, *Vergleichende Betrachtungen über neuere geometrische Forschungen*. Verlag von Andreas Deichert, Erlangen, 1872.

[33] F. Klein, *Über die geometrischen Grundlagen der Lorentzgruppe*. Jahresbericht der Deutschen Mathematiker-Vereinigung 19 (1910).

[34] F. Klein, Über die Differentialgesetze für Erhaltung von Impuls und Energie in der Einsteinschen Gravitationstheorie. *Nachr. D. König. Gesellsch. D. Wiss. Zu Göttingen, Math-phys. Klasse* (1918), 171–189.

[35] G. Mess, Lorentz spacetimes of constant curvature. *Geometriae Dedicata* 126:1 (2007), 3–45.

[36] C. Meusburger, Cosmological measurements, time and observables in (2+1)-dimensional gravity. *Class. Quant. Grav.* 26 (2009), 055006.

[37] C. Meusburger, Global Lorentzian geometry from lightlike geodesics: What does an observer in (2+1)-gravity see?. In *Chern–Simons Gauge Theory: 20 years after*, AMS/IP Studies in Advanced Mathematics Vol. 50, 2011, 261–276

[38] C. Meusburger and K. Noui, Combinatorial quantisation of the Euclidean torus universe. *Nucl. Phys. B* 841 (2010), 463–505.

[39] C. Meusburger and K. Noui, The Hilbert space of 3d gravity: quantum group symmetries and observables. *Adv. Theor. Math. Phys.* 14.6 (2010), 1651–1716.

[40] C. Meusburger and T. Schönfeld, Gauge fixing in (2+1)-gravity: Dirac bracket and spacetime geometry. *Class. Quant. Grav.* 28 (2011), 125008.

[41] C. Meusburger and T. Schönfeld, Gauge fixing and classical dynamical r-matrices in ISO(2,1)-Chern–Simons theory. *Commun. Math. Phys.* 327.2 (2014), 443–479.

[42] C. Meusburger and B. J. Schroers, The quantisation of Poisson structures arising in Chern–Simons theory with gauge group $G \ltimes \mathfrak{g}^*$. *Adv. Theor. Math. Phys.* 7 (2003), 1003–1042.

[43] E. Noether, Invariante Variationsprobleme, *Nachr. D. König. Gesellsch. D. Wiss. Zu Göttingen, Math-phys. Klasse* (1918), 235–257.

[44] N. Reshetikhin and V. Turaev, Invariants of 3-manifolds via link-polynomials and quantum groups. *Invent. Math.* 103 (1991), 547–597.

[45] C. Rovelli, Partial observables. *Phys. Rev. D* 65 (2002), 124013.

[46] E. Witten, Quantum field theory and the Jones polynomial. *Comm. Math. Phys.* 121 (1989), 351–399.

[47] E. Witten, 2+1 dimensional gravity as an exactly soluble system. *Nucl. Phys. B* 311 (1988), 46–78.

[48] C. N. Yang and R. Mills, Conservation of Isotopic Spin and Isotopic Gauge Invariance. *Phys. Rev.* 96(1) (1954), 191–195.

Chapter 12

Invariances in physics and group theory

Jean-Bernard Zuber

Contents

1 Introduction

Let us be honest: most physicists of our time, even theorists, do not have a very clear notion of what Klein's *Erlanger Programm* is about, and this is an understatement... If we read Weyl [32], however,

> According to Klein's Erlanger Program any geometry of a point-field is based on a particular transformation group $\mathfrak{G}$ of the field; figures which are equivalent with respect to $\mathfrak{G}$, and which can therefore be carried into one another by a transformation of $\mathfrak{G}$, are to be considered as the same...

and substitute "physical systems" for "figures", we see that modern physicists, like Molière's character Monsieur Jourdain, who was delighted to learn that he had been speaking prose all his life without knowing, would love to hear that they keep following Klein's program ...

The aim of this chapter is indeed to illustrate how group theory associated with invariances of the geometry or the dynamics of a physical system has pervaded all modern physics and has become of everyday use in the physicist's toolbox. A word of caution, though. As the author of these lines is not a professional historian of science, this chapter will undoubtedly present only a biased and incomplete view of this vast subject.

2 Early group theory in 19th century physics: crystallography

Before the birth of Lie group theory and Klein's Erlangen program, physicists had realized the role of symmetry in Nature and foreseen the importance of group theory in the physical sciences. It had been known for long by artists that 2-dimensional periodic patterns – tilings or wall paper motives, *i.e.* two-dimensional "crystals" – were coming in finite types. There are 17 types of symmetry – 17 space groups in modern terminology – in two dimensions. For a beautiful illustration, see the web site http://en.wikipedia.org/wiki/Wallpaper_group. Crystallographers then set out to tabulate the corresponding structures in 3 dimensions, classifying in turn the point groups, i.e. groups that fix a point of a lattice, the classes of lattices, and the space groups, taking translations into account. This long endeavor kept them busy for the major part of the nineteenth century, with important steps achieved by Frankenheim and by Hessel (32 point groups in 3 dimensions, in the 1830's), by Bravais (14 classes of lattices, circa 1850), Jordan (who emphasized the role of groups), and many others. The program was completed in the early nineties of that century, by Schönflies,[1] Fedorov and Barlow (1891–94), with the classification of the 230 space groups in 3 dimensions, see [2, 22, 11, 17]. The situation is summarized in the following table

Dimension d	Point Groups	Lattices	Space groups
$d = 1$	2	1	2
$d = 2$	10	5	17
$d = 3$	32	14	230

According to H. Weyl [32] "The most important application of group theory to natural science heretofore has been in this field." It is interesting to notice that Weyl wrote this comment in 1928, many years after the birth of relativity – both special and general –, and as he was himself working on the applications of group theory to quantum physics.

Breaking of symmetry If it is important in physical sciences to know the possible types of symmetry, it is maybe even more interesting to understand the way these symmetries may be broken.[2] This was emphasized in a particularly clear way by Pierre Curie, as stated in his principle (1894) [3]: "Elements of symmetry of causes must be found in effects; when some effects reveal some asymmetry, that asymmetry must be found in causes." Or in a more cursive way: "*C'est la dissymétrie qui crée le phénomène.*"

An example is provided by the phenomenon of piezoelectricity, i.e. the creation of an electric (vector) field E in a crystalline material subject to a mechanical stress. The latter is described by a rank-two tensor u; in a linear approximation the electric field

[1] It may be relevant to observe that Schönflies was directed to this problem by Klein, who saw it as a nice illustration of his program.

[2] This is also an interesting issue in art, see [34].

is proportional to u, $E_i = \sum_{jk} \gamma_{i,jk} u_{jk}$, and hence the phenomenon depends on the existence of a non-vanishing rank 3 tensor $\gamma_{i,jk} \neq 0$; if the crystal admits a symmetry by "inversion", (i.e. reflection with respect to a point), $\gamma_{i,jk}$ is changed into $-\gamma_{i,jk}$ under inversion and must vanish, and this rules out piezoelectricity in many crystal classes. Only non-symmetric crystalline classes may give rise to piezoelectricity.

Curie also understood that the breaking of a symmetry under a group G may leave invariance under a subgroup (an "intergroupe" in his terms) H of G, an idea still quite topical. For instance, he classified the possible breaking and subgroups of a system invariant under rotations around an axis, i.e. under the group D_∞ in modern terminology.

Limits of group theory As noticed by M. Senechal [24], "group theory cannot answer a question that seems fundamental today: which shapes tile space and in what way?". That question has of course become highly relevant since the discovery some 30 years ago of quasicrystals. In this new class of materials, rotational order does not extend to large distances and translation invariance is lost. Still diffraction of X-rays leads to patterns of bright spots exhibiting some symmetry. This has led the International Union of Crystallography to redefine the term "crystal" so as to include both ordinary periodic crystals and quasicrystals. According to this new definition, a crystal is "any solid having an essentially discrete diffraction diagram".

3 Special relativity and Lorentz group: Lorentz, Poincaré, Einstein …

Special relativity is often regarded as the first appearance of Lie group theory in modern physics. Let us recall some of the crucial steps, referring the reader to more scholarly sources [19, 4] for further details.

- Lorentz (1892–1904) (after Voigt and FitzGerald, and in parallel to Larmor) discovers what are now called the Lorentz transformations and the resulting contraction of lengths. His purpose is to make the Michelson–Morley experiment consistent with the existence of aether.
- Poincaré (1905) establishes the covariance of Maxwell equations under Lorentz transformations; he also sees that Lorentz transformations together with space rotations leave the form $x \cdot x := x^2 + y^2 + z^2 - c^2t^2$ invariant and form a group, thus giving them their proper geometric meaning, much in the spirit of Klein. In his approach, however, the Lorentz group is not derived from first principles.
- Einstein (1905) starting from two principles – (i) the principle of relativity: physical laws do not depend on the inertial frame of the observer; and (ii) in an inertial frame the speed of light c is an absolute constant of Physics, independent of the uniform motion of the source – *constructs* the Lorentz transformations; he notices as a side remark that Lorentz special transformations (or

"boosts" as we call them now) of *collinear* velocities form a group, "wie dies sein muss" (as they should);[3] he proves that they leave Maxwell's equations invariant, but does not seem to notice or at least does not comment on the fact that they also preserve the form $x \cdot x$.

- Minkowski (1908) introduces "space-time", identifies the Lorentz group as the invariance group of the metric $x_1^2 + x_2^2 + x_3^2 + x_4^2$ with $x_4 = ict$, makes use of the notion of 4-vectors and tensors, and shows the covariant way of writing Maxwell equations. At first Einstein is not impressed by this piece of work, qualifying it as "überflüssige Gelehrsamkeit" (superfluous erudition)! [6]. After starting to work on gravitation, however, Einstein soon realizes the power of tensor methods.

Thus, although Einstein made a real breakthrough in physics and utterly changed our view of space and time by "propounding a new chronogeometry" [4], it seems fair to say that group theory played a very minor role in his work and his lines of thought.

More mathematically inclined people thought otherwise. We have already mentioned Poincaré's and Minkowski's works. Klein (1910) [14] observes: "One could replace 'theory of invariants relative to a group of transformations' by the words 'relativity theory with respect to a group'." For him, Galilean relativity or special relativity were clearly in the straight line of his Program.

4 General relativity... and gauge theories

General relativity is an emblematic case illustrating Klein's program in a differential geometric context. There, following Einstein's vision, one postulates the invariance of the equations of the gravitational field under general coordinate transformations. And by a sort of reverse engineering, one looks for equations knowing the invariance group. This is what was achieved by Einstein and by Hilbert (1915), with the celebrated equation

$$R_{\mu\nu} - \frac{1}{2} R\, g_{\mu\nu} = \kappa T_{\mu\nu}\,, \tag{12.1}$$

with $R_{\mu\nu}$ the Ricci tensor, $R = R^{\mu}_{\mu}$ its curvature, $T_{\mu\nu}$ the energy-momentum tensor, and $\kappa = \frac{8\pi G}{c^4}$ where G is Newton's gravitational constant. Recall that Hilbert derived this equation from the invariant action

$$S = \int \left[\frac{1}{2\kappa} R + \mathcal{L}_{\mathrm{M}}\right] \sqrt{-g}\, d^4x\,, \tag{12.2}$$

with $\mathcal{L}_{\mathrm{M}}$ describing the invariant coupling of gravity to matter, and $T_{\mu\nu} = \partial(\mathcal{L}_{\mathrm{M}}\sqrt{-g})/\partial g^{\mu\nu}$. I shall not dwell more on that subject; it is also treated in Chapter 11 of this volume [16].

[3] This is the only occurrence of the word "Gruppe" in his paper.

Let me rather make a big leap forward in time, and observe that a similar approach was taken in the construction of non-abelian gauge theories. The gauge invariance of electrodynamics had been observed by Weyl (1918) and reformulated later by him into what we now call U(1) gauge invariance [31, 33]. Looking for a generalization to non abelian groups G, i.e. postulating invariance under a certain *infinite-dimensional* group of local, space-time dependent transformations, Yang and Mills (1954) [40] were led to an (essentially) unique solution, with Lagrangian density

$$\mathcal{L} = \frac{1}{2g^2} \operatorname{tr} F_{\mu\nu} F^{\mu\nu}, \qquad F_{\mu\nu} = \partial_\mu A_\nu - \partial_\nu A_\mu - [A_\mu, A_\nu], \tag{12.3}$$

with the gauge field A_μ and its field strength tensor $F_{\mu\nu}$ (a connection and its curvature on a fiber bundle) taking values in the Lie algebra of G or one of its representations. Here and below, ∂_μ stands for $\frac{\partial}{\partial x^\mu}$. $\mathcal{L}$ is invariant under local infinitesimal changes $\delta A_\mu(x) = D_\mu \delta\alpha(x)$, with $D_\mu = \partial_\mu - [A_\mu, \cdot]$ the covariant derivative, and $\delta\alpha \in \operatorname{Lie} G$. A term $\mathcal{L}_{\mathrm{m}}$ may then be added to $\mathcal{L}$ to describe the gauge invariant coupling to matter. This now famous and ubiquitous Yang–Mills theory is the cornerstone of the Standard Model of particle physics, see below.

To summarize, here are two cases (Einstein–Hilbert, Yang–Mills) in which invariances and geometry of space (either real space-time or "internal" space) constrain the dynamics.[4] According to Yang's motto [39], "symmetry dictates interaction."

5 Emmy Noether: invariances and conservation laws

Noether's celebrated paper (1918) [18], presented on the occasion of Klein's academic Jubilee, contains two theorems on group invariance in variational problems. I give a sketch of her results, using modern terminology and notation, and I refer to [15] for a translation of her original article and a detailed and critical reading, see also [13].

Consider a field theory described by an action principle in a, say, 4-dimensional space-time with coordinates $x = (\vec{x}, t \equiv x^0)$

$$S = \int \mathcal{L}(x\,;\, \phi^i(x), \partial\phi^i(x), \ldots) d^4x$$

with S the action and $\mathcal{L}$ the Lagrangian density, a local function of a collection of fields $\{\phi^i\}$ and of finitely many of their derivatives. Assume the invariance of $\mathcal{L}\, d^4x$ (and hence of S) under a Lie group of coordinate and field variations $x \mapsto x'$, $\phi \mapsto \phi'$. Then Noether's first theorem asserts:

[4] "Constrain" does not mean fix uniquely: any smooth function $f(R)$ of curvature in (12.2) gives a diffeomorphism invariant Lagrangian, and likewise, higher order gauge invariant terms could be added to (12.3). Thus in both GR and gauge theories, we need a principle of *minimality* to write the actions (12.2) or (12.3).

Theorem 5.1 (Noether). *An n-dimensional Lie group of invariance of $\mathcal{L}\, d^4x$ implies the existence of n independent divergenceless currents*

$$j_s^\mu = (j_s^0(\vec{x},t), \vec{j}_s(\vec{x},t)), \qquad \text{i.e.} \quad \partial_\mu j_s^\mu \equiv \frac{\partial}{\partial t} j_s^0 - \operatorname{div} \vec{j}_s = 0, \quad s = 1,\dots,n,$$

from which, by Stokes theorem, n independent conservation laws follow

$$\frac{d}{dt} Q_s := \frac{d}{dt} \int d^3x\, j_s^0(\vec{x},t) = \int d^3x \operatorname{div} \vec{j}_s(\vec{x},t) = 0.$$

(The currents are assumed to vanish fast enough at spatial infinity to justify the last step.) Suppose that $\mathcal{L}$ depends only on ϕ and its first derivatives $\partial\phi$. Write coordinate and field infinitesimal variations as $\delta x^\mu = X_s^\mu(x,\phi)\delta a^s$ and $\delta\phi^i = Z_s^i(x,\phi)\delta a^s$, where a^s, $s = 1,\dots n$, are parameters in the Lie algebra, and Einstein's convention of summation over repeated indices is used. Then one finds

$$\begin{aligned} j_s^\mu &= -\frac{\partial \mathcal{L}}{\partial \partial_\mu \phi^i}(Z_s^i - \partial_\rho \phi^i X_s^\rho) - X_s^\mu \mathcal{L} \\ \partial_\mu j_s^\mu\, \delta a^s &= \sum_i \Psi_i\, \delta\phi^i \ \text{ where } \Psi_i := \frac{\delta \mathcal{L}}{\delta \phi^i} := \frac{\partial \mathcal{L}}{\partial \phi^i} - \partial_\mu \frac{\partial \mathcal{L}}{\partial \partial_\mu \phi^i} \\ &= 0 \ \text{ by Euler–Lagrange equations}. \end{aligned} \tag{12.4}$$

In Noether's paper, the converse property, namely that conservation laws imply invariance, is also derived. This first theorem was subsequently generalized by Bessel–Hagen (1921) to the case where $\mathcal{L}\, d^4x$ is invariant up to a total divergence $\delta a^s\, \partial_\mu k_s^\mu\, d^4x$, in which case j_s^μ is just modified by the additional term k_s^μ.

As an example, consider a theory involving a complex scalar field ϕ with Lagrangian $\mathcal{L} = \partial_\mu\phi^* \partial^\mu \phi - V(\phi^*\phi)$, V some arbitrary polynomial potential. The Lagrangian is invariant under the group U(1) of transformations $\phi(x) \to e^{i\alpha}\phi(x)$, leading to a conserved Noether current $j_\mu(x) = i(\phi^*(x)\partial_\mu\phi(x) - (\partial_\mu\phi(x))^*\phi(x))$. The associated conserved U(1) charge may be thought of as an electric (or baryonic or leptonic, etc.) charge.

Thus Noether's first theorem establishes a link between invariances under continuous transformations and conservation laws. This was not a new result in physics. There had been early precursors: Lagrange (1811), Hamilton (1834), Jacobi (1837) had uncovered the fundamental conservation laws of energy, momentum and angular momentum in classical mechanics, but did not make a systematic connection with geometric invariances. This had been elaborated by Schütz (1897) and by other precursors of Noether: Hamel (1904) who introduced the calculus of variations in that context, Herglotz (1911), Engel (1916) and Kneser (1917) who applied it to the 10 conservation laws due to Galilean and to relativistic invariance, see [13, 15]. But E. Noether was the first to give a general and systematic derivation of conservation laws, starting from invariance of an action principle under Lie algebraic transformations.

This important result of Noether had a curious fate. After an initial applause by Klein, Hilbert and others, and some generalization by Bessel–Hagen, came a long freeze. This was caused mainly by the rise of quantum mechanics, which made no use of the Lagrangian formalism. Thus Noether's theorem was essentially forgotten until the early 1950s, when covariant Quantum Field Theory (QFT) developed, causing a revival of interest in the Lagrangian formalism, and Noether's theorem became important again.

In modern QFT, her theorem appears in particular in the guise of *Ward–Takahashi identities* satisfied by the vacuum expectation values of "time-ordered products of fields" – T-products, in short – which are the relevant Green functions. In the latter, the field operators are ordered from right to left according to increasing time, $T\phi_1(y_1)\cdots\phi_n(y_n) := \phi_{\pi_1}(y_{\pi_1})\cdots\phi_{\pi_n}(y_{\pi_n})$, with π a permutation of $\{1,\cdots,n\}$ such that $y^0_{\pi_1} \geq y^0_{\pi_2} \geq \cdots \geq y^0_{\pi_n}$. Take an "internal" symmetry ($X^\mu_s = 0$ in the above notation), consider its Noether currents j^μ_s and the divergence of its time-ordered product with fields $\langle Tj^\mu_s(x)\phi_1(y_1)\cdots\phi_n(y_n)\rangle$. In addition to the explicit divergence which vanishes because of the current conservation, $\partial_\mu j^\mu_s = 0$, there is a contribution coming from the implicit Heaviside functions $\theta(\pm(x^0 - y^0_{i_p}))$ in the T-product. Then one finds

$$\begin{aligned}\frac{\partial}{\partial x^\mu}&\langle Tj^\mu_s(x)\phi_1(y_1)\cdots\phi_n(y_n)\rangle \\ &= \sum_{i=1}^n \delta(x^0 - y^0_i)\,\langle T\phi_1(y_1)\cdots[j^0_s(x),\phi_i(y_i)]\cdots\rangle\end{aligned} \tag{12.5}$$

and the equal time commutator on the r.h.s. is the density of the infinitesimal variation of the field ϕ_j: $[j^0_s(x),\phi_i(y_i)]_{x^0=y^0_i} = Z^i_s(x,\phi_i)\delta^3(\vec{x}-\vec{y}_i)$. These identities lead to very useful relations between different T-products.

In the case the symmetry is not exact but is "softly broken" and one has a *partial conservation* of the current $\partial_\mu j^\mu_s(x) = \chi(x)$, with χ an explicitly known field, the content of the suitably modified identity (12.5) is not void but leads to relations between amplitudes that have been explored in great detail, in particular in the context of weak interactions.

These identities and their various avatars – in particular the Slavnov–Taylor and BRST (Becchi–Rouet–Stora–Tyutin) identities in the framework of gauge theories – play a crucial role at several steps of the study of quantum field theories. They enable one to establish that the renormalization procedure does not jeopardize the symmetries of the original theory; they allow one to prove that conserved currents "do not renormalize" and do not develop anomalous dimensions, thus justifying the notion of universality in "current-current" interactions, see below; they are also used in the derivation of "low energy theorems", see in particular [30].

For completeness, let us mention briefly Noether's second theorem: for an "infinite-dimensional group" of invariance (such as diffeomorphisms in General relativity, or gauge transformations in gauge theories), invariance within a variational principle implies the existence of constraints between the $\Psi_i = \delta\mathcal{L}/\delta\phi^i$, i.e. identities satisfied independently of the Euler–Lagrange equations of motion. Examples are provided by the contracted Bianchi identities in general relativity, $D^\mu G_{\mu\nu} = 0$, where $G_{\mu\nu} = R_{\mu\nu} - \frac{1}{2} g_{\mu\nu} R$, or their analogue $D^\mu D^\nu F_{\mu\nu} = 0$ in gauge theories. Note that, although they are satisfied irrespective of the Euler–Lagrange equations, (12.1) or $\frac{1}{g^2} D^\nu F_{\mu\nu} = J_\mu := \partial \mathcal{L}_{\mathrm{m}} / \partial A^\mu$ respectively, these identities ensure the consistency of the latter, whose right hand sides are covariantly conserved, $D^\mu T_{\mu\nu} = 0$, resp. $D^\mu J_\mu = 0$.

6 Invariances in quantum mechanics

With the triumph of quantum mechanics, a new paradigm appears in the study of symmetries in physical systems. Through the fundamental papers and books of von Neumann and Wigner, Weyl and van der Waerden [29, 32, 35, 28] at the end of the 1920s, representation theory enters Physics. This is particularly well summarized in Wigner's theorem. With any quantum system is associated a Hilbert space $\mathcal{H}$. States of the system are described by vectors Ψ, or more precisely by *rays*, of $\mathcal{H}$ and "observables" A are self-adjoint operators on $\mathcal{H}$. Then Wigner's theorem [32, 35] asserts the following:

Theorem 6.1 (Wigner). *Transformations of a quantum system under a group G are implemented as $\Psi \to U\Psi$, $A \to UAU^{-1}$ with U unitary or anti-unitary and unique up to a phase, satisfying*

$$g, g' \in G \qquad U(g)U(g') = U(g.g')e^{i\omega(g,g')} .$$

Thus $U(g)$ gives a projective (up to a phase) representation of G.

By "anti-unitary", we mean a unitary antilinear operator, a situation which is encountered in the study of the time reversal operator T. Note that the projective nature of the representations is forced upon us by the structure of quantum mechanics: rays rather than vectors are the relevant objects.

Among such transformations, invariances are associated with group actions that commute with the dynamics, i.e. with the Hamiltonian

$$[H, U(g)] = 0 . \tag{12.6}$$

But according to Ehrenfest's theorem, the time derivative of any operator (with no explicit time dependence) is given by its commutator with H, $i\hbar\, \mathrm{d}A/\mathrm{d}t = [H, A]$. Thus (12.6) tells us that any $U(g)$, or any infinitesimal generator of the group action, is conserved: here again, invariances manifest themselves by the existence of conserved quantities. The new feature due to quantum mechanics is that not all conserved quantities are simultaneously observable. If one picks H and a set of commuting operators $U(h)$ – h in a Cartan torus of G if G is a Lie group – eigenstates of those $U(h)$ have conserved eigenvalues, which are, in the physicists' jargon, "good quantum numbers".

For example, consideration of the group of rotations SO(3) shows that its infinitesimal generators (i.e. elements of its Lie algebra) are proportional to the components of the angular momentum $\vec{J}$. The latter is thus quantized by the theory of representations of SO(3). If the system under study is invariant under rotations, one has conservation of $\vec{J}^2$ (the Casimir operator) and of one component, say J_z: their eigenvalues $j(j+1)\hbar^2$ and $m\hbar$ are "good quantum numbers", conserved in the time evolution. States of the system are classified by representations of SO(3) or SU(2), the latter appearing because it gives the projective (up to a sign) representations of the former, through half-integer spin representations.

As a side remark, we also notice that the distinction between discrete and continuous invariances, that was crucial in classical physics, with only the latter leading to conservation laws, fades away. Conservation of parity – to the extent it is conserved – is expressed by the commutation relation $[P, H] = 0$ and implies that the parity of a state is a good quantum number.

This beautiful framework was first applied to the rotation group and its finite subgroups, in conjunction with parity and the symmetric group of permutations. The latter appears in connection with the Pauli principle and the Fermi–Dirac or Bose–Einstein quantum statistics. This resulted in innumerable applications to atomic, molecular and solid state physics: atomic and molecular orbitals, the fine structure of spectral lines of atoms and their splitting in a magnetic or electric field (respectively the Zeeman and the Stark effects), the crystal-field splitting and many other effects were analyzed by group-theoretic methods; selection rules in transitions were shown to be governed by tensor products of representations, etc. See for example [27] for a review, and [23] for an overall presentation of the work of the first actors – Wigner and von Neumann, Heitler and London, Weyl. Early applications to particle physics were exploiting rotation, parity and Lorentz invariance in scattering theory. In the latter context, let us cite Wigner's fundamental work on the representations of the Poincaré group [37]. For a one-particle state, these representations are fully characterized by two real numbers, which describe the mass and the spin of the particle. But more group theory was soon to come in particle physics and we devote the next section to these new symmetries.

As it is often the case when a new theoretical corpus develops, requiring the learning and the practice of an abstract formalism, not everybody accepted happily this irruption of group theory into physics and there was a certain resistance among some physicists. Some even talked about "the group pest"!..., see [38], [28] p. 165, [23], or the prefaces of [32, 25]. In his preface to the 1959 edition of his book [35], Wigner observes: "It pleases the author that this reluctance [among physicists toward accepting group-theoretical arguments] has virtually vanished in the meantime and that, in fact, the younger generation does not understand the causes and the basis for this reluctance."

7 Invariances in particle physics

We have seen above that Noether's reciprocal statement enables one to infer the existence of a symmetry group from conserved quantities. This observation has been beautifully illustrated by the discovery of "flavor groups" in particle physics.

Heisenberg (1932) observing the many similarities of mass and interactions of the two constituents of the nucleus, the *nucleons*, namely the neutron n and the proton p, their electric charge notwithstanding, proposed that they form a 2-dimensional representation of a new SU(2) group of "isotopic spin", or "isospin" in short. This was an extremely fruitful idea, soon confirmed by the discovery (1947) of the π mesons, or pions, coming in three states of charge (π^+, π^0, π^-), and hence forming a 3-dimensional representation of this SU(2) group. Isospin symmetry then predicts relations between scattering amplitudes of nucleons and pions that were well verified in experiments. Later, more instances came with the kaons (K^+, K^0), the Δ resonance $(\Delta^{++}, \Delta^+, \Delta^0, \Delta^-)$ and others, which form representations of isospin $1/2$, $3/2 \ldots$ respectively. This SU(2) group is a symmetry of *hadrons* (i.e. of strongly interacting particles), broken by electromagnetic interactions.

In the sixties, the story repeated itself. In view of the newly discovered "strange" particles, Gell-Mann and Ne'eman (1961) proposed the existence of an SU(3) group of (approximate) symmetry of strong interactions. This "flavor SU(3)" group encompasses the previous isospin group SU(2). The argument leading to SU(3) was that there was experimental evidence of the existence of *two* independent conserved quantities (isospin and *hypercharge* or *strangeness*), hence the group should be of rank 2. Also there were several observed "octets" (8-dimensional representations) of particles of similar masses and same quantum numbers (baryonic charge, spin, parity), and this pointed to the group SU(3) which has an 8-dimensional irreducible representation, namely its adjoint representation. This hypothesis was confirmed soon after by the experimental discovery of a particle Ω completing a 10-dimensional representation, whose mass and quantum numbers had been predicted, and by some other experimental evidence [8]. Associated with the fundamental 3-dimensional representation of SU(3) is a triplet of "quarks", (u, d, s) (for up, down and strange), which according to the confinement hypothesis, should not appear as observable particles in normal circumstances.[5] This Gell-Mann–Ne'eman SU(3) group has been dubbed "flavor" to distinguish it from another "color" SU(3) that appears as the gauge group of "quantum chromodynamics" (QCD), the modern theory of strong interactions. To conclude this discussion, let us stress that the flavor SU(3) group of (approximate) symmetry was more than welcome, in order to put some order and structure in the "zoo" of particles that started to proliferate at the end of the fifties.

This line of thought has proved extremely fruitful, and modern particle physics has seen a blossoming of discoveries structured by the concepts of symmetries and

[5]By "normal circumstances" we mean discarding the extreme conditions of the primordial Universe, immediately after the Big Bang, or of the high-energy heavy ion collisions in the laboratory, where a plasma of unconfined quarks may be created.

group theory. The previous SU(2) and SU(3) groups have been extended to larger flavor groups, in connection with the discovery of new families of particles, with new quantum numbers, revealing the existence of more species (or "families") of quarks.

The role of symmetries is not limited to strong interactions and the other subatomic forces – electromagnetic and weak – are also subject to symmetry requirements. This was hardly apparent in the early Fermi "current-current" theory of the weak interactions, $\mathcal{L}_{\rm F} = -\frac{G_F}{\sqrt{2}} J^{\mu} J_{\mu}$, but then the V-A pattern à la Gell-Mann–Feynman of the current $J = V - A$, the role of the conservation or partial conservation of currents V and A, the Cabibbo angle, etc. were gradually uncovered, see [20, 10] for reviews of these historical developments. This role of symmetries is even more manifest today in the Glashow–Salam–Weinberg model of electroweak interactions, see below.

To look for a group invariance whenever a new pattern is observed has become a second nature for particle physicists.

8 The many implementations of symmetries in the quantum world

When discussing symmetries in contemporary physics, it is common to distinguish space-time symmetries, discrete or continuous, – rotations and Lorentz transformations, translations, space or time reflections, ... – from "internal" symmetries that act on internal degrees of freedom – charge, isospin, etc. While this distinction may be useful, it should not hide the tight interlacing of these two species of symmetries. For instance, one of the fundamental results in QFT is the CPT theorem (Lüders, Pauli and Bell) which asserts that the product of the charge conjugation C by the space reflection or parity P and time reversal T should be an absolute and uninfringed symmetry of Nature. This is established based on fundamental properties like locality and Lorentz invariance that one expects from any decent theory [26].[6]

Another distinction between two big classes of symmetries deals with their "global" or "local" character. The isospin SU(2) or the flavor SU(3) symmetries mentioned above are global symmetries, in the sense that the group element describing the transformation is independent of the space-time point where it applies. In contrast, the diffeomorphisms of GR or the gauge transformations of electrodynamics or Yang–Mills theory are local, with the group (or in infinitesimal form, the Lie algebra) element varying from point to point. As we have seen, that distinction was already clearly perceived by Klein and Noether.

[6]As pointed out by Yang [39], there is a very intriguing sentence in Weyl's preface to the second edition of his book [32], which seems to indicate that as early as 1930, he foresaw some relation between these three transformations.

It turns out that a quantum symmetry may be realized in a multiplicity of ways, namely

- as an *exact* symmetry, *e.g.* in the global U(1) symmetries associated with charge or baryonic number conservation, or in the local gauge invariances of quantum electrodynamics and of quantum chromodynamics (QED, QCD). In the latter, the gauge group is SU(3), and all particles –like quarks and gluons– carrying "color", i.e. a non trivial representation of that SU(3), are confined;
- as an explicitly broken symmetry: this is the case with isospin SU(2) broken by electromagnetism, or flavor SU(3), which is an approximate symmetry, broken by the strong interactions themselves. This is also the case with parity, the space reflection P mentioned above, which is explicitly broken by weak interactions, as discovered by Lee and Yang (1956) and as now implemented in the Standard Model;
- as a spontaneously broken symmetry. This refers to the following situation: in a physical system *a priori* endowed with a certain symmetry, the state of minimum energy, called the ground state or the vacuum depending on the context, may in fact be *non invariant*. This is a very common and fundamental phenomenon, which is familiar from the case of ferromagnetism: in a ferromagnet in its low-temperature phase, the magnetic moments of the individual atoms, although subject to a rotation invariant interaction, pick collectively a direction in which they align on average, thus giving rise to a macroscopic magnetization that breaks the rotation invariance of the whole system. This is accompanied, if the broken symmetry is continuous, by the appearance of massless excitations or particles, associated with the possibility of continuously rotating the ground state at a vanishing cost in energy. These excitations are the *Nambu–Goldstone* particles. In the variant in which the symmetry is only approximate, and in the neighbourhood of a spontaneously broken phase, one expects the would-be Nambu–Goldstone bosons to be not strictly massless but of low mass;
- as a spontaneously broken gauge theory: a global symmetry is spontaneously broken but the resulting theory maintains an exact gauge invariance. Then, and this is the essence of the Brout–Englert–Higgs (BEH) mechanism, the Nambu–Goldstone excitations do not appear as real particles, and instead give rise to additional polarization states of some vector fields and to masses of the corresponding particles. This is a crucial step in the edification of the electro-weak sector of the Standard Model, and the successive discoveries at CERN of massive vector particles (the $W^{\pm}$ and Z^{0}) and lately, of a likely candidate for the remaining massive scalar boson, have corroborated this model;
- anomalously, which means through a breaking of a classical symmetry by quantum effects. Examples are provided by the realization of some chiral symmetries of fermions, which act separately on the left-handed and right-handed components of these particles. Conversely, in the Standard Model of particle physics, where the assignments of representations are different for different chiralities, it is essential that anomalies cancel, see below;

- with supersymmetry: that ordinary Lie groups and algebras could be extended to accommodate anticommuting (Grassmannian) elements has been known and well studied since the seventies. To this date we have not seen any direct manifestation of supersymmetry in the laboratory. But the idea has been so amazingly fruitful in establishing new results and new connections between different fields that it will undoubtedly remain in the physicist's toolbox;
- as quantum symmetries, or "quantum groups", a misnomer for "quantum" deformations of Lie algebras or, more generally, for Hopf algebras. These have not yet manifested themselves in the context of particle physics, but are determinant in the discussion of quantum integrable models and in their applications to many systems of condensed matter physics in low dimension;

... and this list is certainly non exhaustive.

It is truly remarkable that Nature makes use of all these possible implementations of symmetries.

Let us illustrate these various possibilities on a few examples coming from modern physics. Our presentation will be extremely sketchy, as each topic would deserve a separate monograph.

Example 1 "Linear/non linear sigma models" may be regarded as Klein's most direct heirs in the context of QFT. In the simplest possible case, consider a field ϕ defined on $\mathbb{R}^d$ and taking its values in $\mathbb{R}^n$ or in S^{n-1} and write a Lagrangian in the form

$$\mathcal{L} = \frac{1}{2}(\partial\phi, \partial\phi) - V((\phi, \phi))$$

where (,) denotes the O(n) invariant bilinear form. The invariance group of that Lagrangian is obviously O(n), and the field ϕ transforms according to a linear representation or to a non-linear realization, depending on the case $\mathbb{R}^n$ resp. S^{n-1}. According to Noether's (first) theorem, there are $\frac{1}{2}n(n-1)$ independent conserved quantities at the classical level. Using the corresponding Ward identities (12.5), one verifies that the symmetry is preserved by quantum corrections. This was first set up by Gell-Mann and Lévy (1960) in the case $n = 4$, in their investigation of the partial conservation of the "axial current" in weak interactions [7], and involved the fields of pion particles $\pi^\pm, \pi^0$ and of a hypothetical σ, whence the name given to the model; this original model had thus a (softly and spontaneously broken) O(4) symmetry.

This may be generalized to a field ϕ taking its values in $\mathcal{M}$, a Riemannian manifold with isometries. Now in any of these sigma models, the natural questions to ask are

- how is the symmetry realized, as an exact, explicitly broken, or spontaneously broken symmetry?
- how is the symmetry preserved by renormalization? This is where use has to be made of Noether currents and Ward identities;
- what are the physical consequences: are there Goldstone particles, or "almost Goldstone" particles (like the pion of low mass)? is there a dynamical generation of mass? is the theory scale or conformally invariant? and so on, and so forth.

Sigma models have been extensively used with all kinds of manifolds and groups in particle physics and cosmology, in statistical mechanics and solid state physics. For example they appear as effective low-energy theories for various phenomena in condensed matter, describing membranes, surface excitations, order parameters, etc. but also in string theory – again in a low energy limit –, based on ordinary manifolds or generalized geometries à la Hitchin. The study of non compact and/or supersymmetric sigma models is currently a very active subject, for its applications running from condensed matter to string theory.

These sigma models also constitute a mine of mathematical problems. For instance, particular cases with $V = 0$ are studied for their own sake, in Riemannian geometry, under the name "harmonic maps".

Example 2 The Standard Model of particle physics has a symmetry group SU(3)×SU(2)×U(1), with three gauge groups realized in a completely different way.

The SU(3) color (gauge) symmetry of QCD is an exact invariance, and this is believed to be of crucial importance for quark confinement. On the other hand, SU(2)× U(1), the gauge group of weak isospin and weak hypercharge, is spontaneously broken down to an *exact* U(1), the gauge symmetry of ordinary electrodynamics. As mentioned above, a relique of the BEH mechanism at work in this spontaneous breaking should be a spin 0 boson, a good candidate of which has just been observed at CERN.

The absence of anomalies in the Standard Model, crucial for the consistency of the theory, relies on a remarkable matching between families of leptons and of quarks: for both types of particles, three "generations" are known at this time

$$(e, \nu_e),\ (\mu, \nu_\mu),\ (\tau, \nu_\tau) \longleftrightarrow (u, d),\ (c, s),\ (t, b)\,,$$

and anomalies cancel within each generation [1].

On top of the gauge pattern, there are *other* SU(2) and SU(3) groups at work: the flavor SU(2)⊂ SU(3) broken symmetries discussed above. In another vein, a scenario which has been contemplated – and in fact studied in great detail – but does not yet seem to be borne out by experiments is that this Standard Model is in fact a subsector of a larger supersymmetric extension.

Example 3 Quantum integrable systems and Quantum Groups. Consider the spin $\frac{1}{2}$ XXZ quantum chain: this is a quantum system of N spins $\vec{S}_i$ whose interactions are described by the Hamiltonian

$$H = \sum_{i=1}^{N} S_i^x S_{i+1}^x + S_i^y S_{i+1}^y + \Delta\, S_i^z S_{i+1}^z + \text{boundary terms}$$

acting in $(\mathbb{C}^2)^{\otimes N}$. Here, Δ is an anisotropy parameter in spin space. It was first introduced for $\Delta = 1$ by Heisenberg (1928) as a model of ferromagnetism. This is known to be a quantum integrable system after important contributions by Bethe, Lieb and Sutherland, Yang and Yang, Gaudin, Baxter, Faddeev and many others.

For $\Delta = 1$, (and no boundary term), it exhibits SU(2)-invariance. For $\Delta \neq 1$, $|\Delta| < 1$, it has a deformed symmetry $U_q sl(2)$ ("quantum SU(2)"), where $q = e^{i\alpha}$, $\Delta = \cos\alpha$ [21], or an affine quantum $U_q \widehat{sl}(2)$ [12], depending on the boundary conditions. Recent progress on the computation of correlation functions of the XXZ chain and on its connections with problems of combinatorics have been made possible by representation-theoretic considerations.

Other recent advances in the context of integrable gauge theories and the AdS/CFT correspondence also rely to a large extent on representation theory of quantum algebras.

Example 4 Conformal invariance. The last fifteen years of the previous century have witnessed rapid progress in our understanding of quantum field theories in low dimension. In 2 d, conformal invariant field theories (CFTs) have experienced a spectacular development, with a huge number of exact results and applications to critical phenomena and to string theories, thus writing a new chapter of non-perturbative quantum field theory. For the largest part, this progress was made possible by advances at the end of the seventies in the representation theory of infinite-dimensional Lie algebras – Virasoro, affine Lie algebras and their cousins – that are the relevant symmetries of CFTs. For a review, see for example [5]. There, one sees once again the close ties between symmetries, group theory and their physical implications.

9 Conclusions

We have seen that symmetry and group theory play an essential role in modern physics. Their role is:

- to *dictate* the possible form of interactions on geometrical grounds: the cases of general relativity or of gauge theories are exemplary in that respect; but one may also quote non-linear sigma-models, in which the form of the Lagrangian is prescribed by the geometry of the manifold and the isometries play a key role;
- to *predict*: more invariance means less independence, implying relations between different phenomena, selection rules, *a priori* determination of multiplicities, etc., as illustrated by scores of examples in atomic, molecular, solid state and particle physics; and to *organize* a wealth of data, of particles, of phenomena: we have seen that representation theory is instrumental in this undertaking;
- to *protect* in the quantization (and renormalization). Once again, take the example of a gauge theory. Were its symmetries broken by quantum effects (ultraviolet divergences, anomalies), the theory would lose most of its predictive power or even become inconsistent. So we have a self-consistent picture, where symmetry implies constraints (in the form of Ward identities), that in turn guarantee that symmetry is preserved by quantization. This scheme is implemented recursively in the perturbative construction of gauge theories.

The study of groups and of representation theory is now part of the education of a modern physicist. Some domains of representation theory – of superalgebras, of quantum groups and of infinite-dimensional algebras – have developed recently thanks to the incentive of physical applications.

Could a unified theory based on geometry and embracing all fundamental interactions including gravitation be constructed? That was Einstein's dream, this is still regarded today as the Holy Grail by many people, string theorists among others.

We have emphasized the many possible implementations of symmetries in (quantum) physics. We have also stressed that not only the nature of the symmetry group but also the scheme of its breaking, and the residual subgroup of symmetry, are determinant. In that respect, we are still living in the legacy of Klein and Curie...

Acknowledgements. It is a pleasure to thank the organizers, Lizhen Ji and Athanase Papadopoulos, for inviting me to write up this chapter, and Michel Bauer, Matteo Cacciari, Robert Coquereaux, Yvette Kosmann-Schwarzbach and Raymond Stora for several useful suggestions on a first draft of the manuscript.

Bibliography

[1] C. Bouchiat, J. Iliopoulos and P. Meyer, An anomaly-free version of Weinberg's model. *Phys. Lett.* 38B (1972), 519–523.

[2] J. J. Burckhardt, *Die Bewegungsgruppen der Kristallographie*. Birkhäuser, 1966.

[3] P. Curie, Sur la symétrie dans les phénomènes physiques, symétrie d'un champ électrique et d'un champ magnétique. *Journ. de Physique* 3 (1894), 393–415.

[4] O. Darrigol, The Genesis of the theory of relativity. *Séminaire Poincaré* 1 (2005), 57–78; O. Darrigol, 1905: un nouvel élan. In *Einstein aujourd'hui*, eds. M. Leduc and M. Le Bellac, EDP Sciences 2005.

[5] P. Di Francesco, P. Mathieu and D. Sénéchal, *Conformal Field Theory*. Springer 1997.

[6] A. Einstein to V. Bargmann, as quoted in Pais [19], page 152.

[7] M. Gell-Mann and M. Lévy, The axial vector current in beta decay. *Nuovo Cim.* 16 (1960), 705–726.

[8] M. Gell-Mann and Y. Ne'eman, *The Eightfold Way*. Benjamin, 1964.

[9] W. Heisenberg, Über den Bau der Atomkerne. *Z. für Physik* 77 (1932), 1–11.

[10] J. Iliopoulos, Symmetries and the weak interactions, talk given at the Cabibbo Memorial Symposium, Rome 2010, Preprint 2011; `arXiv:1101.3442` [physics.hist-ph].

[11] T. Janssen, *Crystallographic Groups*. North Holland, 1973.

[12] M. Jimbo and T. Miwa, *Algebraic Analysis of Solvable Lattice Models*. (NSF–CBMS Regional Conference 1993), AMS, 1995.

[13] H. A. Kastrup, The contributions of Emmy Noether, Felix Klein and Sophus Lie to the modern concept of symmetries in physical systems. In *Symmetries in Physics (1600–1980)*, ed. M. G. Doncel, Universitat Autònoma de Barcelona, 1987, 113–163.

[14] F. Klein, *Über die geometrischen Grundlagen der Lorentzgruppe*. Jhrber. d. d. Math. Vereinig. 19 (1910) 287. Reprinted in *Gesammelte Mathematischen Abhandlungen*, vol I, Springer, 1921, 533–552.

[15] Y. Kosmann-Schwarzbach, *The Noether Theorems, Invariance and Conservation Laws in the Twentieth Century*. Springer, 2006.

[16] C. Meusburger, Three-dimensional gravity – an application of Felix Klein's ideas in physics. In *Sophus Lie and Felix Klein: The Erlangen program and its impact in mathematics and physics*. Ed. L. Ji and A. Papadopoulos, Zürich, European Mathematical Society (EMS), 2015, 283–306.

[17] L. Michel, Symmetry, invariants, topology. IV Fundamental concepts for the study of crystal symmetry. *Physics Reports* 341 (2001), 265–336.

[18] E. Noether, Invariante Variationsprobleme. *Nachrichten von der Königlichen Gesellschaft der Wissenschaften zu Göttingen* (1918), 235–257.

[19] A. Pais, *Subtle is the Lord . . . , The Science and the Life of Albert Einstein*. Oxford Clarendon Press, 1982.

[20] A. Pais, *Inward Bound*: *Of Matter and Forces in the Physical World*. Oxford University Press, 1986.

[21] V. Pasquier and H. Saleur, Common structures between finite systems and conformal field theories through quantum groups. *Nucl. Phys.* B 330 (1990), 523–556.

[22] F. C. Phillips, *An introduction to Crystallography*. Longman, 1963.

[23] E. Scholz, Introducing groups into quantum mechanics, `arXiv:math/0409571` [math.HO], *Historia Mathematica* 33 (2006), 440–490.

[24] M. Senechal, *Quasicrystals and geometry*. Cambridge University Press, 1995.

[25] S. Sternberg, *Group theory and physics*. Cambridge University Press, 1994.

[26] R. F. Streater and A. S. Wightman, *PCT, Spin and Statistics and All That*. Benjamin, 1964.

[27] M. Tinkham, *Group Theory and Quantum Mechanics*. McGraw-Hill, 1964; Dover, 2003.

[28] B. van der Waerden, *Die gruppentheoretische Methode in der Quantenmechanik*. Springer, 1932.

[29] J. von Neumann and E. Wigner, Zur Erklärung einiger Eigenschaften der Spektren aus der Quantenmechanik des Drehelektrons. *Z. für Physik* 47 (1928), 203–220; *ibid.* 49 (1928), 73–94.

[30] S. Weinberg, *The Quantum Theory of Fields*, vol. II. Cambridge University Press, 1996.

[31] H. Weyl, Gravitation und Elektrizität, *Sitzungsberichte der Königlich Preußischen Akademie der Wissenschaften zu Berlin* (1918), 465–480. Reprinted in *Gesammelte Abhandlungen*, vol. II, K. Chandrasekharan ed., Springer, 29.

[32] H. Weyl, *Gruppentheorie und Quantenmechanik*. Hirzel (Leipzig) 1928); H. Weyl, *The Theory of Groups and Quantum Mechanics*. Dutten (New York) 1931, Dover, 1950.

[33] H. Weyl, Elektron und Gravitation, I. *Z. Physik* 56 (1929), 330–352. Reprinted in *Gesammelte Abhandlungen*, vol III, K. Chandrasekharan ed., Springer, 1968.

[34] H. Weyl, *Symmetry*. Princeton University Press, 1952.

[35] E. Wigner, *Gruppentheorie und ihre Anwendung auf die Quantenmechanik der Atomspektren*. F. Vieweg & Sohn, 1931; E. Wigner, *Group Theory and its Applications to the Quantum Mechanics of Atomic Spectra*. Academic Press, 1959.

[36] E. Wigner, On the Consequences of the Symmetry of the Nuclear Hamiltonian on the Spectroscopy of Nuclei. *Phys. Rev.* 51 (1937), 106–119.

[37] E. Wigner, On Unitary Representations of the Inhomogeneous Lorentz Group. *Ann. Math* 40 (1939), 149–204.

[38] E. Wigner, Oral transcript of an *Interview with Dr. Eugene Wigner by Lillian Hoddeson, Gordon Baym and Frederick Seitz*, (1981), Center for History of Physics of the American Institute of Physics, http://www.aip.org/history/ohilist/4965.html.

[39] C. N. Yang, Hermann Weyl's Contribution to Physics. In *Hermann Weyl, 1885–1985*, Centenary lectures, K. Chandrasekharan ed., Springer, 1986.

[40] C. N. Yang and R. Mills, Conservation of Isotopic Spin and Isotopic Gauge Invariance. *Phys. Rev.* 96 (1954), 191–195.

List of Contributors

Norbert A'Campo Universität Basel, Mathematisches Institut, Rheinsprung 21, 4051 Basel, Switzerland
email: Norbert.ACampo@unibas.ch

Charles Frances Institut de Recherche Mathématique Avancée, Université de Strasbourg et CNRS, 7 rue René Descartes, 67084 Strasbourg Cedex, France
email: frances@math.unistra.fr

Hubert Goenner Institute for Theoretical Physics, University of Göttingen, Friedrich-Hund-Platz 1, 37077 Göttingen, Germany
email: goenner@theorie.physik.uni-goettingen.de

Jeremy J. Gray The Open University, London, UK
email: j.j.gray@open.ac.uk

Lizhen Ji Department of Mathematics, University of Michigan, 530 Church Street, Ann Arbor, MI 48109-1043 USA
email: lji@umich.edu

Catherine Meusburger Department Mathematik, Friedrich-Alexander-Universität Erlangen-Nürnberg, Cauerstrasse 11, 91058 Erlangen, Germany
email: catherine.meusburger@math.uni-erlangen.de

Athanase Papadopoulos Institut de Recherche Mathématique Avancée, Université de Strasbourg et CNRS, 7 rue René Descartes, 67084 Strasbourg Cedex, France
email: papadop@math.unistra.fr

Yuri B. Suris Institut für Mathematik, Technische Universität Berlin, Str. des 17. Juni 136, 10623 Berlin, Germany
email: suris@math.tu-berlin.de

Alexandre Vinogradov Via Roma 27, 40042 Lizzano in Belvedere (BO), Italy
email: vinograd@unisa.it

Sumio Yamada Department of Mathematics, Gakushuin University, Tokyo, 171-8588, Japan
email: yamada@math.gakushuin.ac.jp

Jean-Bernard Zuber LPTHE, CNRS-UMR 7589 and Université Pierre et Marie Curie, 4 place Jussieu, 75252 Paris Cedex 5, France
email: zuber@lpthe.jussieu.fr

Index